Advances in Earth Sciences

Volume - 3

ADVANCES IN EARTH SCIENCES

Volume - 3

Series Editor : **Satish C. Tripathi**

Guest Editors

Arun Kumar
Helder I. Chaminé

SSPH

SATISH SERIAL PUBLISHING HOUSE

403, Express Tower, Commercial Complex, Azadpur, Delhi-110033 (India)
Phone : 011-27672852, Fax : 91-11-27672046
E-mail :info@satishserial.com, hkjain1975@yahoo.com
Website : www.satishserial.com

Published on behalf of :
The Society of Earth Scientists (SES)
C-207, Indira Nagar, Lucknow - 226 016, (U.P.), India
E-mail : india.ses@gmail.com

Published by :
SATISH SERIAL PUBLISHING HOUSE
403, Express Tower, Commercial Complex, Azadpur, Delhi-110033 (INDIA)
Phone : 011-27672852 Fax : 91-11-27672046
E-mail : info@satishserial.com, hkjain1975@yahoo.com

ISBN 978-93-81226-26-1

Cover Photo : *Photomicrographs of Gangotri granite. (a) Showing quartz (qtz), microcline showing cross-hatched twinning (kfs) and plagioclase (pl) under cross polars (sample no. UG 38); (b) Tourmaline (tur) in plane polarized light (sample no. UG 34); (c) Tourmaline (tur) showing zoning in plane polarized light (sample no. UG33); (d) Tourmaline (tur) showing zoning with inclusions of apatite (apt). [Jowhar, Fig. 3, This volume].*

Typeset at: Harminder Singh Kharb *for* Laxmi Art Creation
Printed at: Salasar Imaging Systems, Delhi-35

Preface

Earth scientists, in recent years, have been able to successfully amalgamate researches on different domains for the benefit of mankind and the planet at large. Transgressing research into other disciplines is necessary for strengthening earth system sciences. Book series ***'Advances in Earth Sciences'*** has been established by The Society of Earth Scientists (SES) as a print edition of the e-journal **Earth Science India**- www.earthscienceindia.info. The current volume consists of selected research papers of the Volume- III of the e-Journal. The papers of this volume addresses to all the spheres of the earth.

On behalf of The Society of Earth Scientists, I express sincere thanks to Dr. Arun Kumar, Adjunct Research Professor, Department of Earth Science, Carleton University, Ottawa, Canada and Prof. Helder I. Chaminé, Coordinator Professor, Geotechnical Department, Institute of Engineering of Polytechnic of Porto (ISEP), Portugal, for accepting to edit this volume and for their scientific inputs to improve the content. My Sincere gratitude to Patron of SES Dr. Shailesh Nayak, Secretary, Ministry of Earth Sciences, Government of India and Chairman Sri Rasik Ravindra, Director, NCAOR, Goa for guiding us to the right direction.

I sincerely acknowledge the academic support received from members of Editorial Board and reviewers of the manuscripts in maintaining the optimum quality of the publications. I also extend my thanks to all the earth scientists, who made their scientific contribution to this volume. My colleagues Sri D. D. Bhattacharya, Sri A. C. Pande, Dr. Mukund Sharma, Dr. R. Bali and Dr. S.M Gupta have always extended their helping hand in every activity of the Society and the e-journal. I extend my sincere thanks to them from bottom of my heart. Thanks are also due to Sri H. K. Jain, Partner, Satish Serial Publishing House, Delhi, for bringing out this publication in time. My family always stood behind me while discharging my voluntary duty towards the SES and e-Journal which provided me strength and enthusiasm. However, I am solely responsible for any shortcoming. Hope that I will continue to receive the overwhelming support of earth science community in furtherance of this noble cause.

Satish C. Tripathi
Series Editor

Introduction

Advances in Earth Sciences, volume-III covers research publications addressing to every domain of earth sciences ranging from palaeontlogy, structural geology, geochemistry, geophysical techniques, atmospheric and ocean sciences, thus, filling the gap of a multidisciplinary book. This way scientists are also updated in the advancement of earth sciences in other domains and helping them to incorporate a multidisciplinary dimension in their research programmes and planning. We express our sincere thanks to The Society of Earth Scientists for providing us the opportunity to edit this book.

In the first paper *Joshi and Kotlia* reported for the first time Late Pleistocene fossil lizards from Narmada Basin, Central India. With the help of the jaw fragments and isolated teeth, the material they have assigned to a new species, *Agama schleichi* sp.nov. The presence of lizards indicates dry/arid to semi-arid oscillations within otherwise warm/humid climatic conditions in the Narmada basin around the time of the *Homo erectus,* the only known *Homo* from India. An assemblage of thecamoebians from Early Permian Manjir Formation, Chamba Basin, northwestern Himalaya, Himanchal Pradesh, India is described by *Farooqui et al.* Morphologically these forms are comparable to the extant thecamoebian genera like *Centropyxis, Arcella, Difflugia, Cyclopyxis, Cucurbitella, Amphitrema, Lagenodifflugia* and *Lesquereusia.* The present report from Himalaya is very significant in filling the information gap on Proterozoic records of thecamoebians. *Desai* described biogenic sedimentary structures produced by *Architectonica laevigata* (Gastropod: Architectonicidae) from Mandvi Intertidal Zone, Gulf of Kachchh, western India. These traces are produced during low tides and are concentrated in specific areas within the intertidal zone, and also show substrate preference.

Micro-structural study of the rocks of the Zanskar Shear Zone by *Mukherjee* revealed an oldest top-to-SW shearing of ~22-16 Ma; and subsequent top-to-NE sense of shearing and finally a top-to-NE (down) sense of ductile shearing of ~18-16 Ma. The shape asymmetries as well as the cleavage plane orientation with respect to the C-planes of mineral fish act as reliable shear sense indicators. *Dutta et al.* studied erosional pattern of Majuli island, Jorhat district, Assam, bounded by the river Subansiri to the north and mighty Brahmaputra River to the south. They concluded that the island mass suffered significant rate of erosion than the deposition. *Singh and Bhattacharya* provided field evidences of vertical/subvertical shear zones in the central part of the Bundelkhand massif, central India. They concluded that the nature of the thin crust may be mainly responsible for the formation and preservation of these vertical shear zones. *Vallinayagam and Kumar* reported for the first time occurrence of rhyolite with spheroidal structure in the Neoproterozoic Malani Igneous Suite (MIS) in the Nakora area, Northwestern Peninsular India. Geological setting and the petromineralogical characters of spheroidal rhyolite suggest that at early magmatic processes, the supersaturated liquid crystallized rapidly into relatively differential density silicate liquid phases and growth of crystals lead the formation of spheroidal structure in the studied rhyolite. *Jowhar* provided geochemical characteristics of tourmalines from the Gangotri Granite (23$\pm$0.2 Ma), Garhwal Higher Himalaya. These tourmalines are zoned and from core to rim Mg decreases whereas Fe^{2+}, Ti, Mn and Ca increase. Two-feldspar geothermometer gives temperatures of subsolidus equilibration at about 441-270°C and plagioclase-muscovite gives temperature in the range of 448-339°C.

Samaila and Singh studied porosity loss due to compaction in the Cretaceous Upper Bima Sandstone, Upper Benue Trough, N.E. Nigeria. They concluded that original porosity indicates that the initial porosity of the formation has been reduced to 18.18%. *Bheemalingeswara* suggested a new method for testing of gold in stream sediments by using hot aqua regia and methyl isobutyl ketone solvents (MIBK-AR). The proposed bulk leach extractable gold (BLEG) method is particularly useful in areas where oxide/hydroxide coatings are conspicuous.

The distribution of mangroves along the Red Sea Coast of the Arabian Peninsula was studied using Historical Landsat Enhanced Thematic Mapper (ETM) data of the years 2000 and 2001. In part-I,

Kumar et al. observed depletion of mangroves along the northern coast of western Saudi Arabia due to rapid urbanization and industrialization of the region. In part-II, *Khan et al.* extended study along southern coast of western Saudi Arabia. The distribution of mangroves in both northern and southern Red Sea coast of Saudi Arabia is patchy. However, mangrove growth in south is relatively denser in comparison to north due mainly to the tropical climate, more wadis, higher rainfall and nutrients and less saline waters.

Bansal and Dimri discussed Scaling Spectral Analysis which is a new tool for interpretation of gravity and magnetic data. They concluded that the scaling distribution of the sources is close to the realistic distribution of the sources. The information about the scaling parameter is useful for gridding the data, deriving the lithology and source properties. *Brindha and Elango* Studied bromide in groundwater in parts of Nalgonda District, Andhra Pradesh. They found that only a few groundwater samples had concentration of bromide above permissible limit. Incidences of emergence of springs at surface or in shallow pits reported from discrete locations during the month of January 2005 in the southern Indian peninsular shield area had been an interesting phenomenon. *Muralidharan et al.* attributed the emergence of these springs due to ground movement and squeezing of pores in the saturated zone at shallow depth due the mega thrust Sumatra earthquake.

The relationship between daily meridional winds (v) at 200 hPa taken from NCEP/NCAR, over the domain (15°N to 25°N and 80°E to 95°E) at east coast of India and daily rainfall during the Indian Summer Monsoon (June-September) has been studied by *Dugam*. The persistence of the southerlies meridional at 200 h Pa components of winds in the upper troposphere may be useful for monitoring the large-scale monsoon failures. *Siingh et al.* in their paper reviewed the distribution of thunderstorm/lightning and their association with optical emissions. *Ganguly and Joel* examined long term trend, diurnal and seasonal variations of atmospheric ozone at the Indian Antarctic station Maitri using ground based and satellite data. TOC at Maitri exhibits a clear seasonal variation, which is found to be maximum in January and minimum in September/October while the surface ozone does not exhibit any significant diurnal variation. In order to enhance better understanding and improve the predictive capability for the ocean's role in future climate change scenarios, *Pandey and Singh* compared Estimating the Circulation and Climate of the Ocean, Phase

II (ECCO2) and National Centers for Environmental Prediction (NCEP) assimilation/reanalysis datasets with the TRITON and RAMA mooring buoys observational data at the point 1.5°S, 90°E in the Indian Ocean. They concluded that due to the dull performance of the GODAS data the climate forecast study results can not be very accurate to the real nature, further, performance of the ECCO2 is better than the GODAS/NCEP. In addition to the conventional applications in exploration and geodynamics, gravity methods are now used for numerous geophysical applications, such as modeling of hydrological mass changes, crustal deformation, ice-mass balance in the Polar Regions. *Tiwari* discussed some case studies of recent applications of gravimetry to earth sciences.

Gupta summarised Indian monsoon cycles through the last twelve million years from historical/mythological records to available recent scientific data. *Samui and Sitharam* used Artificial Intelligence (GRNN and LSSVM) techniques for predicting spatial variability of rock depth in Bangalore. This paper gives a powerful model based on LSSVM for the prediction of spatial variability of reduced level. *Olowofela et al.* experimentally studied effect of depth on the thermal signature of buried metallic object. They observed that the burial depth of the buried objects has effect on the amplitude of the temperature at the surface and thus its thermal signature, whereas thickness of the buried objects has a significant effect on its thermal signature.

Arun Kumar
Helder I. Chaminé

Contents

Preface v

Introduction vii

1 First Report of the Late Pleistocene Fossil Lizards from Narmada Basin, Central India 1
Moulishree Joshi and B. S. Kotlia

2 A Thecamoebian Assemblage from the Manjir Formation (Early Permian) of Northwest Himalaya, India 11
A. Farooqui, A. Kumar, N. Jha, A.C. Pande and D.D. Bhattacharya

3 Biogenic Sedimentary Structures Produced by Architectonica laevigata (Gastropod: Architectonicidae), Mandvi Intertidal Zone, Gulf of Kachchh, Western India 23
Bhawanisingh G Desai

4 Microstructures of the Zanskar Shear Zone 31
Soumyajit Mukherjee

5 A Study of Erosion-Deposition Processes around Majuli Island, Assam 51
M. K. Dutta, Swapnali Barman, S. P. Aggarwal

6 Signatures of Archaean E-W Crustal-Scale Shears in the Bundelkhand Massif, Central India: An Example of Vertical Ductile Shearing 65
S. P. Singh and A. R. Bhattacharya

7 First Report of Spheroidal Rhyolite from Nakora area of Malani Igneous Suite, Northwestern Peninsular India 79
G. Vallinayagam and Naresh Kumar

8 Chemistry of Tourmalines from the Gangotri Granite, Garhwal Higher Himalaya 89
T. N. Jowhar

9 Study of Porosity Loss Due to Compaction in the Cretaceous Upper Bima Sandstone, Upper Benue Trough, N.E. Nigeria 107
N.K. Samaila and G.P. Singh

10 Use of Bulk Sample and Hot Acids in Stream Sediment Survey for Gold 119
K. Bheemalingeswara

11 Distribution of Mangroves along the Red Sea Coast of the Arabian Peninsula: Part-I : the Northern Coast of Western Saudi Arabia 133
Arun Kumar, M. Asif Khan and Abdul Muqtadir

12 Distribution of Mangroves along the Red Sea Coast of the Arabian Peninsula: Part-II: The Southern Coast of Western Saudi Arabia 153
M. Asif Khan, Arun Kumar and Abdul Muqtadir

13 Scaling Spectral Analysis: A New Tool for Interpretation of Gravity and Magnetic Data 167
A.R. Bansal and V.P. Dimri

14 Study on Bromide in Groundwater in Parts of Nalgonda District, Andhra Pradesh 183
K. Brindha and L. Elango

15 Excitation of Hard Rock Aquifers in Southern Peninsular India Associated with the 2004 Sumatra Earthquake 193
D. Muralidharan, Rolland Andrade, R. Rangarajan and U. Sathyanarayana

16 Use of Meridional Wind Index at East Coast of India for Monitoring Drought and Flood Monsoon Conditions over Indian Region 205
S. S. Dugam

17 Thunderstorms/Lightning Generated Sprite and Associated Phenomena 211
Devendraa Siingh, Sanjay Kumar and A. K. Singh

18 Long Term Trend, Diurnal and Seasonal Variations of Atmospheric Ozone at Indian Antarctic Station Maitri 243
Nandita D. Ganguly and Vincent Joel

19 Comparison of ECCO2 and NCEP reanalysis using TRITON and RAMA data at the Indian Ocean Mooring Buoy point 253
Vivek Kumar Pandey and Sudhir Kumar Singh

20 On Some Recent Applications of Gravimetry to Earth Sciences 277
V. M. Tiwari

21 Indian Monsoon Cycles Through The Last Twelve Million Years 291
Shyam Murti Gupta

22 Spatial Variability of Rock Depth using Artificial Intelligence Techniques 337
Pijush Samui and T.G. Sitharam

23 Effect of Depth on the Thermal Signature of Buried Metallic Object 351
J. A. Olowofela, O. D. Akinyemi, R. Bello and A. A. Alabi

24 Modelling Electrical Conductivity of Contaminated Nigerian riverbed Sand 361
Olukayode D. Akinyemi and Iyiola Kuforiji

1

First Report of the Late Pleistocene Fossil Lizards from Narmada Basin, Central India

MOULISHREE JOSHI AND B. S. KOTLIA
Department of Geology, Kumaun University, Nainital, 263 002

Abstract

The fossil agamid lizards are reported from a Late Pleistocene locality (ca. 0.74 ka in age) in the Narmada valley, central India. Although the fossil material is scanty, it is described keeping in view that, the fossil lizards are barely studied from south Asia, and secondly, this is the only report of fossil lizards from the Narmada valley. With the help of the jaw fragments and isolated teeth, the material is assigned to a new species, *Agama schleichi* sp.nov. The presence of lizards indicates dry/arid to semi-arid oscillations within otherwise warm/humid climatic conditions in the Narmada basin around the time of the *Homo erectus,* the only known *Homo* from India.

Keywords : Agamid lizards, fossil lizards, Late Pleistocene, Narmada valley

Introduction

In Asia, the first material of the Early Eocene agamid lizards from Kyrgyzstan was made by Averianov and Danilov (1996) who proposed their diversity as characterizing the open landscape. In India, the fossil lizard remains are very scanty. The first report of a fossil lizard (*Varanus sivalensis;* Suborder Lacertilia) was made in 1884 by Lydekker (mentioned in Raghavan, 1991), from an unknown locality in the Siwalik sediments. However, the first well illustrated agamids (*Calotes* and *Uromastyx*) were reported from the Indian Siwalik by Patnaik and Schleich (1998). Recently, Prasad and Bajpai (2009) discovered some new species of agamids from the Early Eocene of western India. Excepting these reports, almost nothing is known from the Indian subcontinent as far as the fossil lizards are concerned. On the other hand, presently, there are over 150 species of lizards in India, majority of which belong to Geckos (*Hemidactylus*). Another widely distributed group is of the garden lizard (*Calotes versicolor*), followed by Chameleon (*Chameleo zeylanicus*). The largest and most conspicuous Indian lizard is *Varanus* with species like *V. salvator*, *V. flavescens* and *V. griseus*, the last being a desert species.

We describe agamids from the Narmada flood plain deposits from where the only known Indian *Homo* has been documented (Sonakia, 1984; de Lumley and Sonakia, 1985; Sankhyan, 1997, 2005) making the valley palaeontologically significant. Agamidae is an old world family of lizards found in Africa, central and southern Asia and Australia. A few forms migrated to northern Asia, Europe and North America during the Eocene. They are generally diurnal, terrestrial or arboreal forms occupying a variety of habitats. The important characters of this family are; premaxilla fused, maxilla forming the lower border of the orbit (jugul in some forms), post frontal absent, cheek teeth acrodont, palatal teeth absent, pleurodont anterior teeth present, splenial reduced or fused with the dentary. The meckelian groove is reduced to a shallow and narrow indentation. Many forms have dentary fangs.

Description of the Locality

At Devakachar site (23^0 23'N; 79^0 17'E), about 120km southwest of Jabalpur (Fig.1a), the sediment sequence is exposed by the Sher river, a tributary of the E-W flowing Narmada river. A 17m thick profile consists of cemented conglomerate, pebbly sand, silt and yellowish clay with a few scattered medium to coarse grained sand lenses (Fig.1b). The sequence belongs to the Hirdepur Formation which palaeomagnetically falls within the Brunhus magnetic chron. The Surajkund Formation which has yielded the *Homo erectus* is older then the Hirdepur Formation (Fig.1c). The fossil horizon, 8.5 m above the base, is a 0.5 m thick coarse pebbly sand bed and is approximated as ca. 0.74 Ma in age (Fig.1c). In addition to the agamid lizards, the collection also consists of murid rodents, fish remains, gastropods and pelecypods.

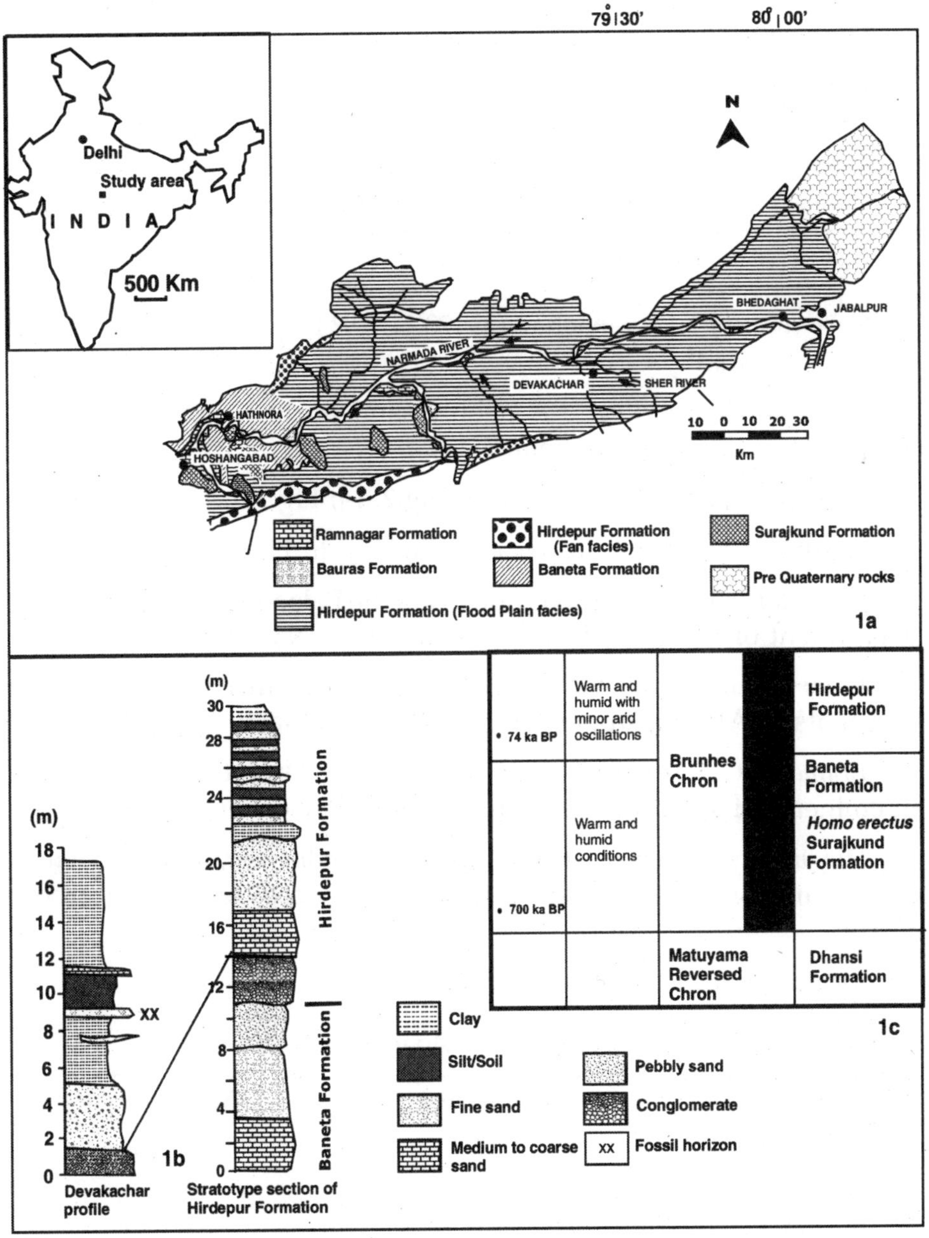

Fig.1: a) Location map showing the study area (modified after Tiwari and Bhai, 1997). b) Lithostratigraphy of the Devakachar section and correlation with the composite sequence (stratotype section of Hirdepur Formation is taken from Tiwari and Bhai, 1997). c) Chronology of a part of the Narmada sequence (compiled from Acharyya and Basu, 1993; Biswas, 1997; Rao *et al.*, 1997; Tiwari and Bhai, 1997; Kotlia and Joshi, 2008).

Systematic	Palaeontology
Class	Reptilia
Order	Squamata
Family	Agamidae Gray, 1827
Genus	*Agama*

Agama schleichi sp. nov.

Type locality : Devakachar (23° 23'N; 79° 17'E), 120km southwest of Jabalpur (Madhya Pradesh).

Fossil horizon and age : A pebbly sand horizon, 8.5m above the base (see Fig.1b) is approximated as ca. 0.74 Ma in age.

Holotype : NAR/L5, a dentary (Fig.2, i-j; Fig.3e).

Paratype : NAR/L4, upper jaw (Fig.2, g-h; Fig.3c).

Repository : The material is housed in the Palaeontology laboratory, Department of Geology, Kumaun University, Nainital.

Referred material : NAR/L1, NAR/L2 and NAR/L3, jaw fragments; NAR/L4, upper jaw; NAR/L5, dentary.

Etymology : Named in the honour of Prof. Dr. Hermann Schleich, an eminent scientist for his contribution in Palaeoherpatology.

Diagnosis : Crown tricuspate, central cusp largest, accessory cusps smaller and well developed, teeth widely spaced, meckelian groove narrow.

Measurements : See Table-1 below

Table-1 : Height (in mm) of the cusps in *Agama schleichi* sp. nov.

Sp. no.	Central cusp	Accessory cusp
NAR/L1	2.15	1.56
NAR/L2	2.15	1.56
NAR/L3	2.3	1.95
NAR/L4	1.95	1.56
NAR/L5	1.95	1.56

Description of the Fossil Material

NAR/L1 (Fig.2a-b; Fig.3a): The single tooth is attached to the jaw bone. It is tricuspate with well developed cusps. The central cusp is the largest while the accessory cusps are comparatively smaller. NAR/ L2 (Fig.2c-d; Fig.3b) is a broken jaw fragment with one complete and other partially broken tooth attached to the jaw bone. The teeth are tricuspate. The central cusp is the most prominent. The accessory cusps are smaller and have a blunt tip. NAR/L3 (Fig.2e-f; Fig.3d) bears two teeth on a fragmentary jaw bone. It is not possible to ascertain whether it is maxillary or dentary. The crown is tricuspate. The central cusp is the most prominent and largest of the three cusps. The meckelian groove can be seen at the base of the teeth. It is narrow but persists throughout the length of the jaw bone. The teeth are widely spaced.

The jaw (NAR/L4, Fig.2g-h; Fig.3c) tapers anteriorly. The crown height increases posteriorly. There are two teeth attached to the jaw. The anterior margin of the jaw bears a single, unicuspate tooth. The posterior teeth are tricuspate and anteriorly inclined. The central cusps are larger than the accessory cusps.

The right maxillary (NAR/L5, Fig.3e) is with five teeth arranged on the jaw bone. The crown height is more or less the same throughout. The central cusps are bigger while the accessory cusps are less developed. Teeth are widely spaced, transversely compressed and swollen at the base. The meckelian groove is continuous at the base.

Comparisons

Agamidae, an old world family of lizards has a geological range extending from the Palaeocene to the Holocene (Estes, 1983). Fossil forms have been recovered from the Palaeocene of Asia, Eocene of Asia, Europe and North America, Oligocene of Europe, Miocene of Asia and Europe, Pliocene of Asia, Pleistocene of Asia, Europe and Australia and Holocene of Europe (Table-2). *Agama* is an African agamid having a depressed body with a dorsal crest and gular sac (Estes, 1983). The crown is tricuspate with a prominent central cusp, smaller and less developed accessory cusps and meckelian grove is present at the base of the teeth.

Agama schleichi sp. nov. is distinct from other genera of family Agamidae. *Clamydosaurus* is a monotypic genus with a prominent erectile frill around the neck. *Mimeosaurus* is Upper Cretaceous agamid which differs from *Agama* in having a strongly developed coarse tubercular sculpture present on the skull roof. *Tinosaurus* differs from *Agama* in having caniniform, widely spaced teeth and open meckelian groove. *Uromastyx*, particularly reported from Indian Siwalik

(Raghavan, 1991; Patnaik and Schleich, 1998) differs from *Agama* in having smaller sized unicuspate teeth, pointed crown and more closely spaced teeth compared to the tricuspate and widely spaced teeth in the later (see Fig. 2 k-l).

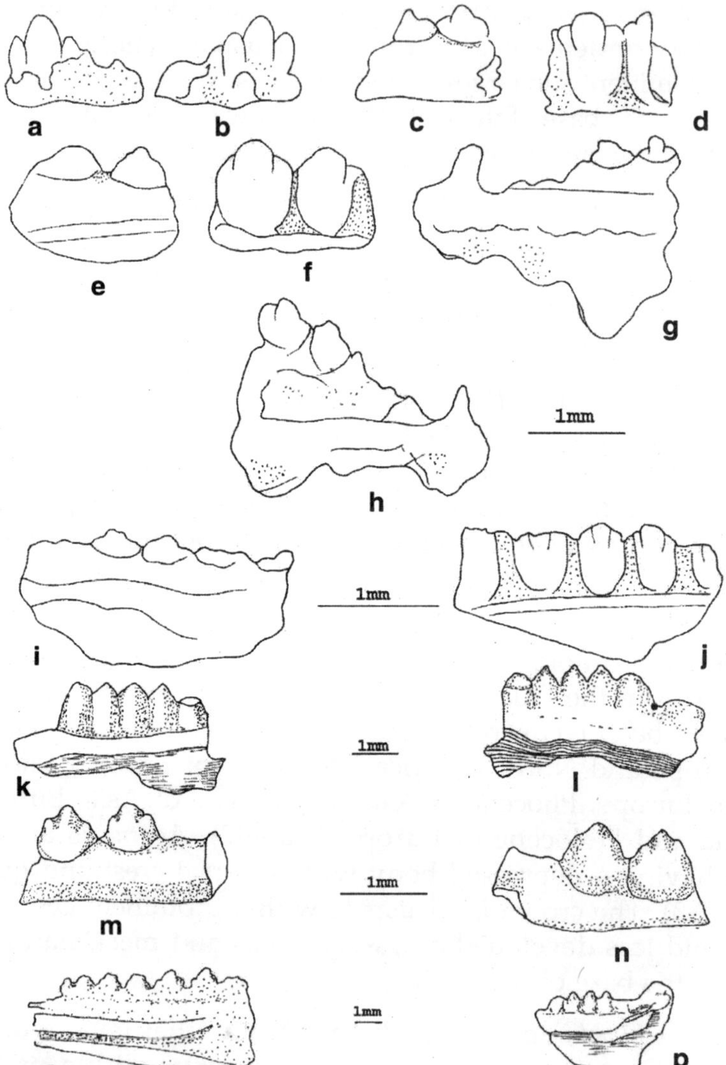

Fig. 2: *Agama schleichi* sp. nov. a-f (NAR/L1-3), labial and lingual views of the jaw fragments; g-h (NAR/L4), upper jaw; i-j (NAR/L5), dentary; k-l, *Uromastyx* sp. (Patnaik and Schleich, 1998); m-n, *Calotes* sp. indet ((Patnaik and Schleich, 1998); o, lingual view of *Tinosaurus stenodon* (Marsh, 1872); p, labial view of *Tinosaurus asiaticus* (Gilmore, 1943).

Fig. 3 : *Agama schleichi* sp. nov. a, b, d, jaw fragments (NAR/L1-3); c, upper jaw (NAR/L4), e, dentary (NAR/L5) (Bar represents 1mm).

Table-2 : Geological range of Agamidae (modified after Estes, 1943).

Age	Asia	Europe	N. America	Australia
Holocene		*Uromastyx*		
Pleistocene	*Agama schleichi* sp. nov. *Agama stellio*	*Agama stellio*		*Clamydosaurus*
Pliocene	*Uromastyx*			
Miocene	*Agamidae*	*Uromastyx*		
Oligocene		*Agama, Uromastyx*		
Eocene	*Tinosaurus, Agama*	*Tinosaurus*	*Tinosaurus*	
Palaeocene	*Tinosaurus*			

Agama schleichi sp. nov. differs from *Agama sinensis*, reported from the Palaeocene of China (Hou, 1974) in having better developed accessory cusps and a smaller sized *Agama galliae* (reported from the Upper Eocene of France by Filhol, 1877) differs from our material in having closely spaced teeth, poorly developed accessory cusps and a bigger size. Although, *A. schleichi* sp. nov. bears close resemblance with *Agama* sp. of Zwick and Schleich (1994) in having a tricuspate crown and presence of meckelian grove at the base, but the teeth of later are more widely spaced on the jaw bone and are larger than the Narmada teeth (see Table 3). *Tinosaurus stenodon* (Marsh, 1872) has tricuspate teeth (Fig. 2o) which are slightly larger in size than *Agama schleichi* sp. nov. Although *Tinosaurus asiaticus* (from the Republic of China; Gilmore, 1943) shares close morphological similarity with *Agama schleichi* sp. nov. (Fig. 2p) but it is much smaller in size (see Table-3) with very closely spaced teeth.

Table-3 : Measurements of height of the cusps in various agamid lizards.

Name of the species	Reference	Height of central cusp (mm)	Height of accessory cusp (mm)
Agama schleichi sp. nov.	Present work	2.10	1.83
Agama sp.	Zwick and Schleich, 1994	3.45	2.40
Agama galliae	Filhol, 1877	1.66	---
Agama sinensis	Hou, 1974	2.00	1.71
Tinosaurus asiaticus	Gilmore, 1943	1.60	1.30
Tinosaurus stenodon	Marsh, 1872	2.30	1.63
Uromastyx sp. indet.	Patnaik and Schleich, 1998	1.71	---

Conclusion

The dental fragments of agamid lizards are reported for the first time from the Narmada valley. *Agama schleichi* sp. nov. is tricuspate with a prominent central cusp and lesser developed accessory cusps. The agamid lizards belong to the Acrodonta, a group which has been suggested to have a Gondwanan origin (Holmes et al., 2009). It seems that the tricuspid agamids appeared in India sometime in early Eocene and diversified in the Pleistocene particularly during the prevalence of semi-arid to arid climatic oscillations in the central India.

Acknowledgements

We are grateful to Prof. Hermann Schleich of Fuhlrott Museum, Germany for helping in identification of the fossil material. Thanks are due to the UGC and CSIR for financial assistance.

References

Acharyya, S. K. and Basu, P. K. (1993) Toba ash on the Indian subcontinent and its implications for the correlation of Late Pleistocene alluvium. Quaternary Research, v. 40, pp. 10–19.

Averianov, A. and Danilov, I. (1996) Agamid lizards (Reptilia, Sauria, Agamidae) from the Early Eocene of Kyrgyzstan. Neus Jh. Geologie und Palaontologie, v. 12, pp. 739-750.

Biswas, S. (1997) Fossil Mammalia of the Quaternary sequence of the Narmada valley: their affinity, age and ecology. J. Geol. Soc. Ind.,Special Publication, v. 46, pp. 91-104.

de Lumley, M. A. and Sonakia, A. (1985) Premiere Dicouverte D'un *Homo erectus* Surte continent Indien a Hathnora, Dans la Mogenne Vallea de La. L'Anthropology, v. 89 (1), pp. 13-61.

Estes, R. (1983) Sauria, Terrestria, Amphisbaenia. Encyclopedia of Palaeoherpatology. Gustav Fisher, Verlag, Stuttgart, New York. 248p.

Filhol, H. (1877) Rescherches sur des phosphorites du Querey. Pt. II. Annals of Science, v. 8, pp. 1-338.

Gilmore, C. (1943) Fossil lizards of Mongolia. Bull. American Museum of Natural History, v. 81 pp. 361-384.

Holmes, R. B., Murray, A. M., Chatrath, P., Attia, Y. S. and Simons, E. L. (2009) Agamid lizard (Agamidae: Uromastycinae) from the lower Oligocene of Egypt. Historical Biology, DOI: 10.1080/08912960903302128.

Hou, L. (1974) Palaeocene lizards from Anhui, China. Vertebrate Palaeontology, v. 12, pp. 193- 202.

Kotlia, B. S. and Joshi, M. (2008) Reconstruction of Late Pleistocene palaeoecology of the Upper Narmada valley (Central India) using fossil communities. Palaeoworld, v. 17, pp. 153-159.

Lydekker, R. (1884) Siwalik and Narbada Equidae. Palaeontologica Indica, v. 2(2), pp. 35-104.

Marsh, O. (1872) Preliminary description of new Tertiary reptiles Parts I and II. American Journal of Science, v. 4, pp. 298-309.

Patnaik, R. and Schleich, H. H. (1998) Fossil micro-reptiles from Pliocene Siwalik sediments of India. Veroffentlichungen aus dem Fuhlrott Museum, v. 4, pp. 295-300.

Prasad, G. V. R. and Bajpai, S. (2009) Agamid Lizards from the Early Eocene of Western India. Palaeontologia Electronica, v. 11(1), pp. 1-19.

Raghavan, P. (1991) Fossil frogs and a lizard from the Basal Pinjor Formation (Sub-Himalaya), Haryana, India. Bulletin of Panjab University, v. 42(1-4), pp. 1-6.

Rao, K.V., Chakraborti, S., Rao, K.J., Ramani, M. S. V., Marathe, S. D. and Borkar, B. T. (1997) Magnetostratigraphy of the Quaternary fluvial sediments. Jour. Geol. Surv. Ind., Special Publication, v. 46, pp. 65-78.

Sankhyan, A. H. (1997) Fossil clavicle of a Middle Pleistocene hominid from the Central Narmada Valley, India. Jour. Human Evolution, v. 32(1), pp. 3-16.

Sankhyan, A. R. (2005) New fossils of Early Stone Age man from Central Narmada Valley. Current Science, v. 88(5), pp. 704-707.

Sonakia, A. (1984) The skull cap of Early man and associated mammalian Fauna from Narmada valley alluvium, Hoshangabad area, Madhya Pradesh (India). Jour. Geol. Soc. Ind., v. 113 (6), pp. 159-171.

Tiwari, M.P. and Bhai, H. Y. (1997) Quaternary stratigraphy of Narmada valley. Geol. Surv. Ind., Special Publication, 46: 33-63.

Zwick, A. And Schleich, H.H. (1994) QuartareReptilifunde von der Grabungsstelle Karain, S. Turkei (Reptilia, Squamata. Courier Forschungs Institute Senckenberg, v. 173, pp. 267-281.

□□□

2

A Thecamoebian Assemblage from the Manjir Formation (Early Permian) of Northwest Himalaya, India

A. FAROOQUI [1], A. KUMAR[2*], N. JHA[2], A.C. PANDE[3] AND D.D. BHATTACHARYA[3]

[1]Birbal Sahni Institute of Palaeobotany, 53, University Road Lucknow, India
[2]Center for Petroleum and Minerals, Research Institute, King Fahd University of Petroleum and Minerals, Dhahran 31262, Saudi Arabia
[3]Geological Survey of India, Lucknow, India

Abstract

A fairly well preserved and diverse assemblage of thecamoebians is reported here from the palynological preparations of the Early Permian Manjir Formation. This Formation is in the Chamba Basin of the northwestern Himalaya in Himanchal Pradesh, India. It has been assigned an Early Permian age on the basis of characteristic palynomorphs (Pande *et al.*, 2004). This study documents for the first time an Early Permian thecamoebian assemblage from the Himalayan region. Morphologically these forms are comparable to the extant thecamoebian genera like *Centropyxis, Arcella, Difflugia, Cyclopyxis, Cucurbitella, Amphitrema, Lagenodifflugia* and *Lesquereusia*. This find supports the current hypothesis of minimal evolution in thecamoebian lineages through geologic time.

Keywords : Thecamoebians, Early Permian, Manjir Formation, Northwestern Himalaya

Introduction

Arcellaceans characterize a special group of testate protists (agglutinated or autogenous) that belong to the Subphylum Sarcodina

(Medioli and Scott, 1988). These have a sac/cap like test, with a simple aperture known as pseudostome through which extrudes pseudopods of the amoeba dwelling inside it (Medioli *et al.*, 1999). Most of the known fossil thecamoebians are agglutinated by calcareous or siliceous grains and/or diatom frustules (Medioli *et al.*, 1990a). They occur in a variety of freshwater habitats (Medioli and Scott, 1983, 1988) and also in marginally brackish environments (Charman *et al.*, 2000; Patterson and Kumar, 2002). They are known from lake sediments from tropics to the Arctic region (see references in Patterson and Kumar, 2002; Boudreau *et al.*, 2005), peat bogs (Woodland *et al.*, 1998; Charman *et al.*, 2000) and salt marshes (Charman *et al.*, 2002; Roe *et al.*, 2002).

Published thecamoebian studies are mainly from the lakes and bogs in Europe and North America. There are relatively fewer thecamoebian studies from the tropical regions (Hoogenraad and Groot, 1940, 1946; van Oye, 1949; Dalby *et al.* 2000, and Sudzuki, 1979; Farooqui and Gaur, 2007). Some of the earliest thecamoebian studies from India include brackish water lakes and ponds of nineteenth century Mumbai, India (Carter, 1856, 1864). Roe and Patterson (2006) studied environmental control on the distribution of thecamoebians in small ponds and lakes of Barbados. Patterson and Kumar (2000, 2002) provide comprehensive reviews of environmental and paleoecological utility of lacustrine thecamoebians in higher latitudes. According to Roe and Patterson (2006), "thecamoebian species can be correlated to a variety of environmental and climatic parameters, including metal and organic pollutant contamination, substrate type, salinity, levels of organics, oxygen concentration, water temperatures, water table fluctuations, humification, changes in intertidal flooding and land use change".

Pre-Quaternary Occurrences of Thecamoebians

Distribution of fossil thecamoebian families from Neoproterozoic through Pliocene is patchy with wide gaps in the fossil record (Medioli *et al.*, 1990a and b, van Hengstum *et al.*, 2007, Kumar, submitted). The oldest thecamoebians are described as Vase Shape Microfossils (VSMs) from the Neoproterozoic rocks of Grand Canyon, Arizona by Porter and Knoll (2000), Porter *et al.* (2003), and Smith *et al.* (2008). There are isolated reports of thecamoebians from Carboniferous Period (Vasicek and Ruzica 1957; Wightman *et al.*, 1994) Mesozoic and Cenozoic Eras (Bradley, 1931; Frenquelli, 1933; Schoborn, 1999; Schmidt *et al.*, 2004; van Hengstum, 2007, Bassi *et al.* 2008). Recently, Kumar (submitted) compiled a list of pre-Quaternary thecamoebian occurrences from all over the world. Despite a long geological history, thecamoebian lineages have shown minimal evolution through geological time and demonstrate close resemblance to their Holocene forms (van Hengstum *et al.*, 2007).

Thecamoebians in Palynological Preparations

Since thecamoebian tests could be secreted (autogenous) and are proteinaceous, they are acid resistant and have been reported in the palynological preparations by Kumar and Patterson (2002), Farooqui and Gaur (2007). Srivastava and Bhattacharya (1998) reported an assemblage of Early Permian palynomorphs from faunal coal balls of Arunachal Pradesh, northeastern Himalaya, India. A specimen referred to as "?Chitinozoa like vesicle" (Plate 1, Fig. 11) is most likely a specimen of *Difflugia*. Likewise, Pande *et al.* (2004) published a palynomorph assemblage from Manjir Formation (Early Permian) from Himachal Pradesh in north India. They too report an "unidentified specimen" (Plate 2, Fig. 6) which is a well preserved specimen of *Centropyxis*. There are significant opportunities for discovering thecamoebian assemblages from the Phanerozoic sediments all over the world that may occur in the palynological preparations. The present paper also deals with Permian thecamoebians observed in the palynological preparations.

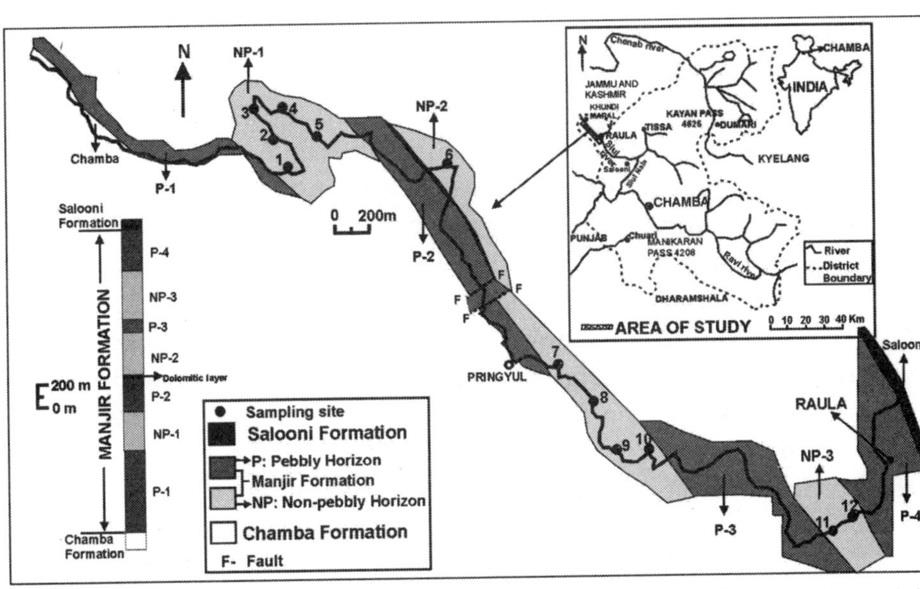

Fig. 1: Location map of Khundi-Maral-Raula Section, Chamba District, Himanchal Pradesh, India.

Stratigraphy and Age of the Manjir Formation

The Chamba Basin (Himachal Pradesh) represents the Tethyan realm of northwest Himalaya and the Manjir Formation is a major stratigraphic unit of this Basin. Most early work on the geology of this region considered Manjir Formation to be of Upper Proterozoic age on the basis of lithology (Rattan, 1973). Later, well preserved and stratigraphically significant palynomorphs comprising chiefly monosaccates (*Parasaccites* and *Plicatipollenites*), non-striate disaccates *Scheuringipollenites, Platysaccus* and striate disaccates (*Striatopodocarpites, Faunipollenites* and *Striatites*) were reported in abundance giving a definite Early Permian age to the Manjir Formation (Pande *et al.,* 2004).

The present study is based on the samples that were collected along the Bhalsu-Sunu Kothi-Khundi Maral Section along the Siul River (Fig.1). The detailed lithostratigraphy is given in Table-1. Lithologically the rocks of Manjir Formation are foliated and moderately metamorphosed and are represented by heterogeneous, poorly sorted pebbly beds separated by non-pebbly horizons. Samples for the present study were taken from the dark grey to black shale/slate separating the pebbly horizons.

Material and Methods

Testate amoebae were isolated from the slate/shale after treatment with hydrochloric acid (HCl) generally used for isolating palynomorphs from hard sediments. After washing the acid, the samples were obtained as a residue in 600 mesh (>15 μ), mounted in glycerol, and slides were scanned for thecamoebians under high power light microscope (Olympus BX-52). The identifications are based on Medioli and Scott (1983), Kumar and Dalby (1998) and Beyens and Meisterfeld (2001). The Manjir formation constitutes three Non-pebbly units separated by pebbly units. The non-pebbly units (NP-1 to 3) yielded thecamoebians and the Early Permian palynomarkers. Five samples each from NP-1, NP-2 and 2 samples from NP-3 were studied. The total count of these in 10 g samples are given in Fig.2.

Table-1 : Lithostratigraphy of the Manjir Formation along Khundi Maral- Raula Section (Siul river), Chamba District, Himanchal Pradesh, India.

Sample Number (Fig.1)	Formation		Lithology	Thickness (m)
	CHAMBA		Fine grained, greyish green quartzite and phyllite	Not measured
	MANJIR FORMATION	Pebbly Unit P-1	Foliated calcareous, matrix-supported pebbly slate/phyllite and quartzite with sparsely distributed granules of transluscent quartz vein in lower part	715
1to 5		Non-pebbly zone **NP-1**	Grey, bluish grey, non-calcareous, slate and phyllite with subordinate siltstone and black shale with occasional limonisation	417
		Pebbly Unit P-2	Grey, bluish grey, calcareous, matrix-supported phyllite/slate and greenish quartzite, medium and small pebble clasts of quartzite, shale and dolomite/dolomitic limestone, stretched clasts conspicuous	315
6-10		Non-pebbly Unit **NP-2**	Black, grayish black,non-calcareous bleached and limonitised pyritiferous shale/ slate and siltstone with minor pockets of dolomite/dolomitic limestone	356
		Pebbly Unit P-3	Foliated calcareous, bluish grey matrix- supported pebbly slate/ phyllite, clasts often have sericitic envelope ranging in size from granular to medium pebble, clasts include quartzite, shale and carbonate	157
11-12		Non-Pebbly Unit **NP-3**	Non-pebbly, non-calcareous, grayish black slate and phyllite with occasional siltstone bands	428
		Pebbly Unit P-4	Foliated calcareous, matrix-supported, bluish grey pebbly horizon, matrix dominantly arenaceous, clasts of quartzite, shale and carbonate range from granular to small-pebble size.	341
	SALOONI		Black pyritised shale, carbonaceous shale, minor siltstone bands	Not measured

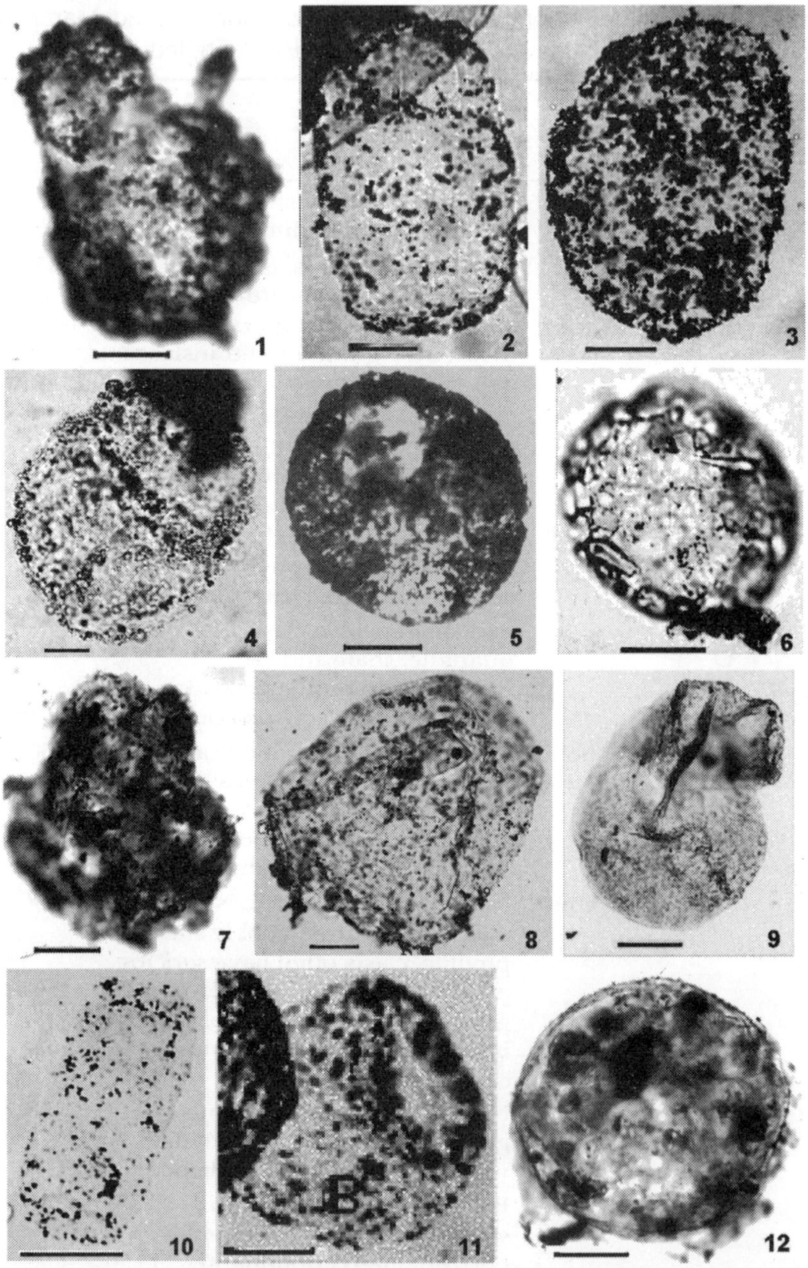

Plate-1: Light Microscopic photomicrographs recovered from Early Permian Khundi-Maral-Raula Section, Chamba District, Himanchal Pradesh.

1. *D.pyriformis*; 2. *Difflugia oblonga*; 3. *Difflugia species*; 4. *D. urceolata*; 5. *Difflugia gramen*; 6. *D. protaeiformis*; 7. *Lagenodifflugia*; 8. *Cucurbitella*; 9. *Lesquereusia*; 10. *Amphitrema flavum*; 11. *Cyclopyxis kahlii*; 12. *Trigonopyxis arcula*.

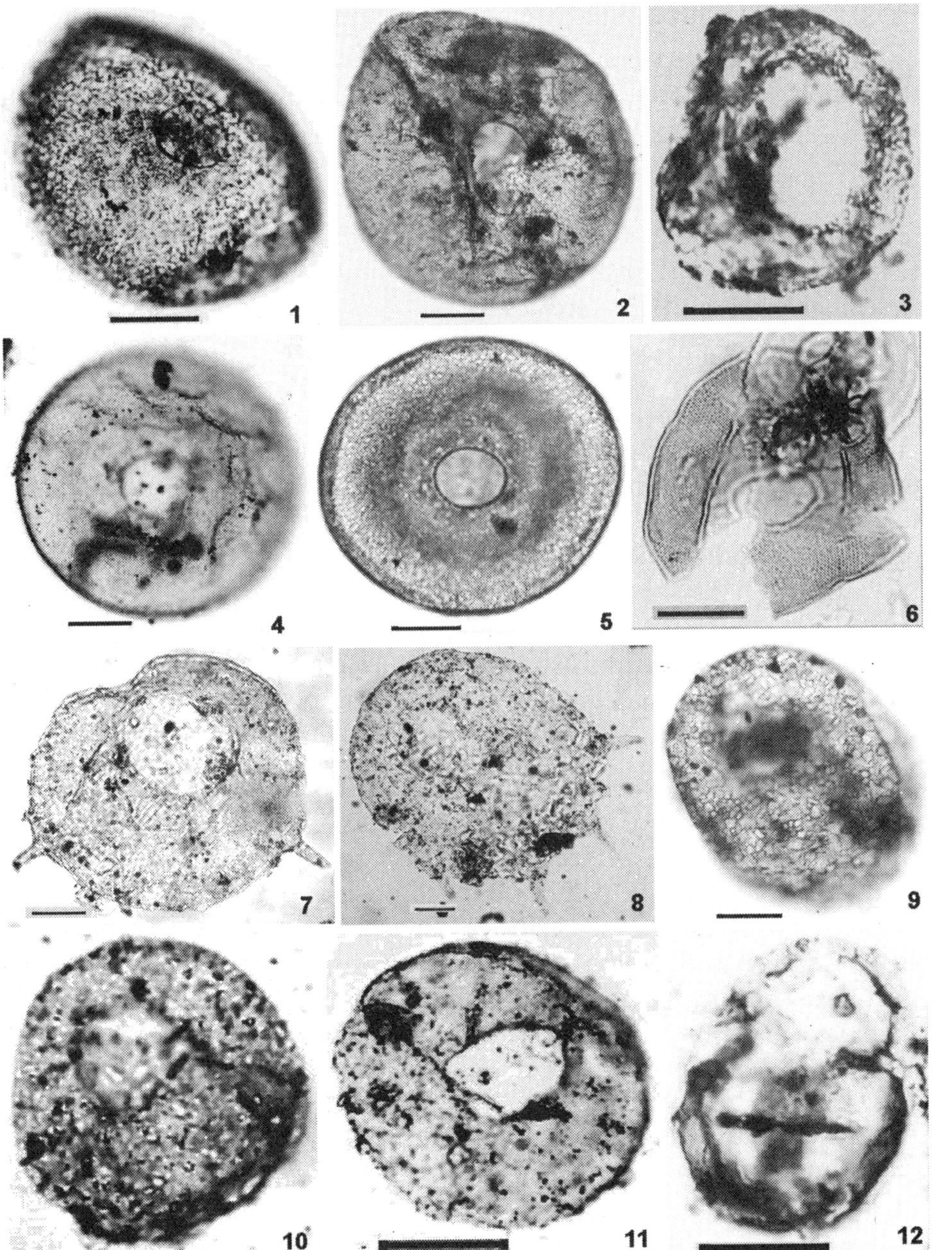

Plate-2: Light Microscopic photomicrographs recovered from Early Permian Khundi-Maral-Raula Section, Chamba District, Himanchal Pradesh.

1. *Arcella discoides*; 2. *A. arenaria*; 3. *A.megastoma*; 4. *A.gibbosa*; 5. *A. artocrea*; 6. *A. discoides*; 7. *Centropyxis aculeata* 'aculeata' ; 8. *C.aculeata* 'spinosa'; 9. *C. hirsuta*; 10 &11. C. *arcelloides*; 12 *C. aerophila*

Results and Discussion

The three non-pebbly units NP-1, NP-2 and NP-3 yielded following thecamoebians. Some of the forms are illustrated in photo plates-1 and 2.

NP-1

Low counts (< 10 specimens) of *Amphitrema flavum, Trinema/ Corythion, Difflugia ovata, D. gramen, D. protaeformis, Cyclopyxis kahlii, Arcella artocrea* and *A. discoides.*

NP-2

The overall thecamoebian count is relatively higher in this unit (10-20 specimens). Both quantitative and qualitative abundance of thecamoebians were recorded. The main forms identified are *Centropyxis aculeata, Arcella artocrea* and *D. protaeiformis*. Other species of *Difflugia* and *Arcella* also occur.

NP-3

Most common forms are *Cyclopyxis kahlii* and *Amphitrema flavum.* Centropyxids, *Cucurbitella* and *Difflugia* also occur.

Oldest geological record of thecamoebians goes back to Neoproterozoic (Porter and Knoll (2000); Porter *et al.* (2003); Smith *et al.* (2008) and Cambrian (Scott *et al.* 2003). Their records are very few and scattered between Neoproterozoic and Tertiary (van Hengstum *et al.*, 2007). Although thecamoebians are well known from Quaternary- Holocene lacustrine environments and peat bogs especially from Europe and North America, the present report on the early Permian thecamoebian assemblage from the Manjir Formation of Himalaya is very significant in filling the information gap on Proterozoic records of thecamoebians. A detailed work is in progress.

Acknowledgements

A. Farooqui and N. Jha thank the Director, Birbal Sahni Institute of Palaeobotany, Lucknow, India, A. Kumar thanks King Fahd University of Petroleum and Minerals, Dhahran, Saudi Arabia, and A.C. Pande and D. D. Bhattacharya thank the DDG, Northern Region, Geological Survey of India, Lucknow for permission to publish this paper. We are grateful to Drs. Prabhat Kumar, Zoology Department, University of Lucknow, Lucknow and David Scott, Dalhousie University, Canada for their valuable suggestions.

References

Bassi, D. E., Fugagnoli, A., Posenato, R. and Scott, D. B. (2008) Testate amoebae from the Early Jurassic of the western Tethys, northeast Italy. Palaeontology, v. 51(6), pp. 1335-1339.

Beyens, L. and Meisterfeld, R. (2001) Protozoa: Testate Amoebae. In: *J. P. Smol, H.J.B., Birks and W.M. Last, (eds.)* Tracking Environmental Change Using Lake Sediments. Vol. 3: Terrestrial, Algal and Siliceous Indicators. Kluwer Acad. Publ., Dordrecht, The Netherlands, v. 3, pp. 121-153.

Boudreau, R.E.A., Galloway, J., Patterson, R. E., Kumar, A. and Michel, F.A. (2005) A paleolimnological record of Holocene climate and environmental change in the Temagami region, northeastern Ontario. J. Paleolimnology, v. 33, pp. 445-461.

Bradley, W. H. (1931) Origin and the microfossils of the oil shale of the Green River Formation of Colorado and Utah. United States Geological Survey Professional Paper 168, 58 p.

Carter, H. J. (1856) Notes on the fresh water Infusoria of the Island of Bombay. No.1. Organization. Annals and Magazine of Natural History, series 2, 18 (105), pp. 221-249.

Carter, H. J. (1864) On fresh water Rhizopoda of England and India. Annals and Magazine of Natural History, series 3, 13, pp. 18-39.

Charman, D., Hendon, D. and Woodland, W. (2000) The identification of peatland testate amoebae. Quaternary Research Association Technical Guide 9, 147 p.

Charman, D., Roe, H. M. and Gehrels, W. R. (2002) Modern distribution of saltmarsh testate amoebae: regional variability of zonation and response to environmental variables. J. Quaternary Science, v. 17(5–6), pp. 387–409.

Dalby, A.P., Kumar, A., Moore, J. M. and Patterson, R. T. (2000) Preliminary survey of Arcellaceans (Thecamoebians) as limnological indicators in tropical Lake Sentani, Irian Jaya, Indonesia. J. Foraminiferal Research, v. 30, pp. 135-142.

Farooqui, A. and Gaur, A. S. (2007) Arcellaceans and pollen/spores of a late Harappan settlement near Porbandar, west coast of India: implications for palaeoecology and environmental monitoring. Current Science, v. 92(7), pp. 992-998.

Frenguelli, G. (1933). Tecamebiani e Biatomee nel Miocene del Neuquen (Patagonia Settentrionale). *Bolletino della Società Geologica Italiana*, v. 52, pp. 33-43.

Hoogenraad, H. R. and Groot, A. A. (1940) Moosbewhnende thekamoebe rhizopoden von Java und Sumatra: Treubia, v. 17, pp. 209–256.

Hoogenraad, H. R. and Groot, A. A. (1946) Thekamoebe sphagnumrhizopoden van Buitenzorg (Java): Biologische Jaarboek, Dodonea (Gent), v. 13, pp. 112–126.

Kumar, A. (ms) Acid resistant Cretaceous Thecamoebians from the Arabian Peninsula: suggestion for palynological study of agglutinated rhizopods. J. Micropalaeontology (revised and submitted).

Kumar, A. and Dalby, A. P. (1998) Identification key for Holocene lacustrine Arcellacean (thecamoebian) taxa. Palaeontologia Electronica, v.1, pp. 1-39

http://palaeo-electronica.org/1998_1/dalby/issue1.htm

Kumar, A. and Patterson, R. T. (2002) Dinoflagellate cyst assemblages from Effingham Inlet, Vancouver Island, British Columbia, Canada. Paleogeography, Plaeoclimatology, Paleoecology, v. 180, pp. 187-206.

Medioli, F.S. and Scott, D.B. (1983) Holocene Arcellacea (Thecamoebians) from Eastern Canada: Cushman Foundation for Foraminiferal Research, Special Publication no. 21, pp. 1-63.

Medioli, F.S. and Scott, D.B. (1988) Lacustrine thecamoebians (mainly Arcellaceans) as potential tools for palaeolimnological interpretations. Paleogeography, Paleoeclimatology, Paleoecology, v. 62, pp. 361-386.

Medioli, F.S., Scott, D.B., Collins, E., Asioli, A. and Reinhardt, A. (1999) The thecamoebian bibliography. Palaeontologia Electronica, v.3, pp. 1-161.

http://palaeo-electronica.org/1999_1/biblio/issue1_99.htm.

Medioli, F.S., Scott, D.B., Collins, E.S. and Wall, J.H. (1990a) Thecamoebians from the Early Cretaceous deposits of Ruby Creek, Alberta (Canada). In: *C. Hemleben, M.A. Kaminski, W. Kuhnt and D.B. Scott (eds.)* Proceedings of the NATO Advanced Study Institute on Paleoecology, biostratigraphy, paleoceanography and taxonomy of agglutinated foraminifera. NATO ASI Series. Series C: Mathematical and Physical Sciences. 327: 793-812, D. Reidel Publishing Company. Dordrecht-Boston, International.

Medioli, F.S., Scott, D.B., Collins, E.S. and Wall, J.H. (1990b) Fossil thecamoebians: present status and prospects for the future. In: *C. Hemleben, M.A. Kaminski, W. Kuhnt, and D.B. Scott (eds.)*, Proceedings of the NATO Advanced Study Institute on Paleoecology, biostratigraphy, paleoceanography and taxonomy of agglutinated foraminifera. NATO ASI Series. Series D: Mathematical and Physical Sciences. 327, pp. 813-839, D. Reidel Publishing Company. Dordrecht-Boston, International.

Pande, A. C., Bhattacharya, D. D., Jha, N., Misra, R. S. and Chandra, S. (2004) Discovery of Early Permian palynomorphs from the Manjir Formation, Chamba District, Himachal Pradesh. J. Geol. Soc. Ind., v. 63, pp. 665-669.

Patterson, R. T. and Kumar, A. (2000) Use of Arcellacea (Thecamoebians) to gauge levels of contamination and remediation in industrially polluted Lakes. Environmental Micropaleontology, v. 15, pp. 257-278.

Patterson, R.T. and Kumar, A. (2002) A review of current testate rhizopod (thecamoebian) research in Canada. Palaeogeography, Palaeoclimatology, Palaeoecology, v. 180, pp. 225-251.

Porter, S. A. and Knoll, A. (2000) Testate amoeba in the Neoproterozoic Era: evidence from vase-shaped microfossils in the Chuar Group, Grand Canyon. Paleobiology, v. 26 (3), pp. 360-385.

Porter, S., Meisterfeld, R. and Knoll, A. (2003) Vase-shaped microfossils from the Neoproterozoic Char Group, Grand Canyon: a classification guided by modern testate amoebae. J. Paleontology, v. 77(3), pp. 409-429.

Rattan, S. S. (1973) Stratigraphy and sedimentation of the Chamba area, western Himachal Pradesh. Himalayan Geology, v. 3, pp. 231-248.

Roe, H. and Patterson, R. T. (2006) Distribution of Thecamoebians (Testate Amoebae) in small lakes and ponds, Barbados, West Indies. J. Foraminiferal Research, v. 36(2), pp. 116-134.

Roe, H. M., Charman, D. and Gehrels, W. R. (2002) Fossil testate amoebae in coastal deposits in the UK: implications for studies of sea-level change. J. Quaternary Science, v. 17 (5–6), pp. 411–429.

Schmidt, A. R., Schönborn, W. and Schäfer, U. (2004) Diverse fossil amoebae in German Mesozoic amber. Palaeontology, v. 47(2), pp. 185-107.

Schönborn, W., Dörfelt, H., Foissner, W., Krienitz, L. and Schäfer, U. (1999) A fossilized microcenosis in Triassic amber. J. Eucaryotic Microbiology, v. 46(6), 571-584.

Scott, D.B, Medioli, F.S. and Braund, R. (2003) Foraminifera from the Cambrian of Nova Scotia: the Oldest Multichambered Foraminifera. Micropaleontology, v. 49, pp. 109-126.

Smith, H. G., Bobrov, A. and Lara, E. (2008) Diversity and biogeography of testate amoebae. Biodiversity Conservation, v. 17, pp. 329–343.

Srivastava, S. C. and Bhattacharya, A. P. (1998) Early Permian microfossils in faunal coal balls from Arunachal Pradesh, India– phytogeographic and palaeoenvironmental significance. Geophytology, v. 26 (2), pp. 75-82.

Sudzuki, M. (1979) Marine interstitial testacea from Plau Pinang, Malaysia. Anot. Zool. Japan, v. 52, pp. 50-53.

van Hengstum, P. J., Reinhardt, E. G., Medioli, F. S. and Gröcke, D. R. (2007) Exceptionally preserved late Albian (Cretaceous) Arcellaceans (Thecamoebians) from the Dakota Formation near Lincoln, Nebraska, USA. J. Foraminiferal Research, v. 37(4), pp. 300-3008.

Van Oye, P. (1949) Rhizopodes de Java: Bijdragen tot de Dierkunde, v. 28, pp. 327–352.

Vasicek, M. and Ruzicka, B. (1957) Namurian Techamoebina from the Ostraxa-Karxina coal district. Sbornik Naradniho Museav Praze, Reda B, Prirodni Vedy Acta Musei Nationalis Pragae. Series B, Historia Naturalis, v. 13, pp. 333-340.

Wightman, W.G., Scott, D.B., Medioli, F.S. and Gibling, M.R. (1994) Agglutinated foraminifera and thecamoebians from the Late Carboniferous Sydney Coalfield, Nova Scotia: Paleoecology, Paleoenvironments and Paleogeographical implications. In: *J.H. Calder and M.R. Gibling (eds.)* The Euroamerican coal province; controls on tropical peat accumulation in the Paleozoic. Palaeogeography, Palaeoclimatology, Palaeoecology, v. 106, pp.187-202.

Woodland, W. A., Charman D. J. and Sims, P. C. (1998) Quantitative estimates of water tables and soil moisture in Holocene peatlands from testate amoebae. The Holocene, v. 8, pp. 261-273.

❑❑❑

3

Biogenic Sedimentary Structures Produced by *Architectonica laevigata* (Gastropod : Architectonicidae), Mandvi Intertidal Zone, Gulf of Kachchh, Western India

BHAWANISINGH G DESAI
School of Petroleum Technology, Pandit Deendayal Petroleum University, Raisan Village, Gandhinagar-382009

Abstract

Biogenic structures of *Architectonica laevigata* (Gastropod: Architectonicidae) were studied along the lagoon of the Modwa spit area, Mandvi, Gulf of Kachchh. The traces are bilobed in cross-section, forming complex meandering trails, while their burrows are closely packed small holes about 1-2 mm in diameter. The traces are produced during low tides and are concentrated in specific areas within the intertidal zone, and also show substrate preference. These traces and burrows are intimately associated with each other, which reflect their behaviour modification to avoid steep thermal gradients and desiccation stress during low tide conditions in the lagoon and intertidal area. Similar traces are comparable with the ichnogenera *Gyrochorte* and *Scolicia,* which are abundant from Precambrian times to recent. Such traces, if preserved, serve as a good indicator for paleo-environmental conditions.

Keywords : Biogenic structures, Gastropod, Architectonicidae, traces, burrows, Gulf of Kachchh.

Introduction

The intertidal zone is considered to be a "stressed" area (Vermiji, 1978), and those organisms dwelling in this zone are considered to be "stress tolerant species" (*sensu stricto* Vermeij, 1978). The organisms often undergo intense environmental and physiological stress, leading to behavioural switching (Faulkes, 2005) and behaviour modification. The stress factors include the thermal gradient and desiccation, along with competition. The behaviour of any organism is reflected in the biogenic sedimentary structures produced by them within the sediment or on surface. Biogenic sedimentary structures are the dominant features of modern as well as ancient soft-sediment intertidal environments (Patel and Desai, 2009). In the ancient rock record, morphological patterns preserved in the form of traces, burrows, and borings are good indicators of the behavioural strategy adopted by organisms during changing conditions (Desai, 2003). Biogenic structures made by intertidal gastropods are particularly interesting as they give insights into the behaviour of stress-tolerant species that are adapted to behavioural switching. Some of these structures are rarely preserved as they are made at the sediment-water interface and are commonly destroyed by erosion.

The soft-sediment intertidal zone of the Mandvi area, Western India (Fig. 1) registers varied behavioural activities and their resultant traces produced by the gastropod *Architectonica laevigata* Lamarck, 1816 (Architectonicidae). *A. laevigata* is a common intertidal gastropod of the Gulf of Kachchh (Hornell, 1909; Gopalkrishnan, 1970), having broad and conical shells that vary in size from minute to 50 mm or larger, with a varying umbilicus width. The protoconch of the Architectonicidae grows upside down, with the umbilicus facing upwards and hence appearing to be coiled sinistrally, although the animal is organized dextrally (Bieler, 1993).

A. Laevigata shows a great variety of behavioural adaptation at the sediment-water interface, which includes substrate-specific locomotion, dwelling and feeding activities (Desai, 2003). Two basic interrelated behavioral activities are studied, namely burrowing and locomotion. Both activities are commonly observed in modern environments and are governed by stress acting in the intertidal zone during tidal fluctuations. Like other intertidal organisms, *A. Laevigata* takes refuge in and makes locomotion excursions into the surficial sediments during low tide.

Area of Investigation

The Mandvi area (Fig. 1a) has a semi-diurnal tide-affected coastline situated at the southernmost tip of the Kachchh Peninsula. The intertidal zone here has a soft substrate with sediments ranging from medium sand to fine silt along with different proportions of clay and mud. Geomorphologically, there is a dominance of ridge-and-runnel systems (RRS) in the area (Patel *et al.,* 2001). The western part of the study area is dominated by RRS, giving way to a tidal flat environment towards the east. It also comprises two small intertidal lagoons, one situated at Rawal Pir and the other at Modwa. The present study was carried out in the Modwa spit lagoon, which is shallow and elongated in nature and connected to the open sea towards the southeast. During low tide, it partially drains, leaving about 50% of the lagoonal area exposed. The remaining 50% of the lagoon area is comprised of thick algal mats. The sediments are fine-grained, with a peat layer 5-7 cm below the sediment-water interface. This zone is also inhabited by other benthic organisms such as crustaceans, polychaetes, bivalves and nemertea along with the abundant gastropod *Architectonica laevigata.*

Observed Habitat of *Architectonica*

Architectonica is found in a soft substrate of fine sand in the aerobic environment, 1-2 mm below the sediment-water interface (Fig.1b). It communicates with the external environment through a pair of highly active siphons by which it performs all vital physiological activities. It extends a siphon from the aperture upward towards the sediment-water interface, which creates a definite pattern on the surface. Miller (1974) suggested that gastropods living in sandy substrates have diverse locomotion modes and possess a long foot, often moving in a discontinuous way with the foot totally or partially buried.

The gastropod communities comprise about 500-600 individuals per m^2 area. The life position of each individual inside the burrow is with its long axis parallel to the bedding and its aperture nearly vertical. It was observed that the animal excavates a tunnel by its slender foot in an oblique manner, thus the opening is located on the surface slightly away from the shell and linked to the former by a tiny inclined tunnel 5-8 mm in length. Field observations during various tides on the Modwa spit suggest that the gastropods do not leave the burrow during high tides, unless they are eroded and relocated elsewhere by stronger tides. However, during receding tides they do come to the surface and move about, thus constructing locomotion traces. Several trails with blunt, burrowed ends were observed in the intertidal zone just after the area became exposed during low tide. These traces were abundant (up to

25 individuals per m²) and of varied lengths, often crossing each other. However, according to the density of individual gastropods, less than 0.05% actually produces traces. The production of traces thus seems to be the exception rather than the rule.

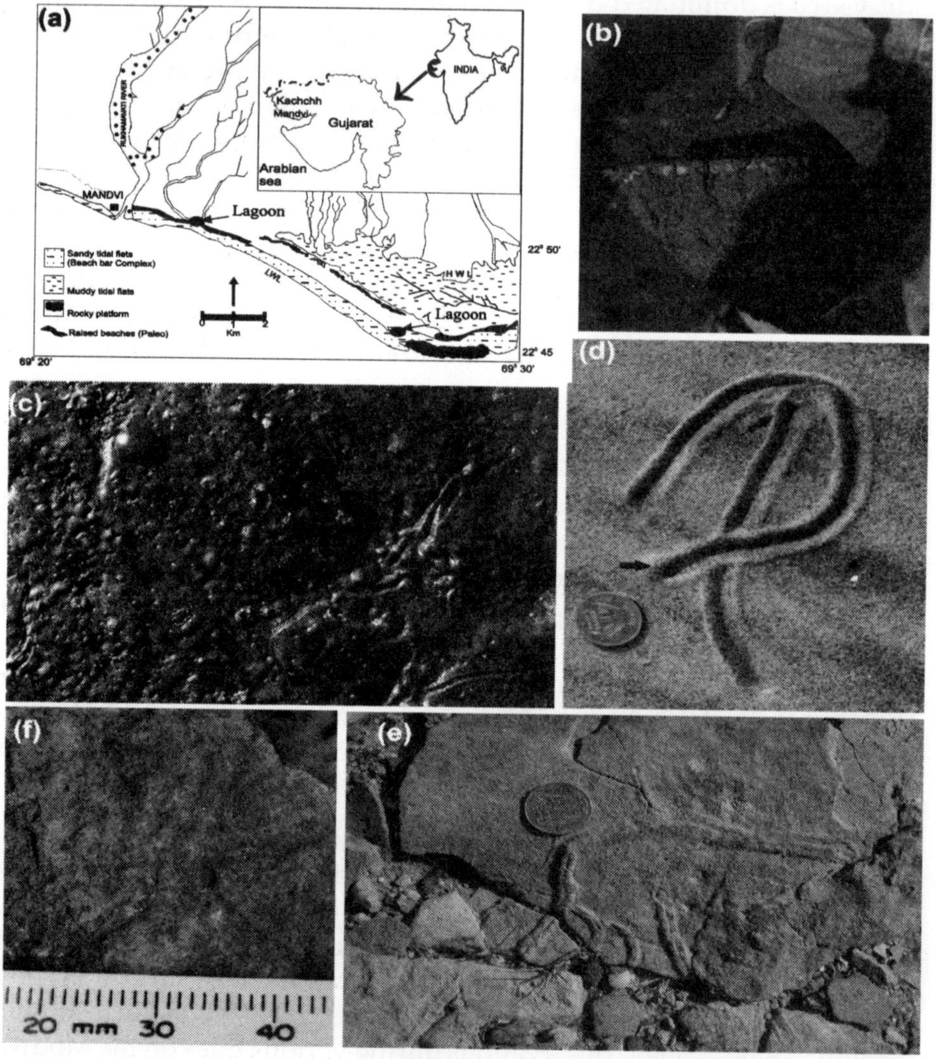

Fig. 1: (a) Location map of the Mandvi intertidal zone; (b) *Architectonica laevigata* in life position; (c) Burrows of *A. Laevigata*. Red cap is 15 mm in diameter; (d) V-shaped locomotion trace of gastropod during low tide. Arrow indicates raised rim of temporary burrows. Coin is 24 mm in diameter; (e) Trace fossil Gyrochorte from the Jurassic (Jhuran Formation) of the Jara Dome in Kachchh. Coin is 24 mm in diameter; (f) trace fossil, cf *Archaeonassa*- a bilobed trace from the early Cambrian Nigalidhar section of Himalayas.

Morphology of Trails and Burrows

Trails

These gastropods produce generally smooth, bilobed trails (Fig. 1d), with a prominent V-shaped morphology in cross-section. As they move about on the sediment-water interface, loose sediments along the lateral margins of the trail collapse outward after their passage. The resultant morphology of the trails is shallow (<1 cm), narrow and bilobed without any transverse ridges. They show rims of about 3-4 mm thick, with a total width of about 16-18 mm, the latter remaining constant throughout the trail. The trails show a low order of meandering and are often found criss-crossing each other. The end of each trail is marked with a semi-circular, rimmed depression. The substrate preferred for locomotion is slightly firm, medium-grained sediments. The depressions in the end of the trail are morphologically different from the burrows. These depressions may or may not be connected to the actual burrows. The orientation also differs significantly. In case of depression, the axis of orientation is coinsides with the trail and is at very small angle with respect to substrate.

Burrows

Cylindrical, unlined burrows, with circular cross section, showing surficial expression as closely packed small holes of about 1-2 mm diameter are often seen all along the lagoon of the Modwa spit area (Fig. 1c). Each hole is associated with a near-vertical, cylindrically shaped, straight burrow of nearly 1-3 mm in length. The end of the burrow coincides with the aperture of the shell of *Architectonica laevigata*.

Locally, the gastropods also make temporary inclined burrows close to the surface at the end of the trail, as discussed earlier. The surficial expression of the burrows is in the form of elevated semicircular mounds about 1-2 mm high. The substrate preferred for burrowing is loose, fine-grained sediments.

Discussions

The present study describes modern *Architectonica laevigata* traces formed in a lagoon/intertidal setup. From a paleo-ecological point of view, the narrow, V-shaped meandering trail, if preserved, would be classified as *Repichnia* (locomotion traces). However, if the burrows were to be preserved, it would resemble small *Skolithos*.

The surfacial trace made by *Architectonica laevigata* in the lagoon of the Mandvi area, if preserved in the rock record, may be identical to the ichnogenera *Gyrochorte* or *Archaeonassa* (Fig. 1e, f). *Gyrochorte* is a common and abundant trace fossil of Mesozoic deposits in Kachchh (Patel *et al.*, 2008). These traces, which show a low order of meandering and cross-over structures, are frequently observed in the Jurassic Jara Dome (Desai *et al.*, 2008) and can also be interpreted to have formed in an intertidal environment. Similarly the ichnogenus *Psammichnites,* a gastropod locomotion trace, is significant in Precambrian deposits of the Himalayas (Tal Group, Mussoorie area). *Psammichnites,* a Palaeozoic trace is also similar to the present form, but is interpreted to be constructed by hydraulically operated push-and-pull action of Gastropod (Seilacher, 2007). Fenton and Fenton (1937) also described similar bilobed trails, Ichnogenus *Archaeonassa* and their associated burrows from the recent sand and compared them with Cambrian occurrences of similar forms from British Columbia. Badve and Bhonsle (1987) studied similar burrowing habits of *Umbonium* in intertidal zones. However, in the case of *Umbonium* the burrows are inclined, whereas *Architectonica* burrows are nearly vertical.

Several studies have shown that the locomotion of intertidal dwelling gastropods is related to rhythmic tidal fluctuations, whereby these gastropods are adapted to rhythmic zonal migration synchronised with the tides (Naylor, 1987; Chelazzi *et al.,* 1987).

Architectonica is considered to be among the fastest burrowing gastropods, although its shell morphology does not reveal any burrowing behaviour (Savazzi, 1994). Two distinct types of activites were studied here, namely burrowing and locomotion. In the former activity, the nature of the burrows is in the form of short siphon tubes, suggesting that the gastropod is a suspension feeder. The close association of these gastropods suggests that they are non-aggressive and non-competitive, but still opportunistic.

Studies have suggested that in intertidal areas with extensive tidal oscillations, organisms are active during one phase and inactive during the other phase of tidal activity (Naylor, 1987). In the present study, burrowing/feeding activity is interpreted to be related to high tide, thus enabling the gastropods to suspension feed. During low tide the intense thermal gradient and desiccation might instigate the gastropods to undertake excursions, during which meandering locomotion traces are produced. The burrow at the end of the trace is interpreted to be a temporary resting burrow made in order to avoid the thermal gradient.

Conclusion

Architectonica, the fastest burrowing gastropod, inhabits intertidal stress zones. Their activities are governed by tidal fluctuations.

1. During high tides, burrows are stabilized and the animal does not come out of its burrow, being a suspension feeder.
2. The dense burrow systems indicate non-aggressive, non-competitive behaviour.
3. During low tides, the gastropods leave the burrow and undertake short random excursions.
4. Due to extreme exposure and to avoid desiccation, they make inclined temporary shelters at the end of the traces, which have raised rims.

Acknowledgement

Author is grateful to Director, School of Petroleum Technology for providing opportunity. Part of the work was done at M. S. University of Baroda during the doctoral work for which author is thankful to Dr. Satish J. Patel (MSU).

References

Badve, R. M and Bhonsle, S. P. (1987) Burrowing of gastropod *Umbonium* Link and its implication for ichnofossil study. Current Science, v. 56(22), pp. 1164-1166.

Bieler, R. (1993) Architectonicidae of the Indo-Pacific (Mollusca: Gastropoda). *Abhandlungen des Naturwissenschaftlichen Vereins in Hamburg* (G. Fischer Verlag; Stuttgart), v. 30, 376p.

Chelazzi, G., Focardi, S. and Deneubourg, J-L. (1987) Analysis of Movement patterns and orientation mechanisms in intertidal chitons and Gastropods. In: *G. Chelazzi, and M. Vannini*, (eds.) Behavioural adaptation to intertidal life, NATO ASI Series, Series A, Life science, v. 151, pp. 173-184.

Desai, B. G., Patel, S. J., Shukla, R. and Surve, D. (2008) Analysis of Ichnoguild and their significance in interpreting Ichnological events,: A study from Jhuran formation (Upper Jurassic), Western India. J. Geol. Soc. Ind., v. 72, pp. 458-466.

Desai, B.G. (2002) Animal-sediment relationship of the two benthic communities (crustaceans and polychaetes) in the intertidal zone around Mandvi, Gulf of Kachchh, Western India. Ph.D. Thesis, M.S. University of Baroda, Vadodara, 231p.

Faulkes, Z. (2005) Mechanisms of Behaviour switching. J. Comp Physiology A : Neuro, sens, neural and Behav. Physio., v. 191 (3), pp. 197-199.

Fenton, C. L. and Fenton, M. A. (1937) *Archaeonassa*: Cambrian snail trail and burrow. Amm. Mid. Natur., v. 18(3), pp. 454-456.

Gopalkrishnan, P. (1970) Some observation on the shore ecology of the Okha coast. J. Mar. Biol. Ass. Ind., v. 12, pp. 15-34.

Hornell, J. (1909) Report of the Government of Baroda on the marine zoology of Okhamandal. London.1

Miller, S. L. (1974a) Adaptive design of locomotion and foot form in prosobranch gastropods. J. Exp. Mar. Biol. Ecol., v. 14, pp. 99-156.

Miller, S. L. (1974b) The classification, taxonomic distribution, and evolution of locomotor types among prosobranch gastropods. Proc. Malac. Soc. Lond., v. 41, pp. 233-261.

Naylor, E. (1987) Clock-controlled behaviour in intertidal animals. In: *G. Chelazzi, and M. Vannini,* (eds.) Behavioural adaptation to intertidal life, NATO ASI Series, Series A, Life science, v. 151, pp. 1-14.

Patel, S. J. and Desai, B. G. (2009) Animal Sediment Relationship of the crustaceans and polychaetes in the intertidal zone around Mandvi, Gulf of Kachchh, Western India. J. Geol. Soc. Ind., v. 47, pp. 233-259.

Patel, S.J., Desai, B.G. and Bhatt, N.Y. (2001) Neotectonic evolution of the coastal landforms between Jakhau and Mundra, Gulf of Kachchh, Western India. Bull. Ind. Geol. Assoc., v. 34, pp. 221-232.

Patel. S.J., Desai, B.G., Vaidya, A.D. and Shukla, R. (2008) Middle Jurassic Trace Fossils from Habo Dome, Mainland Kachchh, Western India. Jour. Geol. Soc. India, v. 71, pp. 345- 362.

Savazzi, E. (1994) Adaptations to burrowing in a few recent gastropods. Historical Biology, v. 7(4), pp. 291-311.

Seilacher, A. (2007) Trace fossil Analysis. Springer, 226p.

Vermiji, G. J. (1978) Biogeography and Adaptation: Patterns of marine life. Haward University Press, Cambridge, 332p.

❑❑❑

4

Microstructures of the Zanskar Shear Zone

SOUMYAJIT MUKHERJEE
Department of Earth Sciences
Indian Institute of Technology Bombay, Mumbai-400 076

Abstract

Thin-section studies of the rocks of the Zanskar Shear Zone led to document a top-to-SW ductile shearing; a top-to-NE shearing and finally a top-to-NE (down) shearing. Shape asymmetry and cleavage orientations of mineral fish reliably give the latter two shear senses. Although some of the fish reveal different phases of shearing, their specific orientations indicates simple shear. Type-1 flanking microstructures indicate the earliest top-to-SW shear. Boudins denote extension along the main foliation. Intrafolial folds demonstrate a top-to-NE shearing. A top-to-SW brittle shear is displayed by asymmetric trapezoid-shaped minerals. The brittle planes cut across rigid minerals. As in other shear zones, micas are found to be most susceptible to deformation.

Keywords : Microstructure, Zanskar Shear Zone, Ductile shearing, Deformation, Himalaya

Introduction

The Zanskar Shear Zone (ZSZ), the northern boundary of the Higher Himalaya in Kashmir, comprises mainly of mylonitized gneisses, leucogranites and migmatites (Fig. 1) where the grade of metamorphism varies from lower greenschist to upper amphibolite facies (Herren, 1987). Note that in Fig. 1, the term 'High Himalaya' of Searle *et al.* (1988) is replaced by 'Higher Himalayan Shear Zone'. The granitic rocks of the area belong to Miocene Period (reviews by Godin *et al.*, 2006; and Yin, 2006). Structural geology and tectonics of this shear zone have been discussed in detail and updated by a number of workers, *e.g.* Herren (1987), Patel *et al.* (1993), Steck *et al.* (1993), Dèzes (1999), Dèzes *et al.* (1999), Walker *et al.* (1999) and Mukherjee and Koyi (in press, 1). The ZSZ characterizes a normal way up isograds (see review in Walker *et al.*, 2001). Dèzes *et al.* (1999) estimated the vertical displacement (minimum amount) and the net slip of the ZSZ to be 12 ± 3 km and 35 ± 9 km, respectively. They also deduced the thickness of the shear zone to be 1 km and observed that the leucogranite dykes are sheared parallel to the foliation planes. Walker *et al.* (1999) estimated the 'total slip' (net slip?) to be 40-60 km. Foliation planes dip 20-30^0, the outcrop width of the ZSZ is about 5 km (Walker *et al.*, 1999) and the shear zone can be traced along the Himalayan trend for about 80 km (Herren, 1987). Herren (1987) and Walker *et al.* (1999) estimated the throw of the ZSZ to be ~ 15-20 km. Herren (1987) noted a top-to-NE sense of extensional ductile shearing from the ZSZ. Subsequent work by Patel *et al.* (1993) and Dèzes *et al.* (1999) revealed a relict top-to-SW shear sense too. Mukherjee and Koyi (in press, 1) recently documented three phases of ductile deformation in micro-scale on a plane perpendicular to the northeasterly dipping main foliation and parallel to the northeasterly plunging stretching lineation. These three phases are (i) a top-to-SW sense of compressional shearing ('shearing-1'), (ii) subsequent top-to-NE sense of extensional shearing ('shearing-2'), and (iii) a latest top-to-NE (down) sense of extensional ductile shearing ('shearing-3'). Out of these, shearing-1 and 3 were first documented from field by Patel *et al.* (1993) and subsequently by Dèzes (1999) and Dèzes *et al.* (1999). In addition, Mukherjee and Koyi (op. cit.) showed (i) different varieties of micro-boudins parallel to the main foliation indicating a brittle-ductile extension parallel to the foliation, (ii) symmetric duplexes of high-grade minerals not indicative of any brittle shear sense, (iii) asymmetric duplexes of low-grade minerals indicating a top-to-SW sense of brittle shearing, (iv) northeasterly dipping sharp brittle normal micro-faults, (v) transgranular fractures (or T-fractures) that cut the primary shear planes at high angles, and (v) extensive grain boundary migrations. The angular relations measured under microscope by Mukherjee and Koyi (in press, 1) is summarized in Table-1.

Walker *et al.* (2001) considered the timing of extrusion of the HHSZ in the Zanskar section to be 21.5-19.5 Ma. On the other hand, compiling existing geochronological data, Mukherjee and Koyi (in press, 1)

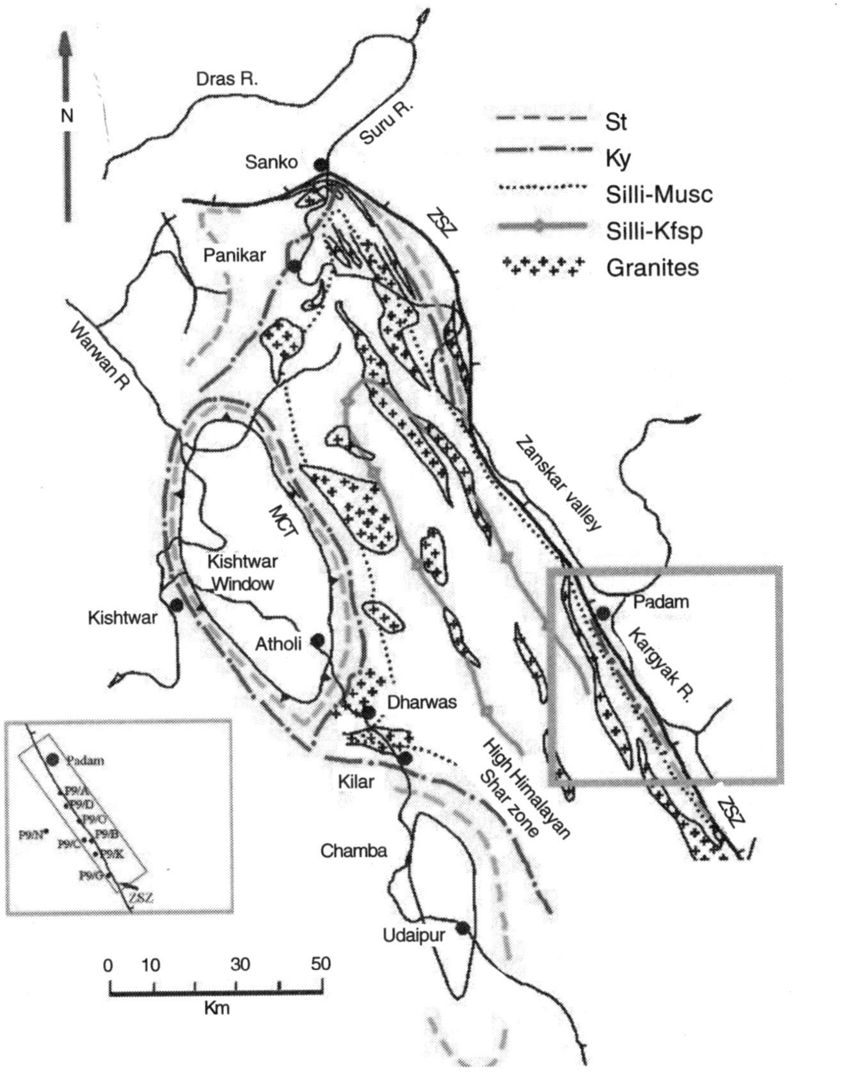

Fig. 1 a: Geological map of the Higher Himalayan Shear Zone in the Zanskar–Chamba–Kulu area, reproduced from Searle *et al.* (1988). Isograds are shown. The studied thin-sections are from rocks within the Zanskar Shear Zone SE to the locality Padam. **b.** Locations of rock samples SE to Padam.

interpreted the top-to-SW sense of shearing in the ZSZ to be a ~22-16 Ma event and as a delayed product of India-Eurasia collision. The extensional ductile shearing was interpreted to be of a shorter span of ~18-16 Ma. During ~18-16 Ma was in fact the span of combined activation of top-to-SW and –NE shearing. Brittle fractures were considered as perturbing events to the southwesterly brittle shearing. Recently channel flow model (Beaumont *et al.*, 2001) has evolved as an almost unanimously accepted extrusion mechanism of the HHSZ (see review in Mukherjee, 2005). Mukherjee and Koyi (in press, 1) applied a combined simple shear and channel flow model of extrusion of the HHSZ in the Zanskar section, which is an elaboration of Grujic *et al.'s* (2002) model proposed for the Bhutan Himalaya. This work aims at detail micro-structural studies of the ZSZ rocks.

Table-1: Micro-structural data of the ZSZ as compiled from Mukherjee (2007) and Mukherjee and Koyi (in press, 1). Symbols: top-to-NE (down) ductile shearing: 'shearing-3', top-to-NE ductile shearing: 'shearing-2', top-to-SW ductile shearing: 'shearing-1'. All measurements were done on the XZ thin-sections.

S. No.	Shearing-3		Shearing-2		Shearing-1
	Angle between S & C planes	Angle between axial trace & C-plane	Angle between axial trace & C plane	Angle between S &C planes	Angle between S &C planes
1-5	47^0	70^0	13 to 21^0	30^0	4 to 8^0
6	Mineral fish with aspect ratios: 2 to 4.8				
7	Mineral fish: angles between long axes & ductile shear plane: 0-32^0				
8	Angle between ductile shear planes of shearing-1 and shearing-3: 25-53^0				
9	Primary shear planes for shearing-1 and -2 coincide				
10	Angles between P- and Y-brittle shear planes vary up to 35^0				
11	The primary brittle shear plane (Y-plane) and the primary shear planes of shearing-1 and -2 coincide				
12	Angle between Y- and secondary synthetic brittle shear R-plane ~ 15^0				
13	Angle between T-fractures and Y-plane: 69-86^0				
14	Angle of V in V-pull-aparts: 10-27^0				
15	Aspect ratios of boudins: 3.3-4.2				

Microstructures

Mineral fish

Mineral fish are lozenge-shaped porphyroclasts and single crystals in fine-grained matrices in mylonitized rocks (ten Grotenhuis *et al.* 2002; also see Mukherjee and Pal, 2000). Mineral fish (Figs. 2a-c; 3a-d; 4a-d; 6c; 8d; 12b) are the most common ductile shear sense indicators in the ZSZ in micro-scale. In the studied thin-sections, quartz, muscovites, biotites, feldspars, garnets and staurolites show the characteristic fish shape. Unlike Passchier and Trough's (2005) view that quartz fish are rare in shear zones, they are rather abundant in the ZSZ (Figs. 3c, 4a, 4d, 6c). Where the mineral fish touch the shear planes that bound them, the shear strain is maximum and equidistant from a pair of C-planes the shear strain is minimum. The C-planes are straight and envelope (or, are tangent to) the mineral fish. These planes are defined by very fine-grained recrystallized minerals. The C-planes seem to be irregularly spaced (Figs. 2d, 3d). Three broad morphologies of mineral fish were observed- (i) sigmoid (Figs. 2a-c, 3a-d, 4a-d); (ii) lenticular (Fig.5c); and (iii) parallelogram (Figs. 7c, 9d). Sigmoid-shaped fabrics show progressive curving near the bounding C-planes. Either these three shapes were defined by single minerals (Figs. 2a-c, 3a, 4b, 4d, 5c, 6c, 8d) or as aggregate of minerals (Figs. 2d, 3b-d, 4a, 4c) that are usually micas. Two senses of ductile shearing are documented from mineral fish- a top-to-NE (Fig. 2a, 2d, 3a-d, 4a, 4c-d, 5d, 6c, 8a, 12b-d) *i.e.* just in the opposite direction to that of the previous top-to-SW sense, and a top-to-NE (down) (Figs. 2a-d, 3a, 4b, 6a, 6c-d, 12b-d). The cleavages of mineral fish are in all cases tilted to the primary shear plane same as that of the fish itself. Thus, the cleavage planes of mineral fish alone can act as reliable shear sense indicators.

Sometimes mineral fish are found to be affected by late brittle shearing (Fig. 6c). Individual minerals, in composites that define the fish, are not indicative of any sense of shearing. These aggregates can be considered as 'decussate texture' (Fig. 6b). For example, the possible shape of a parallelogram grain of mica could be a rectangle such as shown by dashed lines in Fig. 6b, and that grains 'p' and 'q' might have concealed parts of them giving rise to the parallelogram shape. In some cases rigid quartzo-feldspathic minerals are wrapped by incompetent micas that overall defines the fish shape (Figs. 2b, 3d, 4c, 12c). A very peculiar example of a train of muscovite fish nucleated inside a host of muscovite grain (Fig. 11c) was noted. The nucleated muscovite fish display the top-to-SW sense of shearing. In some cases, lenticular fish, presumably products of the prior top-to-SW sense of shearing underwent late brittle top-to-NE shearing (Fig. 11a). In other words, the latter shear sense although most of the times acted ductilely on minerals, at times it also deformed minerals in a brittle manner. Strongly swerved cleavages of mica are found to be in sympathy with the shearing of the clast (*e.g.* Fig. 3a). As in Fig. 2d, the ratio of thicknesses of shear zones formed in micro-scale are 14.3:6.3:1:13. On the other hand in Fig. 3d, the ratio of distances (white lines) between the pair of C-planes to that bounding the clast is 1:15. One of the C-planes could be remarkable straight than the other one. Minerals recrystallized near the C-planes of porphyroclasts define tiny shear zones. The ductile shear planes in parallel arrays could be defined by intense grain size reduction (Fig. 3d).

Many times the same mineral fish are affected by shearing of two phases *i.e.* a top-to-NE and a top-to-NE (down) (Figs. 2a-b). Near their tips, the fish may be kinked as also revealed from their sharply folded cleavage planes (Fig. 8d). In a rare example, the C-planes that bound the top-to-NE shear fabric were found to be non-parallel (Fig. 4a). Extensive migration of boundary of grains from matrix into fish might

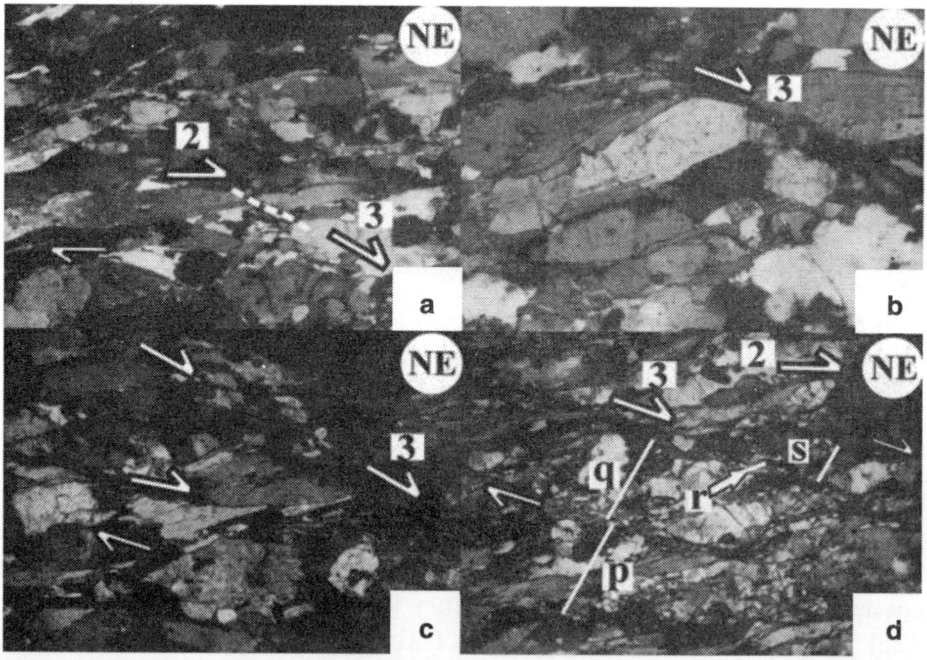

Fig. 2: a. Single sigmoid muscovite fish shows a top-to-NE shear (shearing-2). Top-to-NE (down) shear (shearing-3) brought one of the fish tips at gentler angle to the shear plane of '2'. (Cross-polarized light; photo width: 2 mm.) **b.** A single sigmoid alkali feldspar fish showing a top-to-NE (down) shear (shearing-3). The boundary of this grain is partly occupied by biotites. (Cross-polarized light; Photo width: 1 mm.) **c.** A top-to-NE (down) shear (shearing-3) is indicated by imperfect parallelogram and sigmoid shaped mica fish. (Cross-polarized light; Photo width: 1 mm.) **d.** A top-to-NE (down) shear (shearing-3) is indicated by an aggregate of porphyrclastic nearly sigmoid shaped alkali feldspars. (Cross-polarized light; Photo width: 2 mm.).

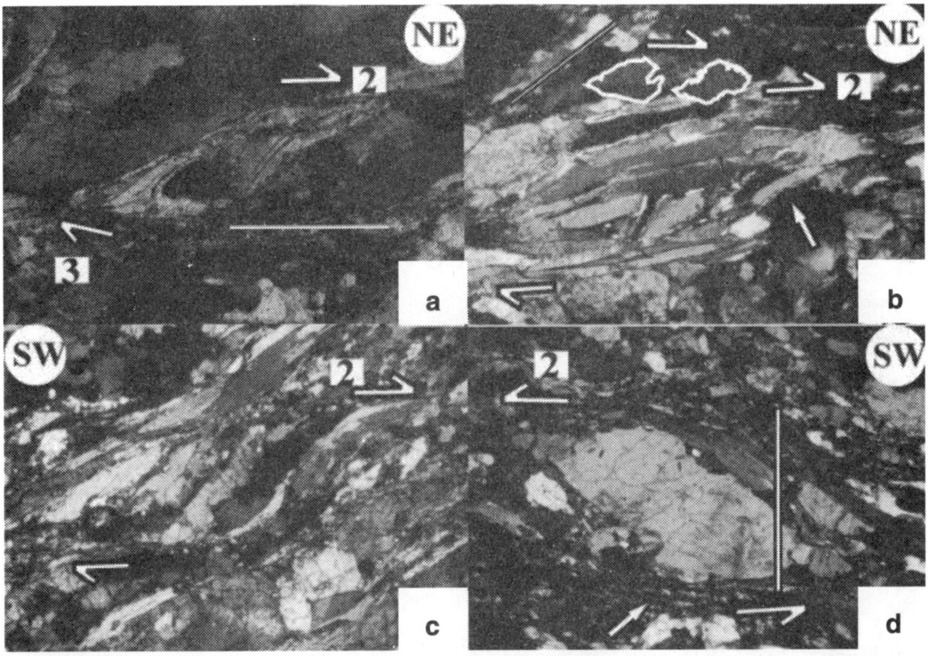

Fis. 3: a. A sigmoid mica fish affected by a top-to-NE (shearing-2) and top-to-NE (down) shear (shearing-3). Strongly swerved cleavages of mica is in sympathy with the shear. The C-plane is traced (white line). (Cross-polarized light; Photo width: 4 mm.). **b.** A composite of bent mica minerals of nearly rectangle shapes and with a haphazard orientation overall defines a lenticle shape and gives a top-to-NE (shearing-2) sense of shearing. (Cross-polarized light; Photo width: 4 mm.). **c.** A stack of mica and quartzo-feldspathic minerals showing a prominent top-to-NE (shearing-2) sense of shearing. (Cross-polarized light; Photo width: 4 mm.). **d.** A weakly sigmoid or a parallelogram shaped alkali feldspar shows a top-to-NE shear (shearing-2). Micas (right hand side) and finer matrix minerals (left hand side) of feldspars were sheared. (Cross-polarized light; Photo width: 4 mm.).

lead to loss of the fish shape almost beyond recognition. Inclusion of alkali feldspar inside sigmoid mica fish are also noted (Fig. 3a). Such more competent inclusions were found to be undeformed. Extensively recrystallized minerals in the matrix define the S-fabric of Bèrthe *et al.* (1979). Some of the mineral fish possess mouths at their corners (*e.g.* Fig. 4d) probably indicating boudinage from one fish to the adjacent one. Pronounced top-to-NE (down) sense of shearing lowered the local orientation of the mineral fish to the northeasterly dipping main foliation planes. Secondary shearing in some cases led to stair stepping of the fish tails. Pressure solution might be a reason of sigmoidality of some of the fish (*e.g.* Bell and Cuff 1989). A consistent northeast dipping orientation of the primary foliation plane that bound the mineral fish indicate that they certainly were the products of simple shearing and not by pure shear on clasts of significantly different viscosity than the matrix as proposed by Treagus and Lan (2000).

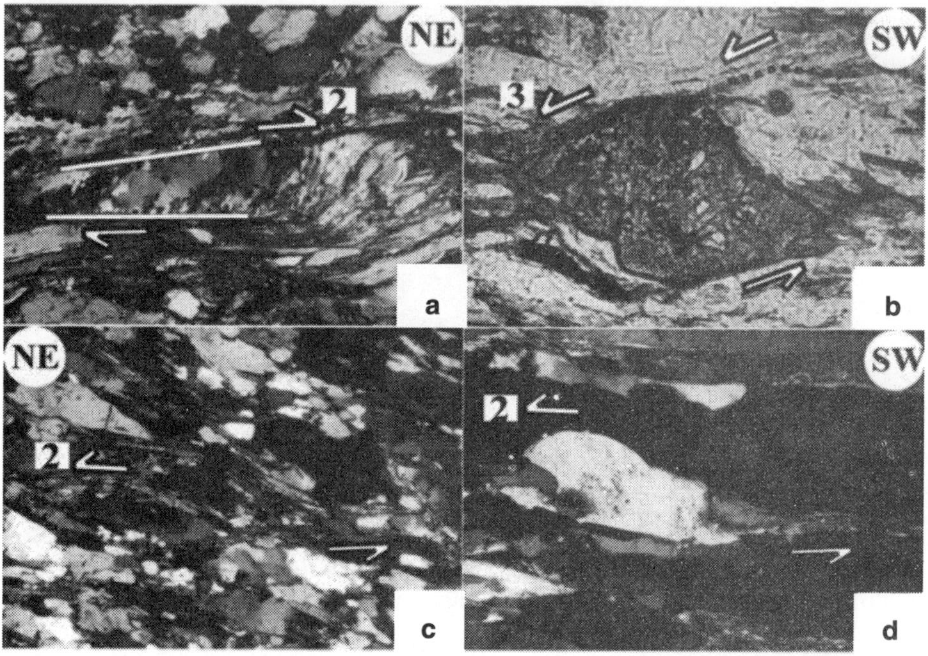

Fig.4: a. Sigmoid matrix define the S fabric: top-to-NE shear (shearing-2). The C-planes are non-parallel (white lines) and are at an angle of 8^0. Below the dashed blue line, quartz took part in shearing. (Cross-polarized light, Photo width: 4 mm.). **b.** A porphyroblastic garnet fish shows a top-to-NE (down) shear sense (shearing-3). Very fine, elongate, discrete biotites also define this shear (dashed blue line). (Plane polarized light; Photo width: 2 mm.). **c.** Discontinuous discrete mica warping larger matrix minerals define a sub-sigmoid/lenticular geometry. A top-to-NE shear ('shearing-2') is displayed. (Cross-polarized light. Photo width: 4 mm.). **d.** A single sigmoid quartz fish with a notch at one of the corners display a top-to-NE shear ('shearing-2'). (Cross-polarized light; Photo width: 2 mm.).

Flanking Microstructures

'Flanking structures' is deflection of planar or linear fabric elements in a rock alongside a cross-cutting object (Passchier, 2001). Following the same trend, Mukherjee and Koyi (2009) defined flanking microstructures in micro-scales where nucleated minerals defined the cross-cutting elements (CEs) and deflected cleavages and margins of

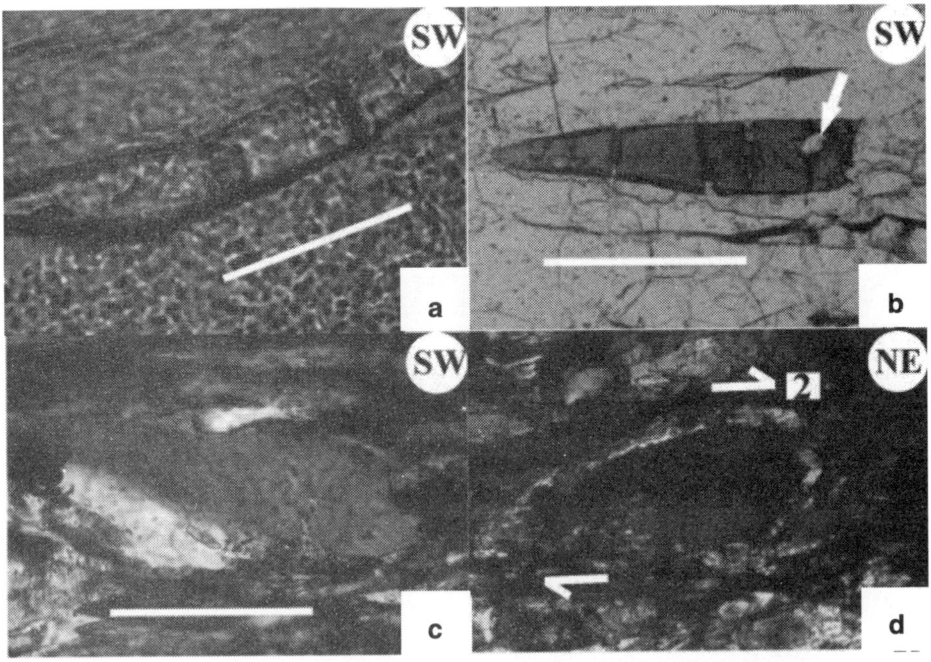

Fig. 5: a. A parallel pull-apart or a boudin of sillimanite grain. A white line shows the orientation of the main foliation deciphered from elsewhere in the thin-section. (Plane polarized light; Photo width: 1 mm.). **b.** A series of pull-apart structures within originally a single tourmaline grain inside quartzo-feldspathic matrix. (Plane polarized light; Photo width: 1 mm.). **c.** A rare spindle-shaped single grain of quartz indicating the tectonic force to be insufficient to completely break apart. (Cross-polarized light; Photo width: 1 mm.). **d.** A rootless intrafolial fold of quartz grain with thick hinge zone and thin unequal limbs dipping towards NE. A relict top-to-SW ('shering-1') is indicated. (Cross-polarized light; Photo width: 1 mm.).

the host grains the host fabric elements (HEs). The former elements cross-cuts the latter planes. The CE minerals that are parallelogram shaped and are inclined to the C-planes indicate ductile shearing and were categorized into the type-1 flanking microstructures. On the other hand, same sense of drag of the CE across the HE minerals indicate preferential growth of the latter and the HE-CE composite together was classified as the type-2 flanking microstructures. Both the type-1 (Fig. 7b-c, 8c, 9c, 10b-c) and the type-2 varieties (Fig. 7a) of these structures were observed in the studied thin-sections. The type-1 structures indicate that the host grain's cleavage planes efficiently acted as ductile C-planes (Fig. 7c). Few of the HE-CE contacts are marked with zones of haziness (Figs. 9b). Inclination and the usual sigmoid and parallelogram shapes of the HE of the type 1 structures indicated a top-to-SW sense of ductile shearing that Mukherjee and Koyi (in press, 1) adjudged to be a 18-16 Ma event. This indicates that the nucleations of those HE minerals were certainly no younger than 16 Ma. The rationale of deciphering the top-to-SW as the earliest deformation in the present micro-structural study is that it is least abundant and occurs as a relict.

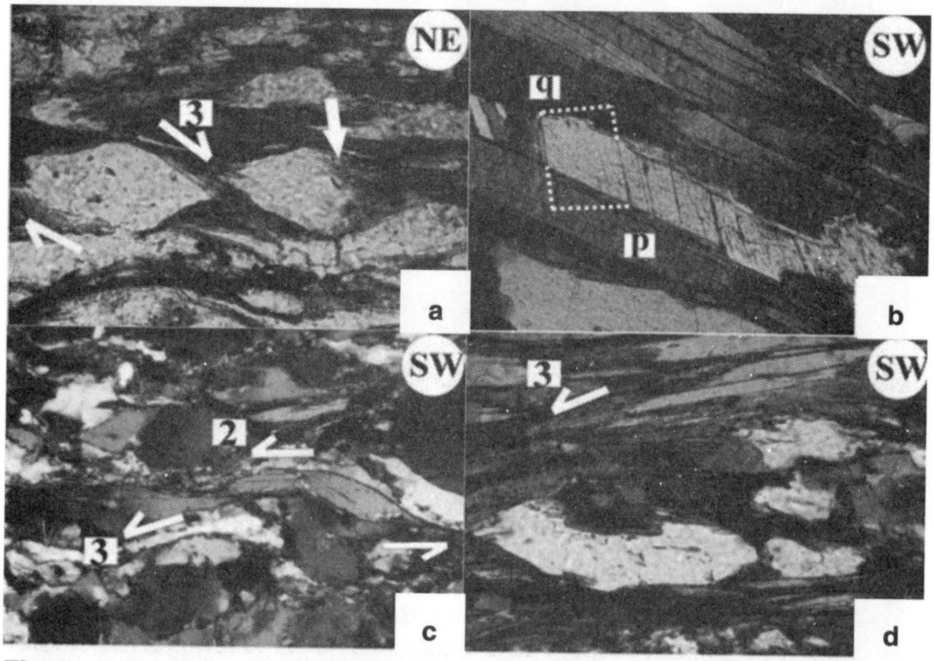

Fig. 6 : a. A pair of single nearly sigmoid quartz fish showing a top-to-NE (down) (shearing-3) sense of shearing. (Plane polarized light; Photo width: 2 mm.). **b.** Bounded by layers of micas, a mica grain assumes a parallelogram shape. (Cross-polarized light; Photo width: 1 mm.). **c.** A single sigmoid-shaped mica fish displays a top-to-NE shear ('shearing-2', white half arrows). Pronounced top-to-NE (down) shear of recrystallized quartz is noted. (Cross-polarized light; Photo width: 2 mm.). **d.** A top-to-NE (down) shear ('shearing-3') has affected foliation micas. (Plane polarized light; Photo width: 2 mm.).

Pull-apart structures and boudins

Pull-apart structures were originally described in micro-scale by Hippereitt (1993) but were latter found also in hand specimen scale by Mukherjee (in press). Singh (1999) provided a comprehensive review on pull-aparts. Mukherjee and Chakraborty (2007) reported the passive folding of foliations inside the opening of the pull-apart minerals in the ZSZ to be of class 3 folds. In the present study, high-grade metamorphic minerals displayed sets of parallel pull-apart (nearly same as boudinage) over a number of parallel zones of breakage (Figs. 5a-b, 11b). Quartz at the V opening is recrystallized (arrow in Fig. 5b). These breakage lines defining the openings are at high angle to the main foliations. At very high magnification, the breakage is found to have taken place along curved planes (Fig. 11b). Further, spindle-shaped isolated boundins of rigid minerals such as quartz are noted. The long axes of the boudins are parallel to the main foliation. These indicate intense brittle-ductile extension parallel to the foliation plane.

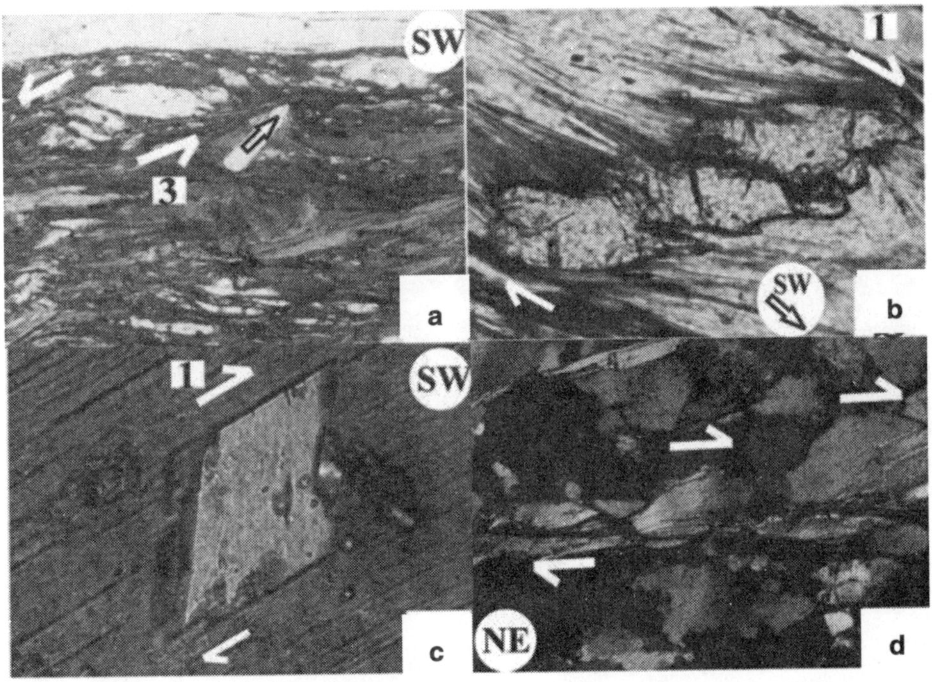

Fig. 7: a. A top-to-NE (down) shear ('shearing-3') is displayed by sigmoid alkali feldspars. A type 2 flanking micro-structure is defined by an elongated grain of alkali feldspar (CE). (Plane polarized light; Photo width: 2 mm.). **b.** A type-1 flanking microstructure defined by nearly sigmoid staurolite as the Host fabric element (HE) and deflected biotites with opposite senses of drag across the HE as the CE. (Plane polarized light; Photo width: 1 mm.). **c.** A parallelogram-shaped muscovite nucleated within a biotite host mineral. The former shows a top-to-SW shear ('shearing-1'). (Plane polarized light; Photo width: 1 mm.). **d.** A top-to-SW brittle shear is displayed by asymmetric duplex of muscovite grain that also defines the foliation plane. (Cross-polarized light; Photo width: 2 mm.).

Rootless intrafolial folds

Intrafolial folds of quartz are quite common structures in the ZSZ rocks in micro-scale (Fig. 5d). Their limbs and axial traces dip toward SW, are sub-parallel, and indicate a top-to-NE sense of shearing. The enveloping surfaces of these folds define the ductile shear C-planes of Passchier and Trouw (2005). The fold limbs are unequal in length. Their hinges are rounded and thicker than the limbs. Rootless folds indicate extreme shearing leading to breakage of the limbs from the train of fold.

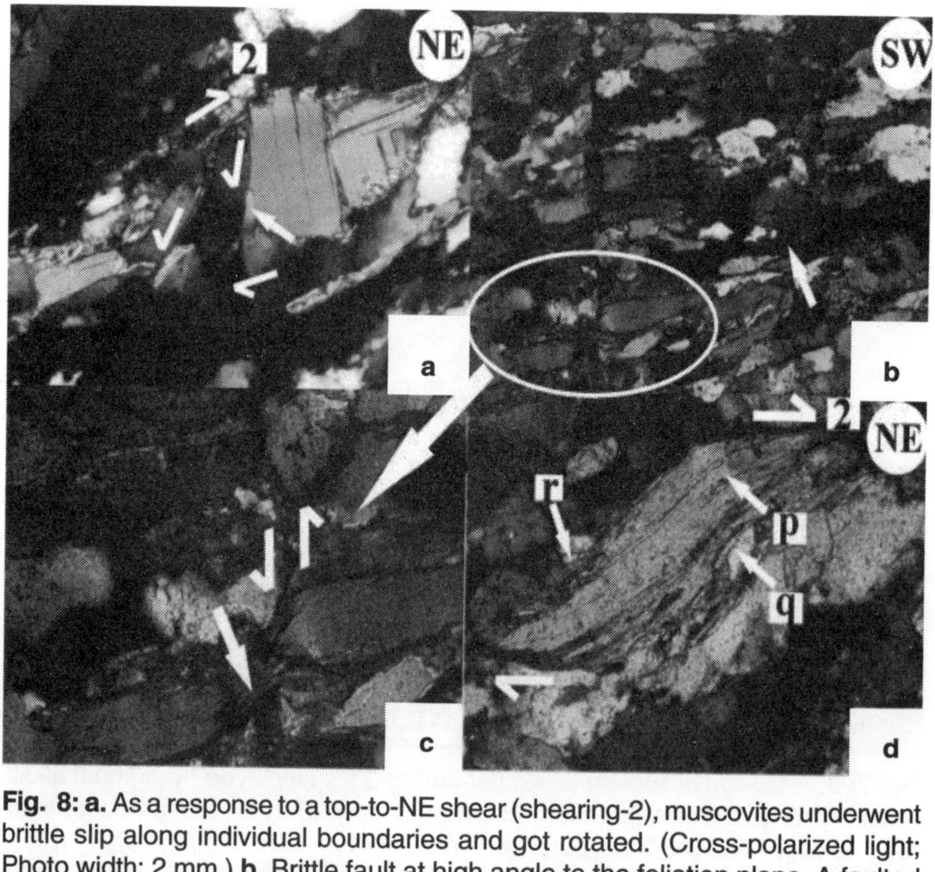

Fig. 8: a. As a response to a top-to-NE shear (shearing-2), muscovites underwent brittle slip along individual boundaries and got rotated. (Cross-polarized light; Photo width: 2 mm.) **b.** Brittle fault at high angle to the foliation plane. A faulted biotite grain at either side of the fault plane act as a marker (ellipse). (Cross-polarized light; Photo width: 0.5 mm.). **c.** Biotite is zoomed and reveals brittle-ductile deformation, normal faulting and an s-type flanking micro-structure. (Cross polarized light; Photo width: 0.5 mm.). **d.** A single sigmoid mica fish. Its crest region (arrow 'p') is kinked. A top-to-NE shear ('shearing-2') is displayed. (Cross polarized light; Photo width: 1 mm.).

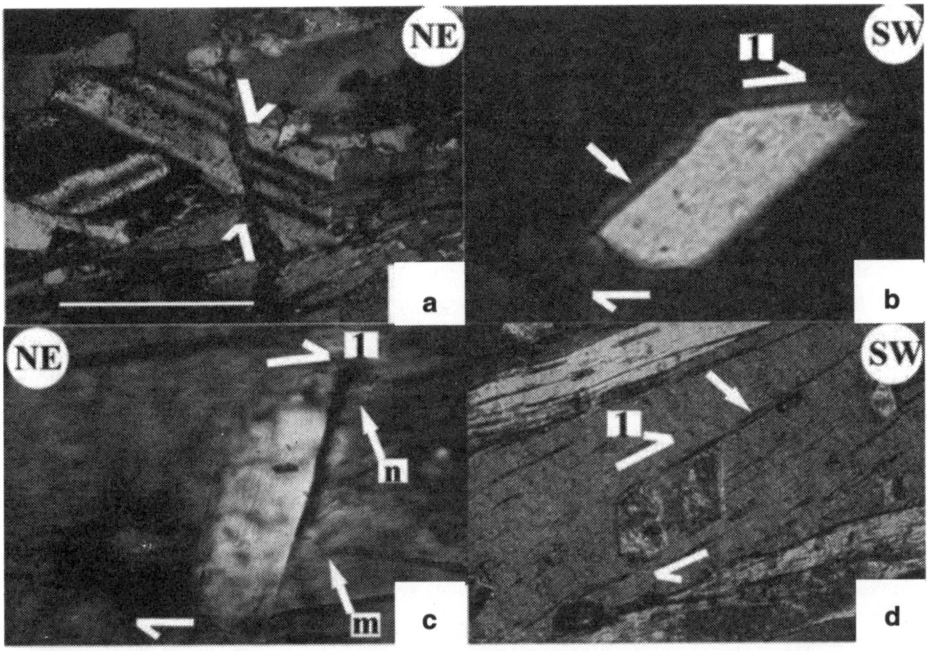

Fig. 9: a. A brittle normal fault in micro-scale (Cross-polarized light; Photo width: 0.5 mm.). **b.** A nearly parallelogram-shaped muscovite grain nucleated inside a biotite host mineral. The shape of the nucleated mineral indicates a top-to-SW shear ('shearing-1'). (Plane polarized light; Photo width: 0.5 mm.). **c.** A nearly parallelogram-shaped muscovite grain nucleated inside a biotite host mineral. Top-to-SW sense sheared ('shearing-1'). (Plane polarized light; Photo width: 0.5 mm.). **d.** A pair of adjacent nearly parallelogram-shaped muscovite grain nucleated inside biotite host mineral. Top-to-SW sheared ('shearing-1'). (Cross polarized light; Photo width: 1 mm.).

Other micro-structures

A top-to-SW sense of brittle shearing along Y-planes is indicated by asymmetric nearly trapezoid-shaped minerals- most commonly micas (Figs. 7d, 12a). These trapezoids define the P-planes (as per the terminology of Passchier and Trouw, 2005). Mukherjee (2007) and Mukherjee and Koyi (in press 1, in press 2) previously recognized such shapes of grains as characteristic of brittle shear sense from a number of shear zones in the Himalaya. In some cases the brittle plane sharply cut across rigid minerals in the matrix (Fig. 12a). The brittle shear plane dips towards NE and is parallel to the ductile shear C-plane. In few cases, minerals are sharply faulted across it without any decipherable dragging (*e.g.* Fig. 9a). In one example, imperfect book-shelf gliding of muscovite grains indicates inter-granular slip under an overall top-to-NE sense of shearing (Fig. 8a). Northeasterly dipping sharp normal faults cut across the pre-existing ductile shear planes (Fig. 8b-c, 9a). At high magnification, individual biotite grain margins as well as their cleavage planes were found to be brittle-ductilely affected by these fault planes (Figs. 8c). On the other hand, more competent minerals such as plagioclase feldspar underwent sharp slip without any drag effect. Fault gouge is expected in such deformation (Passchier and Trouw, 2005) but were not encountered in the studied thin-sections. In the present micro-structural study, micas are found to be most vulnerable to deformation presumably because of their lower competence. Nearly straight limb but round hinged folds of dense intergrowth of biotite and sillimanite was noted within the foliation plane (Fig. 10d) but whether these belong to the pre-Himalayan D_1 or late phase D_3 folds (as per the compilation by Mukherjee and Koyi in press, 1) remained indeterminate.

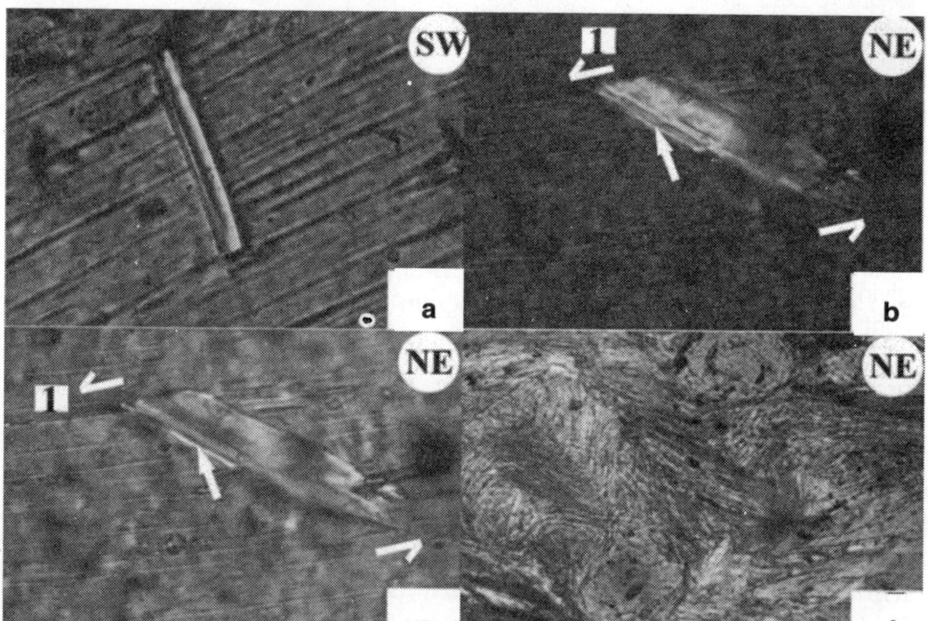

Fig. 10: a. An elongated unsheared muscovite grain nucleated within another muscovite grain. (Plane polarized light; Photo width: 1 mm.). **b & c:** A nearly parallelogram-shaped muscovite grain nucleated inside biotite host mineral. The shape of the nucleated mineral indicates a top-to-SW shear ('shearing-1'). (Plane and cross polarized lights, respectively; Photo width: 0.5 mm.). **d.** Kinked aggregate of biotites and sillimanites that define the main foliation. (Plane polarized light; Photo width: 1 mm.).

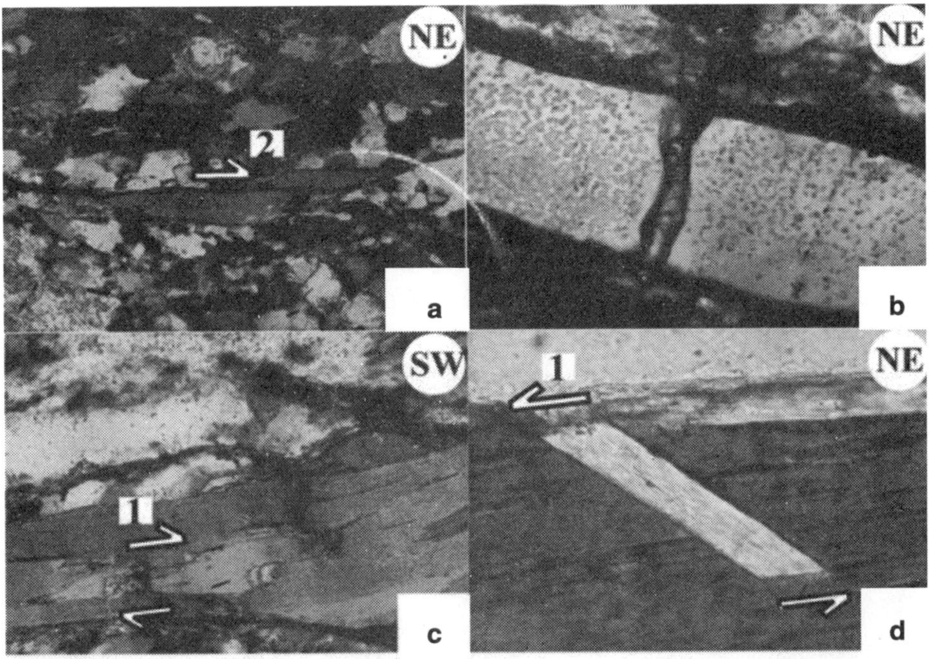

Fig. 11: a. Sharp slip of a single muscovite grain along a brittle fault plane to a top-to-NE sense. (Cross-polarized light; Photo width: 1 mm.). **b.** Zoomed photo of parallel pull-apart reveals that the opening is actually curved. (Cross polarized light; Photo width: 0.5 mm.). **c.** Inside a muscovite host mineral, two nucleated muscovite grains of unequal sizes underwent a top-to-SW shear ('shearing-1') giving rise to their asymmetric shapes and notches (mouths).(Cross-polarized light; Photo width: 1 mm.). **d.** A parallelogram-shaped muscovite nucleated within a biotite host mineral. A top-to-SW shear ('shearing-1') is displayed by muscovite grain. (Plane polarized light; Photo width: 0.5 mm.).

Conclusion

Micro-structural study of the rocks of the Zanskar Shear Zone revealed an oldest top-to-SW shearing of ~22-16 Ma; and subsequent top-to-NE sense of shearing and finally a top-to-NE (down) sense of ductile shearing of ~18-16 Ma. The effect of combined simple shear and channel flow mode of deformation and extrusion are well manifested by micas that act as the commonest shear sense indicators. The shape asymmetries as well as the cleavage plane orientation with respect to the C-planes of mineral fish act as reliable shear sense indicators. Few fish are affected by shearing of different phases. Specific orientation of fish indicates they are produced dominantly by either a top-to-SW or a –NE simple shear. The top-to-SW sense of shear is also well demonstrated by the type-1 flanking microstructures defined by sheared nucleated minerals as the cross-cutting elements and cleavages and grain boundaries of host minerals as the host fabric elements. Rigid minerals are boudinaged by pronounced pulling along the main foliation. Rather frequent (rootless) intrafolial folds with axial planes dipping towards southwest indicate a top-to-NE sense of shearing. A top-to-SW brittle shear sense is displayed by asymmetric trapezoid-shaped duplexes of micas. The brittle Y-planes, which are parallel to the preexisting ductile C-planes, at places are so pervasive that they cut rigid minerals. Amongst all the minerals, micas are most vulnerable to ductile and brittle deformations. Rare micro-structures that are possibly not of any regional significance are (i) intense folding of sillimanite-biotite woven foliations, and (ii) haphazard growth of minerals that overall define the main foliations.

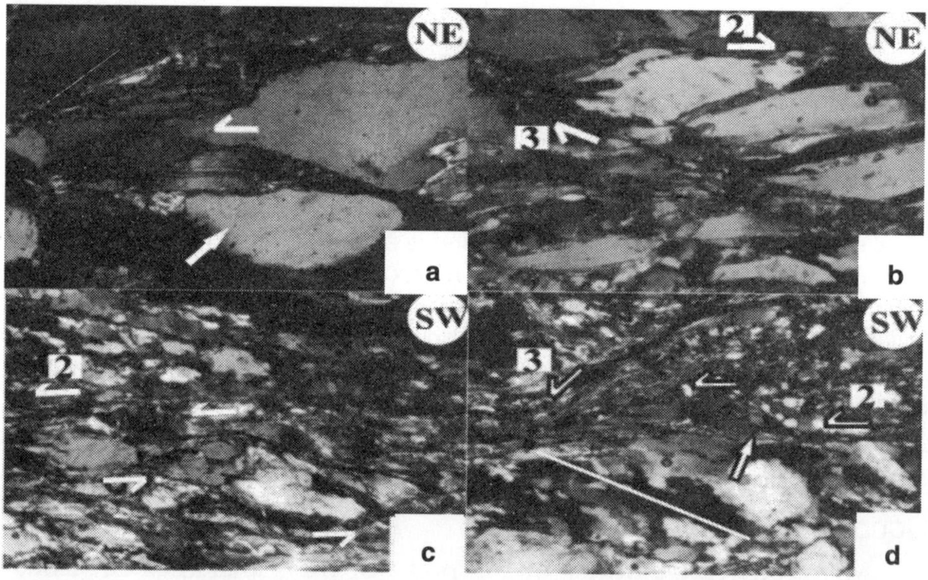

Fig. 12: a. A top-to-SW shear is indicated by a muscovite grain that overrode another grain of muscovite. The brittle fault plane sharply cuts an otherwise rounded grain of a rigid grain of quartz (arrow) (Cross-polarized light; Photo width: 1 mm.). **b.** Top-to-NE (down) ('shearing-3') and a top-to-NE shear ('shearing-2') are deciphered from sigmoid-shaped single larger muscovites and alkali feldspar. (Cross-polarized light; Photo width: 1 mm.). **c.** A top-to-NE shear (shearing-2) is indicated by an aggregate of quartzo-feldspathic minerals (Cross-polarized light. Photo width: 1 mm.). **d.** A top-to-NE (down) shear (shearing-3) is indicated by an aggregate of fine-grained quartzo-feldspathic minerals. A top-to-NE shear ('shearing-2') is deciphered from recrystallized and oriented (white line) quartz in the bottom half of the photo (Cross-polarized light, Photo width: 1 mm.).

Acknowledgements

Author acknowledges Indian Institute of Technology Bombay's Seed Grant: Spons/GS/SM-1/2009. Emeritus Professor Christopher J. Talbot (Uppsala University) provided thorough internal review. Positive criticism and critical comments of an anonymous reviewer is acknowledged.

References

Beaumont, C., Jamieson, R.A., Nguyen M.H. and Lee, B. (2001) Himalayan tectonics explained by extrusion of a low-viscosity crustal channel coupled to focused surface denudation. Nature, v. 414, pp. 738-742.

Bell, T.H. and Cuff, C. (1989) Dissolution, solution transfer, diffusion versus fluid flow and volume loss during deformation/metamorphism. J. Meta. Geol. v. 7, pp. 425-447.

Bèrthe, D., Choukroune, P. and Jegouzo, P. (1979) Orthogneiss, mylonite and non-coaxial deformation of granite: the example of the south Armorican shear zone. J. Struct. Geol., v. 1, pp. 31-42.

Dèzes, P.J. (1999) Tectonic and Metamorphic evolution of the central Himalayan domain in southeast Zanskar (Kashmir, India). Ph.D. Thesis. University of Lausanne, Switzerland. 160p.

Dèzes, P.J., Vannay, J.C., Steck, A., Bussy, F. and Cosca, M. (1999) Synorogenic extension: Quantitative constraints on the age and displacement of the Zanskar shear zone. Geol. Soc. Am. Bull., v. 111, pp. 364-374.

Godin, L., Grujic, D., Law, R.D. and Searle, M.P. (2006) Channel flow, extrusion and exhumation in continental collision zones: an introduction. In: *R.D. Law and M.P. Searle (eds.)* Channel flow, extrusion and exhumation in continental collision zones. Geol. Soc. London, Spec. Pub., v. 268, pp. 1–23.

Herren, E. (1987) Zanskar Shear Zone: northeast southwest extension within the Higher Himalaya (Ladakh, India). Geology, v. 15, pp. 409-413.

Hippertt, J.F.M. (1993) 'V' pull-apart microstructures: a new shear sense indicator. J. Struct. Geol, v. 15, 1394–403.

Lister, G.S., and Snoke, A.W. (1984) S-C Mylonites. Journal Struct. Geol, v. 6, pp. 617-638.

Mukherjee, S. (2005) Channel flow, ductile extrusion and exhumation of lower-middle crust in continental collision zones. Current Sci, v. 89, pp. 435–436.

Mukherjee, S. (2007) Geodynamics, deformation and mathematical analysis of metamorphic belts of the NW Himalaya. Unpublished Ph.D. thesis. Indian Institute of Technology Roorkee. 267 p.

Mukherjee, S. (in press) Macroscopic 'V' pull-apart structure in garnet. Photo of the Month. J. Struct. Geol.

Mukherjee, S. and Chakraborty, R. (2007) Pull-apart micro-structures and associated passive folds. In: *J. Aho (ed.)* Annual Transactions of the Nordic Rheology Society v. 15, pp. 247-252. 16th Nordic Rheology Conference, Stavanger, Norway.

Mukherjee, S. and Koyi, H. A. (2009) Flanking Microstructures. Geol. Mag, v. 146, pp. 517-526.

Mukherjee, S. and Koyi, H.A. (in press, 1) Higher Himalayan Shear Zone, Zanskar Section-Microstructural Studies and Extrusion Mechanism by a Combination of Simple Shear & Channel Flow. Int. Journal Earth Sci., online published, doi: 10.1007/s00531-009-0447-z.

Mukherjee, S. and Koyi, H.A. (in press, 2). Higher Himalayan Shear Zone, Sutlej section: structural geology and extrusion mechanism by various combinations of simple shear, pure shear and channel flow in shifting modes. Int. Journal Earth Sci., online published, doi: 10.1007/s00531-009-0459-8.

Mukherjee, S. and Pal, P. (2000) Tectonic structures of the Karakoram metamorphic belt, its significance in the Geodynamic Evolution. Unpublished Report. Summer Undergraduate Research Award. University of Roorkee.

Passchier, C.W. and Trouw, R.A.J. (2005) Microtectonics. Springer Verlag, Heidelberg. p.p. 1-366.

Patel, R.C., Singh, S., Asokan, A., Manickavasagam, R.M. and Jain, A.K. (1993) Extensional tectonics in the Himalayan orogen, Zanskar, NW India. In: *P.J. Treloar, M.P. Searle* (*eds.*) Himalayan Tectonics, Geol. Soc. Lond. Spec. Publ., v. 74, pp. 445-459.

Searle, M.P., Cooper, D.J.W. and Rex, A.J. (1988) Collision tectonics of the Ladakh-Zanskar Himalaya. In: *R.M. Shackleton, J.F. Dewey and B.F. Windley (eds.)* Tectonic evolution of the Himalayas and Tibet. Philosophical Transaction of the Royal Society, London, v. A326, pp 117–150.

Singh, K. (1999) Pull apart microstructures in feldspar from Chail Thrust Zone, Dhauladhar Range, Western Himachal Pradesh. In: *A. K. Jain and R.M. Manickavasagam (eds).* Geodynamics of the NW Himalaya. Gondwana Research Group Memoir, v. 6; pp. 117–23. Osaka.

Steck, A., Spring, L., Vannay, J.C., Masson, H., Stutz, E., Bucher, H., Marchant, R., and Tièche, J.C., (1993) Geological transect across the Northwestern Himalaya in eastern Ladakh and Lahul (A model for the continental collision of India and Asia): Eclo. Geol. Helv, v. 86, p. 219.

Treagus, S.H. and Lan, L. (2000) Pure shear deformation of square objects, and applications to geological strain analysis. J. Struct. Geol, v. 22, pp. 105-122.

Walker, J. D., Martin, M. W., Bowring, S. A., Searle, M. P., Waters, D. J. and Hodges, K. V. (1999) Metamorphism, melting and extension: age constraints from the High Himalayan slab of southeast Zanskar and northwest Lahul. Geol., v. 107, pp. 473-495.

Walker, J. D., Searle, M. P. and Waters, D. J., 2001. An integrated tectonothermal model for the evolution of the Higher Himalaya in western Zanskar with constraints from thermobarometry and metamorphic modeling. Tectonics, v. 20, 810-833.

Yin, A. (2006) Cenozoic tectonic evolution of the Himalayan orogen as constrained by along-strike variation of structural geometry, extrusion history, and foreland sedimentation. Earth Sci. Rev., v. 76, pp. 1-131.

□□□

5

A Study of Erosion-Deposition Processes around Majuli Island, Assam

M.K.DUTTA[1], SWAPNALI BARMAN[2], S.P. AGGARWAL[2]
[1]Department of Civil Engineering, Jorhat Engineering College, Jorhat, Assam
[2]Indian Institute of Remote Sensing (National Remote Sensing Centre)
4, Kalidas Road, Dehradun

Abstract

Majuli, the largest inhabited river island bounded by the river Subansiri to the north and mighty Brahmaputra River to the south, is one of the subdivisions of the Jorhat district, Assam. Erosion of the island is a continuous process since historical times and posses a significant concern. The present approach have been made to study the erosion-deposition processes with the help of data resource generated from the Survey of India (SOI) toposheets and Indian Remote Sensing (IRS) satellite imagery spanning the period from 1966-1975 to 2008 in a Geographical Information System (GIS) environment. The observation has revealed a dramatic change in reduction of land area of the Majuli Island. The land area as evidenced till 1966-1975, 1998 and 2008 were 706.14, 578.38 and 484.34km^2 respectively. The total average annual rate of erosion and deposition covering the entire period were 8.76 km^2/yr and 1.87 km^2/yr. It can clearly be attributed that the island mass suffered significant rate of erosion than the depositional processes.

Keywords : River island, Erosion- deposition process, Majuli Island, Brahmaputra river, Subanisri river, Assam

Introduction

The geographical extent of the study area (Fig.1) is 26°45' N- 27°12' N latitude and 93°39'E- 94°35'E longitude with mean height of 84.5 m above MSL. Majuli, bounded by the river Subanisri on the northwest, the Kherkatia Suti (a spill channel of the river Brahmaputra) in the northeast and the main Brahmaputra river on the south and the south west is regarded as the largest fresh water inhabited island of the world. Water yield of the Subansiri is 0.076 cumec/km^2 (Goswami, 1998). The Brahmaputra is a classic example of a braided river consisting of a network of interlacing channels with unstable bars and islands (locally known as chars). As the flow begins to rise with the onset of the monsoon, most of the islands are submerged and the river then flows in more or less single channel. The most striking feature is the continuous shift of the thalweg (deep channel) from one location to another within the bankline, its movement being high in the rising stage (May to August), most erratic during the falling stage (September to October) and minimal in the low flow stages (Goswami, 1998). The mechanism of braiding may be attributed as the affect of excessive sediment load, large and variable flow, erosion-prone banks and the rapid aggradation of the channel (Goswami, 1991, Leopold *et al.*, 1964).

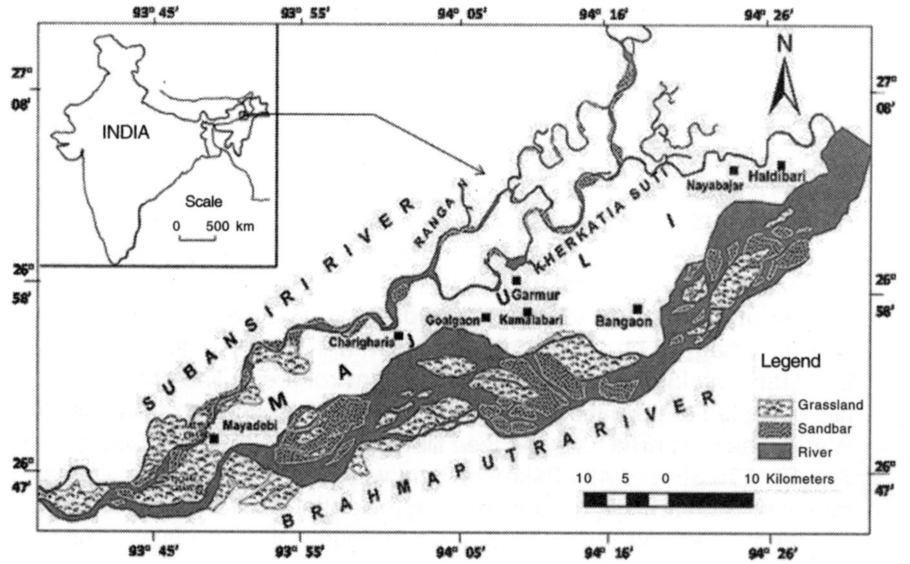

Fig.1: Location map of the study area.

Majuli is seriously affected by erosion of the Brahmaputra and the Subansiri rivers (Fig. 2, 3 and 4). The extreme braided nature of the Brahmaputra coupled with silt and sand strata of the banks is the main cause of erosion. Erosion in this area was not much before the

1950 earthquake of magnitude 8.6 Richter scale but became active thereafter and attained serious dimension after the 1954 flood. In 1987, Majuli suffered the most severe flood having lost 50,000 cattle and crop (Centre for Natural Disaster Management, Assam Administrative Staff College, Jawaharnagar, Khanapara). Porcupines are being used to control erosion only at a few places, but it has not been found to be effective (Fig.5). The area of the island has been reduced from 706.14 km^2 in 1966-1975 to 578.38 km^2 in 1998 and to 484.34 km^2 in 2008. Space Application Centre (SAC) and Brahmaputra Board (1996) jointly studied the river erosion problem of Majuli Island and identified the areas of the island which have undergone changes along the bankline due to erratic behavior of the river. Brahmaputra Board (1997) prepared a report where the area of the island was mentioned to be 925 km^2 in 1971. The erosion is a serious problem which threats the existence of Majuli-one of the oldest cultural heritage site of Assam. Mani *et al.* (2003) studied the erosion affect around Kaniajan village in south Majuli using satellite data of 1991, 1997 and 1998 and reported a loss of 1900 ha during 1991 to 1997 and 845 ha during 1997 to 1998. Sarma and Phukan (2004) gave a comprehensive account on the origin and geomorphological changes including erosion and deposition in Majuli Island. Kotoky *et al.* (2005) studied the erosion and deposition of the island from 1914 to 1998 and revealed that the extent of erosion and deposition was not same for the period 1914–75 and 1975–98. The aim of this present study is to evaluate the migration of the rivers Brahmaputra and Subansiri, the locations of erosion and deposition and the rate of erosion and deposition in the island starting from 1966-1975 to 2008.

Fig. 2: Bank erosion by river Brahmaputra (at Bengena Ati).

Fig. 3: River bank under process of active erosion.

Fig. 4: Bamboo trees remained in the river after the bank has been shifted due to erosion.

Fig. 5: Existing Porcupines used to check the bank erosion.

Methodology

The present approach employs Survey of India (SOI) toposheets (1966-1975) and Indian Remote Sensing (IRS) satellite imagery (1998 and 2008) to investigate spatial changes over available period of time. Such information is considered valuable in providing information for a period of more than 33 years, which is often beyond the scope of empirical observation. The Survey of India toposheet nos. 83F/5,6,9,13; I/4,8; J/1, 5 of 1966-1975 (scale 1:50,000) covering the study area were scanned, georeferenced and mosaiced and used to prepare the base map. The IRS 1B LISS II satellite imagery of 1998 and IRS P6 LISS III digital image of 2008 are registered to the base map using a set of Ground Control Points (GCPs) in ERDAS IMAGINE 9.1 software. Thematic maps of different periods were prepared on 1:50.000 scale and were integrated using ArcView GIS. The bank lines for different periods *i.e.* 1966-1975, 1998 and 2008 of the Brahmaputra and Subansiri Rivers, which limits the south and north banks of the Majuli Island, were digitized and integrated with the help of Arc View GIS. The migration as well as the erosion and deposition caused by the rivers were studied by dividing the whole stretch of the island into 5 km long blocks. Then for each block the extents of migration as well as the amount of erosion and deposition are determined. Ground truth observations were made to prepare the interpretation key and Garmin Global Positioning System (GPS) was used to locate the latitude, longitude and altitude of the study sites. Geomorphic attributes of the flood plain, morphology of the channel and banks, and erosion/deposition activities have been evaluated from the toposheets, imageries and after proper field checking were used for interpretative use of the present study.

Results and Discussions

Bank-line Migration

The sequential analyses of geomorphological maps (Fig.6) for the period from1966-1975 to 2008 have been utilized to study the bank line migration for the stretch under study. The channel configuration of 1966- 1975 was considered as the base line for evaluation of bank line migration during the period under study.

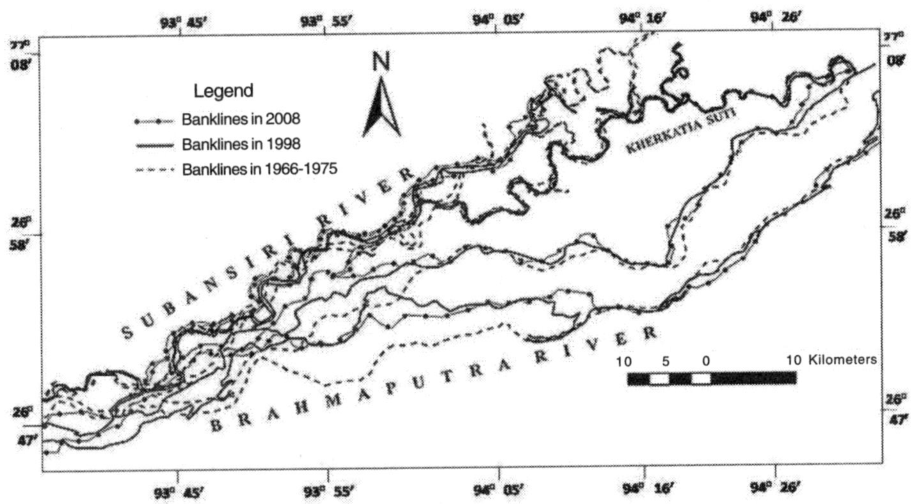

Fig. 6 : Banklines of Rivers Brahmaputra and Subansiri in different periods.

Migration of the Subansiri

a) 1966-1975 to 1998: The maximum migration was observed near Bohumari and Nunibari for a distance of about 3.12km towards south. The Subansiri migrated towards north for 0.64 km at the western end of Majuli causing deposition. Within 20 to 40 km towards east of the island, the river recedes away towards north for 2.02 km at Majar Chapori and for 2.86 km at Kumaliagaon causing deposition. For the rest of the distance from 35 to 80 km the river migrated towards south with a maximum of 1.25 km at Chaporigaon and 1.21 km at Jengraimukh causing erosion (Fig. 7a).

b) 1998-2008: The major change that had occurred during this period was that the Subansiri river had shifted its course towards east and met Brahmaputra about 15 km ahead from its earlier confluence. Thereafter, the river migrated towards north about

1.58 km at Majar Chapori which shows deposition of sediments to the island. In the block within 25 km to 35 km the river migrated towards south with a maximum of 2.71 at Aunati Jangal and with 0.8 km at Kumalia gaon resulting in heavy erosion of the land area of Majuli Island. Next from 40 km to 80 km the river migrates towards north with a maximum of 0.55 at Khengmiligaon.

Fig.7a clearly shows that the river Subansiri has migrated for a distance of more than 3 kms towards south at Bohumari and Nunibari. These two places also evidenced highest rate of migration during the period (Fig.7b) of total observation.

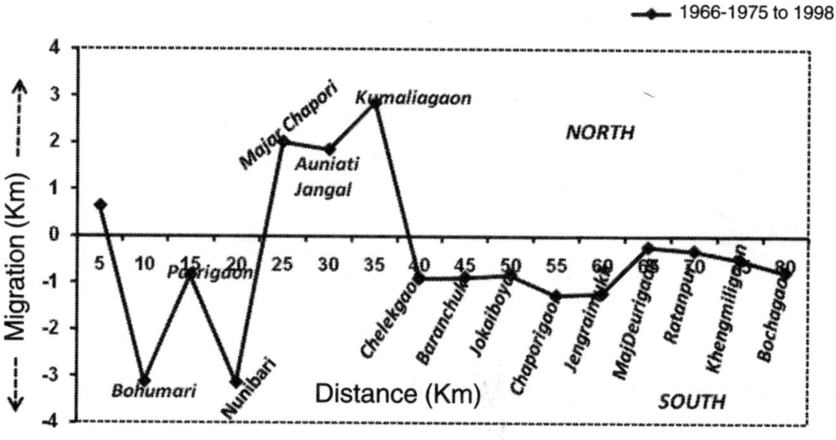

Fig. 7a: Bankline Migration of Subansiri River during 1966-1975 to1998.

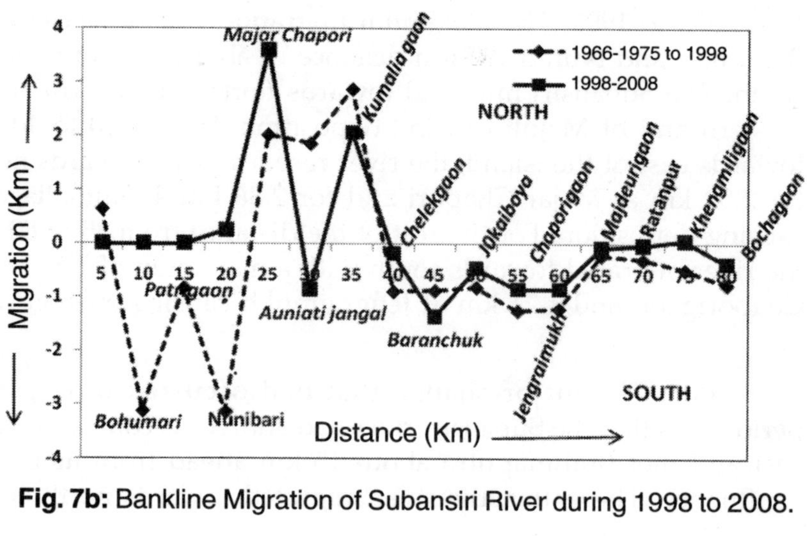

Fig. 7b: Bankline Migration of Subansiri River during 1998 to 2008.

Migration of Brahmaputra

a) 1966-1975 to 1998: During the period of 1966-1975 to 1998, a significant change was observed in the migration of bank line of river Brahmaputra. The migration of bank lines during this period is presented in Fig.7c. Heavy erosion of the island was caused by Brahmaputra river during this period as the river migrates towards north at most of the places. The river channel had migrated for a distance of about 6.49 km near Mayadebi, 5.24 km near Manikmukh kuamargaon, 1.73 km at Bengena Ati and 4.24 kms near Khorapargaon (Fig. 7c).

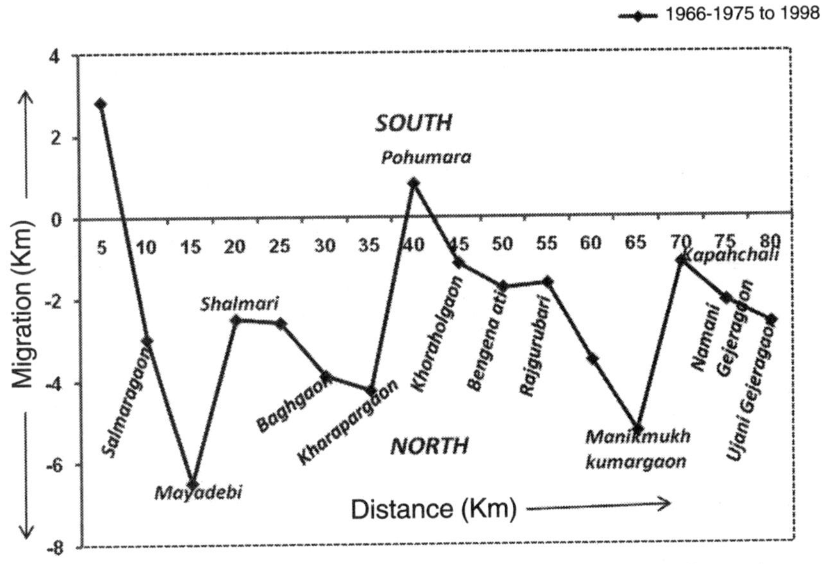

Fig. 7c: Bankline Migration of Brahmaputra River during 1966-1975 to 1998.

b) 1998-2008: During the period of 1998 to 2008 it was evidenced that the Subansiri River had shifted its course and joined the Brahmaputra 15 km ahead of its earlier point of debouch in the western part of the Majuli Island. This results in erosion of Majuli Island and thereby decreasing the total area. The next stretch of 65 km mainly showed migration towards the north. The river migrated towards north about 1.53 km near Khorapargaon, 1.14 km near Bengena Ati, 1.53 km near Rajgurubari and 2.5 km near Ujani Gejargaon. In between, the river had migrated towards south at Khoraholgaon and Kapahchali for about a distance of 1.25 kms and 1.8 kms respectively with reference to the 1998 position of the river bank line.

Analysis on total observation covering the period from 1966-1975 to 2008, it was clearly visible that the river Brahmaputra has migrated for a maximum distance of about 6 km towards north near Mayadebi and Khorapargaon areas. However, near Bengena Ati and Ujani Gejeragaon the bank line was migrated towards north of a distance of about 2.87km and 5.1 km respectively. From the present investigation it can clearly be attributed that there have been a continuous shift in the bank line of the Brahmaputra River channel indicating significant erosion affect at many places of the Majuli Island (Fig. 7d).

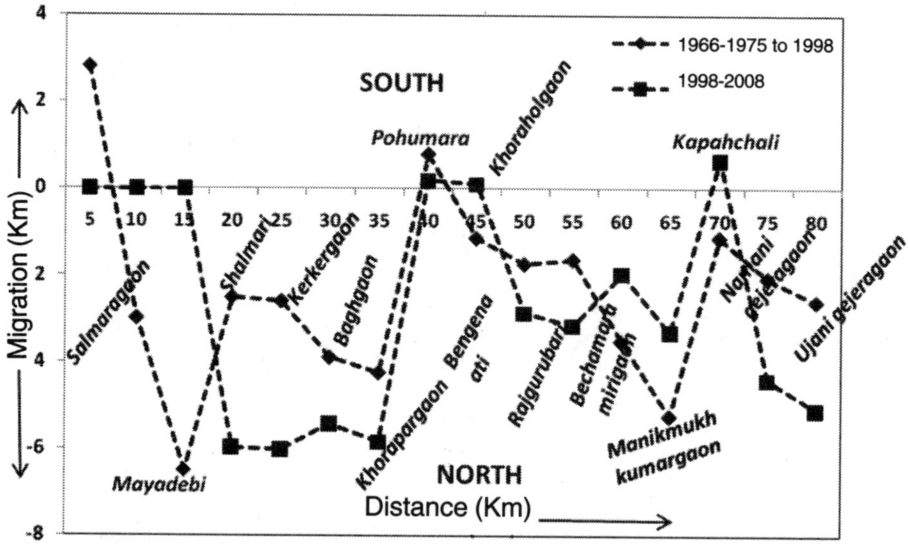

Fig. 7d: Bankline Migration of Brahmaputra River during 1998 to 2008.

Erosion/Deposition Activities around Majuli Island at Different Periods

The activity of erosion deposition processes around Majuli Island with reference to the Subansiri (North) and Brahmaputra (South) Rivers have been evaluated for two successive periods *i.e.* from 1966-1975 to 1998 and 1998 to 2008. In the light of generated information an interpretative discussion have been attempted in the preceeding sections.

Activities related to Subansiri River on the north

a) During the period from 1966-1975 to 1998: The stretch up to 20km from the western end of Majuli Island suffered erosion with maximum at Bohumari and Nunibari of about 8.83 km^2 and 10.77 km^2 respectively. Thereafter, there was deposition of about 6.07 km^2 at Major Chapori and 12.26 km^2 at Kumolia gaon. The remaining steretch under study have shown normal processes of erosion and deposition. The total average annual rate of erosion and deposition during this period were 1.23 km^2/yr and 1.17 km^2/yr (Fig.8a)

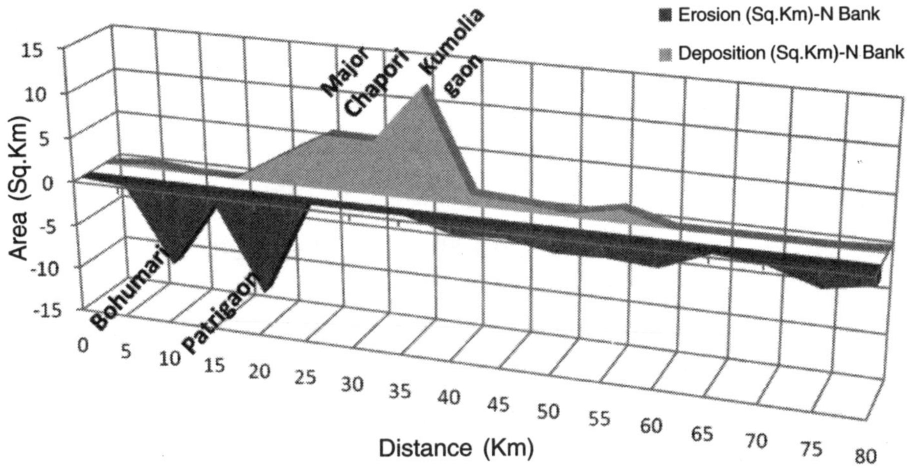

Fig. 8a : Erosion/deposition activities on the north bank of Majuli Island during 1966-1975 to 1998.

b) During the period from 1998 to 2008: The erosion/deposition process in the later period of observation *i.e.* 1998 to 2008 have shown a dissimilar behaviour than the earlier period of observation. The island mass faces an extreme rate of erosion during the period than the depositional processes. The area near Nunibari has faced hgh intensity of erosion measuring about 17.82 km^2. At Auniati Jangal, the erosion was about 5.74 km^2 and and a maximum intensity of erosion was observed near Michhamara measuring about 24.1 km^2. Deposition was not very promnent with only 3 km^2 at Majar Chapori and a maximum of 3.05 km^2 at Kheng mili gaon. It is clear from the observation that the bank on north of the Majuli Island was eroded away by the Subansiri River. The average annual rate of erosion and deposition were 5.05 km^2/yr and 1.22 km^2/yr respectively (Fig.8b).

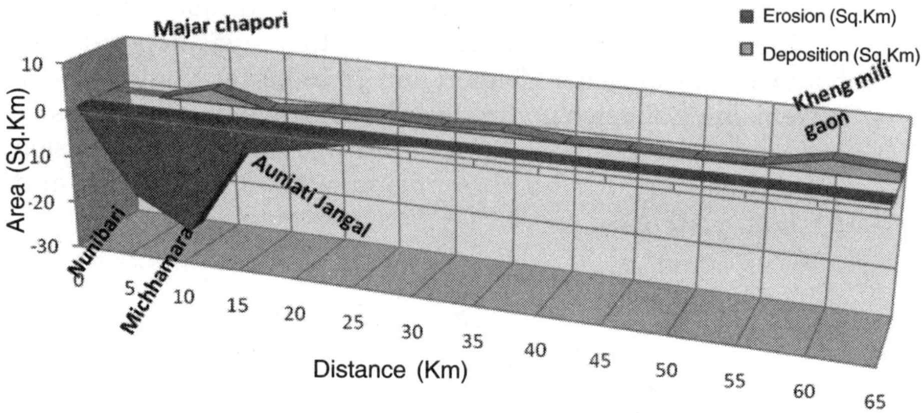

Fig. 8b : Erosion/deposition activities on North bank of Majuli Island during 1998-2008.

Activities related to Brahmaputra River on the south

a) *During the period from 1966-1975 to 1998* : The stretch under study on the south bank of the Majuli Island underwent extensive erosion. The areas around Mayadebi and Baghgaon have evidenced maximum intensity of erosion up to 25.44 km^2 and 18.94 km^2 respectively. However, the stretch from 45 to 50 km from western side of Majuli near Bengena Ati the bank also faced an intensity of erosion about 5.45 km^2. The areas around Manimukh-Kumargaon and Ujani-Gejargaon towards the eastern end of the island faces an erosion activity up to a maximum of 10.64 km^2 and 9.52 km^2 respectively. The westren end of the island and at Chelekhgaon have evidenced depositional activities measuring 9.08 km^2 and 2.45 km^2 respectively. The average annual rate of erosion and deposition during this period on the souther bank of the Majuli Island were 4.73 km^2/yr was 0.42 km^2/yr respectively (Fig.8c).

b) *During the period from 1998 to 2008* : The period represent increasing intensity of erosion due to the shifting of bankline of the Brahmaputra river towards the north. The initial stretch of 20 km from the western side suffered erosion with maximum of 16.58 km^2 near Kerkergaon. Shalmari has evidenced erosion of 10.52 km^2 whereas at Baghgaon the river washed away an area of 7.49 km^2. In the eastern end of the island Namani Gejera gaon and Ujani Gejera gaon measured erosions of 9.07 km^2 and 8.15 km^2 respectively.Near the Bengena Ati Satra an erosion of 3.65

km² has been observed. Deposition occurred in the stretch from 30 to 60 kms with a maximum of 4.04 km² at Manimukh Kumaragaon and 2.03 km² at Khorahol gaon. The annual erosion was 6.51 km²/yr and the annual deposition was 0.94 km²/yr (Fig.8d).

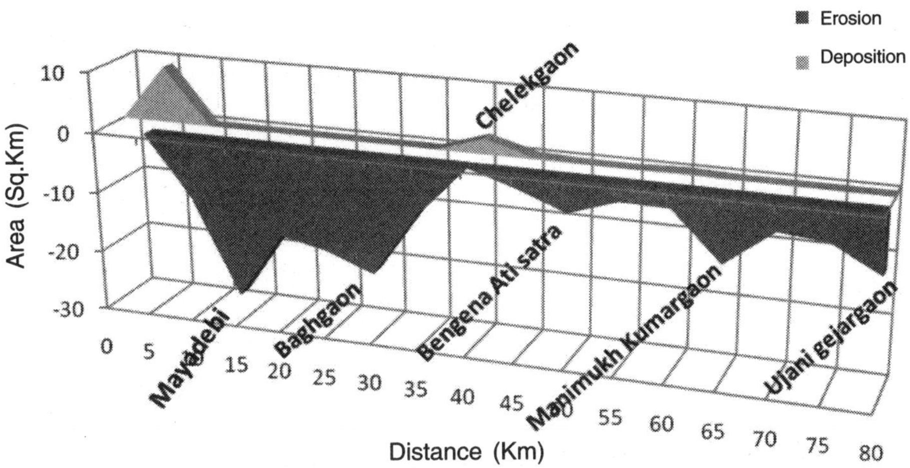

Fig. 8c : Erosion/ deposition activities on the South bank of Majuli Island during1966-1975 to 1998.

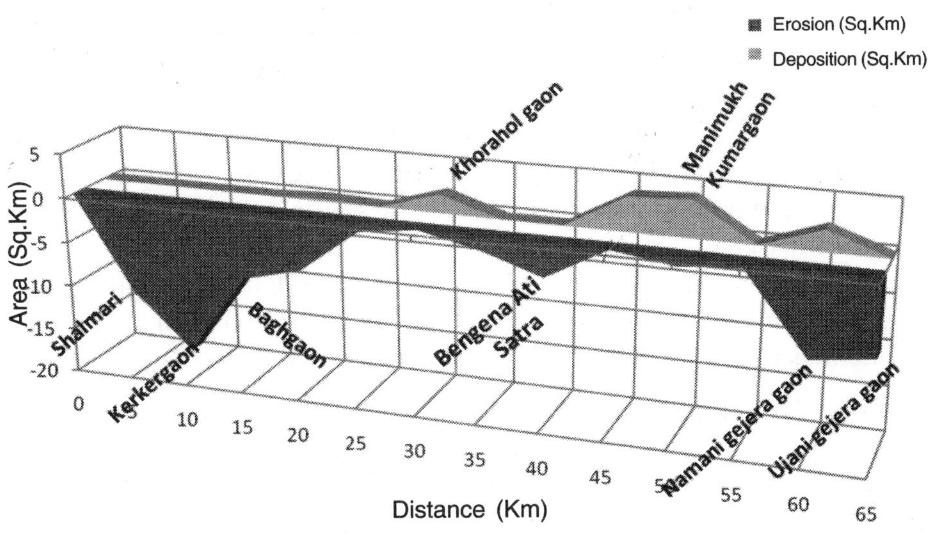

Fig. 8d : Erosion/ deposition activities on South bank of Majuli Island during 1998-2008.

The area of the Majuli Island and area added (deposition) or washed away (erosion) as evidenced from the study during different periods (1966-1975, 1998 and 2008) are as follows:

Area in 1966-1975 = 706.14

Erosion (66-75 to 98) = 171.91

Deposition (66-75 to 98) = 44.15

Area in 1998 = (706.14-171.91) + 44.15
= 578.38 km^2

Area in 1998 = 578.38 km^2

Erosion (98-08) = 115.58

Deposition (98-08) = 21.54

Area in 2008 = (578.38-115.58) + 21.54
= 484.34 km^2

Rate of average annual erosion and deposition

The average annual erosion and deposition are given by the total area of erosion and deposition divided by the period of years. It gives the amount of erosion or deposition in each year. The rate of average annual erosion and deposition per unit length is given by average annual erosion or deposition divided by the stretch in kilometers. Spanning the period over 1966-1975 to 2008 the study attributed that the total average annual erosion and deposition covering the entire period were 8.76 km^2/yr and 1.87 km^2/yr respectively. Comparatively higher erosion potential was confined to the south bank of Majuli Island than the north bank. The south bank of Majuli Island showed a higher rate of average annual erosion per unit length of 0.08 km^2/km by the Brahmaputra River than 0.046 km^2/km of the north bank of the island by Subansiri (Table-1).

Also the annual rate of erosion from 1998-2008 was much higher than that of the period 1966-1975 to 1998. 1966-1975 to 1998 showed an annual rate of erosion per unit length of 0.074 km^2/km and 1998-2008 showed a much higher erosion rate of 0.178 km^2/km. Thus, the present erosion rate was much more than that of the previous years. Thus heavy erosion threatens the existence of Majuli.

Table-1 : Erosion and deposition along the stretch of Majuli at different time periods.

Period	River	Average Annual Erosion (km²/yr)	Average Annual Deposition (km²/yr)	Rate of Average Annual Erosion per unit length (km²/km)	Rate of Average Annual Deposition per unit length (km²/km)
1966-1975 to 1998	Subansiri	1.225	1.166	0.015	0.015
	Brahmaputra	4.73	0.422	0.059	0.0052
	Total	5.955	1.588	0.074	0.0202
1998-2008	Subansiri	5.05	1.22	0.078	0.0188
	Brahmaputra	6.51	0.94	0.100	0.0145
	Total	11.56	2.16	0.178	0.0333
Mean	8.76	1.87	0.126	0.0267	

Conclusion

The fresh water river island Majuli have suffered a significant rate of erosion since historical time. Observation on the bank line migration spanning the period from 1966-1975 to 2008 an extensive rate of migration was observed. The bank line of the Brahmaputra Rivers has migrated for a distance of about 6 km towards north at Khorapargaon and 5.1 kms near Ujani Gejeragaon. The bank line of the Subansiri River has shown its migration up to a distance of more than 3km towards south specifically around Bohumari and Nunibari areas.

The study has clearly attributed that the total average annual rate of erosion and deposition covering the entire period were 8.76 km^2/yr and 1.87 km^2/yr respectively. The south bank of the Majuli Island shows higher rate of annual erosion per unit length of 0.08 km^2/km by the Brahmaputra River than the Subansiri River on the north amounting of 0.047 km^2/km. The areas around Majar Chapori, Mayadebi, Baghgaon and Bengena Ati have suffered extensive intensity of erosion.

The average annual rate of erosion per unit length from 1998-2008 was much higher than that of the period 1966-1975 to 1998. An average annual rate of erosion per unit length of 0.074 km^2/km was observed during the period 1966-1975 to 1998 where as higher average rate of erosion rate of 0.178 km^2/km was observed during the period 1998-2008. The present erosion rate was more than double that of the previous years. Thus, the erosion of the world's largest inhabited river island Majuli should be considered as a national problem and should be addressed in time to stabilize ecological balance.

Acknowledgement

The authors thank Department of Science and Technology (DST), Govt. of India for supporting the research. We are grateful to Indian Institute of Remote Sensing (IIRS), Dehradun and Jorhat Engineering College (JEC), Assam for providing facilities to carry out the project works.

References

Brahmaputra Board (1997) Report on the Erosion Problem of Majuli Island, Brahmaputra Board, Guwahati.

Goswami, D.C. (1991) Braiding of the Brahmaputra river channel in Assam: A fluvio-geomorphological enquiry. Northeast Geographer, v. 23 (1&2), pp. 52-60.

Goswami, D.C. (1998) Fluvial Regime and Flood Hydrology of Brahmaputra River, Assam. In: V.S. Kale (ed.) Flood Studies in India, Memoir 41, Geological Society of India, pp. 53-75.

Kotoky, P., Bezbaruah, D., Baruah, J. and Sarma, J. N. (2004) Nature of bank erosion along the Brahmaputra river channel, Assam, India. Current Science, v. 88(4), pp. 634-640.

Leopold, L.B., Wolman, M.G. and Miller, J. P. (1964) Fluvial Processes of Geomorphology. W. Freeman, San Francisco, 522 p.

Mani, P., Kumar, R. and Chatterjee, C. (2003) Erosion Study of Part of Majuli River Island Using Remote Sensing Data. J. Indian Soc. of Remote Sensing, v. 31, pp. 11-18.

SAC and Brahmaputra Board (1996) Report on bank erosion on Majuli Island, Assam: a study based on multi temporal satellite data. Space Applcation Centre, Ahmedabad and Brahmaputra Board, Guwahati.

Sarma J. N. and Phukan, M. (2004) Origin and some geomorphological changes of Majuli Island of the Brahmaputra River in Assam, India. Geomorphology, v. 60, pp. 1-19.

❑❑❑

6

Signatures of Archaean E-W Crustal-Scale Shears in the Bundelkhand Massif, Central India : An Example of Vertical Ductile Shearing

S. P. SINGH[1] AND A. R. BHATTACHARYA[2]
[1] Department of Geology, Bundelkhand University, Jhansi
[2] Centre of Advanced Study in Geology, University of Lucknow, Lucknow

Abstract

The study incorporates a documentary evidence of vertical/subvertical shear zones in the central part of the Bundelkhand massif, central India. The shear zones are aligned E-W and are characterized by the occurrence of mylonites that show vertical/subvertical foliation. The shear zones involve a variety of metamorphic and metavolcanic rock types. These shear zones appear to be the oldest and pre-tectonic in the context of early Proterozoic Bundelkhand granitoids. It is possible that the thin nature of the Archaean crust was mainly responsible for the formation of these vertical shear zones.

Keywords : Bundelkhand complex, Vertical shear zone, Mylonites, Ductile shearing, Archaean crust.

Introduction

Large scale shear zones are known from several orogenic belts of the world (Knipe, 1989; Burg, 1999). Most of them are of low to moderate angles. Vertical/subvertical shear zones are, however, rare in the world (Westaway and Kusznir, 1993; McCaig and Knipe, 1990;

Neves *et al.*, 1996) and their report from the Indian subcontinent is negligible to absent. The present work documents possibly the first occurrence of vertical shears from the Indian shield. In the world, occurrence of vertical shear zones is relatively uncommon; some examples include shear zones of the Spanish Pyrenees (Burg, 1999; Brun and Burg; 1982). The present report of E-W trending, vertical shear zones in the Bundelkhand complex of Central India (Fig. 1) should therefore have significant bearing for Archaean tectonics and evolution of the Archaean crust.

Regional Geological Setup

The northern part of Central India exposes five major geological divisions:(1) Bundelkhand massif, (2) Early Proterozoic sedimentary basins (Bijawar and Gwalior groups),(3) Middle to Late Proterozoic sedimentary basins (Vindhyan Supergroup), (4) Malwa/Deccan Trap with inter- and intra-trappean sediments, and (5) Quaternary deposits of marginal alluvial plain.

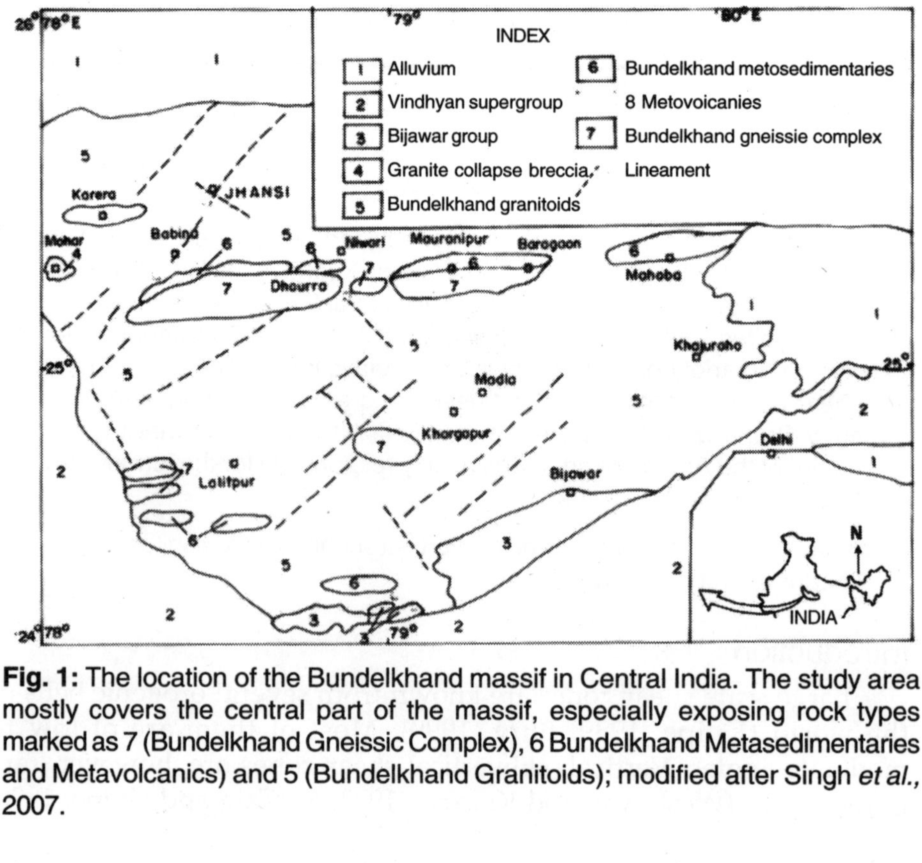

Fig. 1: The location of the Bundelkhand massif in Central India. The study area mostly covers the central part of the massif, especially exposing rock types marked as 7 (Bundelkhand Gneissic Complex), 6 Bundelkhand Metasedimentaries and Metavolcanics) and 5 (Bundelkhand Granitoids); modified after Singh *et al.*, 2007.

The Bundelkhand massif covers about 26,000 sq. km. in Central India and occurs as elliptical body located between lat. 24^0 30' N and 26^0 30' N and long. 77^0 30' E and 81^0 50′E. The massif is mostly dominated by Palaeoproterozoic granitoids known as Bundelkhand Granitoids (Singh *et al.*, 2010) that are described as pink granite, grey granite, biotite granite and granite porphyry (Basu, 1986; Sharma, 1998). The NE-SW trending quartz reefs and NW-SE trending massive dykes impart large scale linear fabrics in the massif (Fig.1 & 2). The massif represents the culmination of a long period of ancient structural, tectonic and petrologic cycles that may have been operative from Archaean to Early Proterozoic. Singh *et al.*, 2007 reviewed the geology of Bundelkhand and suggested that the Bundelkhand Gneissic Complex (BnGC) is the oldest component that comprises high grade metamorphics reaching up to granulite facies (Singh and Dwivedi, 2009) that is followed by low grade metamorphic rocks of green schist facies of the Basement metamorphic complex, known as the Bundelkhand metasedimentaries and metavolcanics (BMM). These low and high grade metamorphites have angular relationships with each other and are characterized by polyphase deformation. The Bundelkhand granitoid is found to be intrusive and is supposed to have developed in volcanic arc environment (Mondal *et al.*, 2002). The Bundelkhand region has thus experienced cycles of sedimentation, igneous activities, granitization, metamorphism and tectonic deformation. The prolonged deformation history since the Archaean times has produced complex structures in the internal domain of the massif that have left numerous signatures on various scales.

Some of the studies on the structure and tectonics of the Bundelkhand massif including shear fabric (*e.g.* Roday *et al.*, 1995; Singh *et al.*, 2007; Bhattacharya, 1986), shows mainly three types of shears, *viz.* E-W shear, NE-SW and NW-SE shears (Fig. 2). The E-W shear zones are mainly confined to the central part of the Bundelkhand massif. The field evidences suggest the E-W shear to be the oldest and pre-tectonic in the context of early Proterozoic Bundelkhand granitoids, while the latter two types are relatively younger and post-Bundelkhand granitoids. The E-W shear has been found to be superimposed by the NE-SW and NW-SE shears as is clearly noticed at several places, *e.g.* Shivgarh and Pahunj.

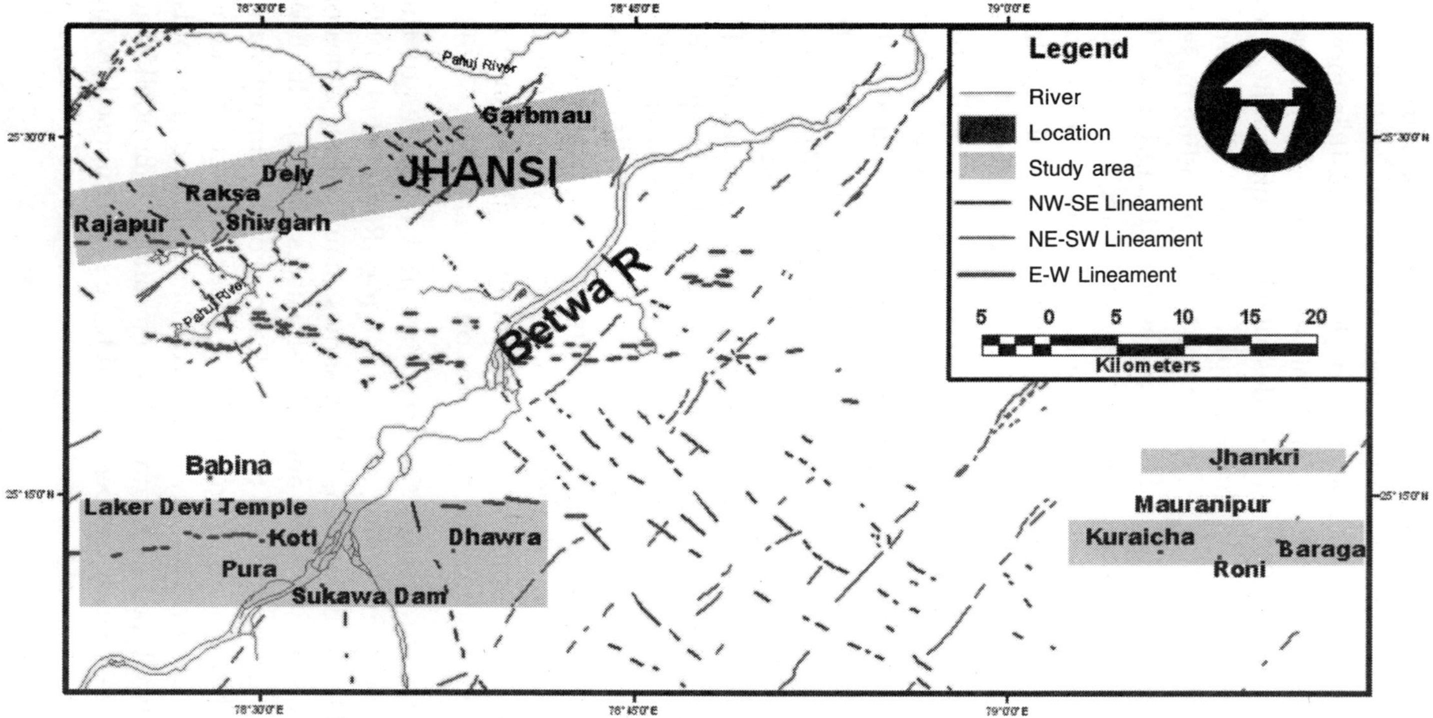

Fig. 2 : The pattern of shear zones in the Bundelkhand massif. The region is characterized by the presence of dominantly three types of shear zones: E-W, NE-SW and NW-SE. The E-W trending, vertical shear zones constitute the subject matter of the present study and the areas where these shear zones are exposed have been shown. It may be noted that in addition to the E-W shear zones shown in this map, there are many such shears of relatively smaller extent that could not be shown because of the scale of the map, however, were studied because of their typical features, *e.g.* Jhankri, Kuraicha, Roni, etc.

Geology of the Shear Zones

Our recent structural studies revealed the occurrence of a number of kilometre-scale, isolated, ductile shear zones in the central part of the Bundelkhand massif (Fig. 2). These shear zones trend E-W; variation from this general trend is rare and is confined to about 10^0 only. The shear zones contain mylonitic rocks, and a striking feature is the vertical to subvertical attitude of the mylonitic foliation of the shear zone rocks. However, the rock types/lithogical associations vary with the individual shear zones.

Petrographically, the mylonites of the shear zones look much similar to each other (Fig. 3). The rocks generally show a good deal of pulverization and grain-size reduction with protoliths of feldspar, quartz and quartzo-feldspathic aggregates. The grain size reduction is mainly attributed to dynamic recrystallization of the original quartz and feldspar grains due to which the originally large, strained grains released their internal strain energy by the formation of smaller, strain-free, clear grains. Thus, with continued or prolonged process of dynamic recrystallization, the rock continued to become fine grained including the formation of ultramylonites. All these processes operate at great depths and under highly ductile conditions. Most of the shear zones rocks are ultrmylonites. Since the rocks of the Bundelkhand shear zones show a long history of recrystallization and deformation, the ultramylonites occasionally show evidences of late recrystallization (Figs. 3 A, B) in the form of development of large grains some of which contain pre-existing smaller grains also that look like inclusions. The latter (*i.e.* pre-existing smaller grains) may also represent some 'blebs' left out during the recrystallization process. Since these shear zones have rose up in the crust, some of them have also preserved evidences of brittle, *i.e.* near-surface, deformation in the form of cataclasites (Fig. 3 C). Since the occurrences of vertical shear zones are very rare in the world, we still do not have any well worked-out mechanism for their present-day occurrence. However, an attempt has been made under 'Discussion'.

These shear zones extend over a length of about 150 km from the east of Koti in the west to Mauranipur in the east (Fig. 2) and occur as scattered outcrops. Individual shear zones have a strike length up to 5 km. and a width up to 300 m. North of this belt also, a series of E-W trending shear zones are prominently developed, while towards the south, such shear zones are rather poorly developed.

Field photographs of the rocks of a few representative shear zones have been shown in Fig. 4. In general, these shear zones are oriented E-W with variation between N 80^0 E and N 80^0 W. Detailed studies of these shear zones have been carried out in Pahunj, Dely, Raksa, Shivgarh, Rajapur, Babina, Badera, Sukanwa, Dhaura, Jhankri, Kuraicha-Mauranipur, Roni and the areas adjoining to all these localities. Some field characteristics of these shear zones are highlighted below.

In *Pahunj* (25^0 27′46″: 78^0 31′42″), about 1 km west of Jhansi, the host rock is pink granite that is mylonitized. Some lenses of pink granite does not show mylonitic foliation as well as signatures of any E-W shearing. This shear zone is also affected by a NNW-SSE trending fault along which the mylonites are brecciated.

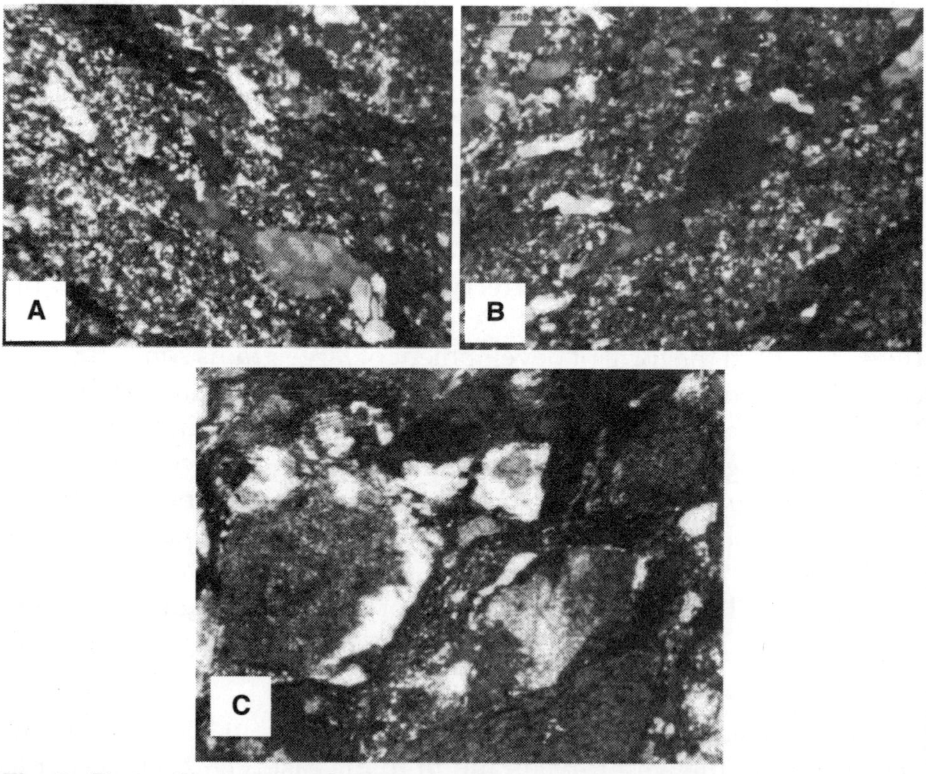

Fig. 3: Photomicrographs of some common rocks of the shear zones. A & B- Ultramylonite showing high degree of grain-size reduction because of dynamic recrystallization. Locality: A-Dely, B-Dhaura; C- Cataclasite showing angular grains set in a fine grained matrix. A few larger, angular grains occur in the fine matrix. Locality: Babina.

At *Dely* (25^0 27′25″: 78^0 30′85″), the width of the shear zone is 500 to 600 m. The shear zone rocks are mainly mylonitic gneisses showing vertical foliation striking N80^0E- S80^0W. The adjoining gneisses with foliation dipping 75^0 to 65^0 towards S and ptygmatic folding is commonly noticed.

North of *Raksa* (25^0 26′ 96″ : 78^0 29′70″), the hornblende- and biotite-bearing gneissic rocks are mylonitized with vertical/subvertical mylonitic foliation showing dips up to 80^0 towards N. Copper mineralization occurs along the E-W shear zone.

In the *Shivgarh* area, the mylinitic foliation shows vertical attitude (Fig. 4 a) with E-W orientation. This E-W shear zone is superimposed by at least three distinct younger lineaments or possibly shears zones: (1) A NE-SW trending shear zone represented by quartz reef showing dextral sense of shear, (2) A N-S trending shear zone represented in the form of epidote veins and quartzo-feldspathic veins with minor mineralization. This N-S trending shear is a relatively rare feature in the Bundelkhand terrain. (3) A NW-SE trending shear zone represented by the dolerite dikes. Thus, the E-W shear seems to be the oldest followed (possibly) by the NE-SW and NW-SE shears.

Near *Dhaura* (25^0 10′ 9″: $78^0$31′ 67″), rhyolites are exposed showing strong mylonitic foliation which is vertical (Fig. 4 b) with E-W trend. In a nearby hillock, about 500 m south of this spot, the shear zone is seen intruded by massive pink granite.

Near *Rajapur* (25^0 25′ 75″: 78^0 22′ 62″), the shear zone is represented by quartzo-feldspathic gneisses and the mylonitic foliation is vertical with E-W strike.

South of *Babina* (on the Sukunwa road), the E-W shear zones with ultramylonites (Fig. 4 c) are exposed at least in three localities (25^0 13′ 10″ : 78^0 27′ 94″; 25^0 12′ 08″ : 78^0 28′ 06″ and 25^0 12′ 08″ : 78^0 25′ 49″).

At *Badera* (25^0 12′ 36″: 78^0 29′ 60″), the mylonitic gneisses show vertical foliation striking E-W. Within the shear zone, a variety of folds - mostly sheath folds, sheath-like folds, shear folds and pack of similar folds, are developed. Common occurrence of sheath folds suggests dominance of a rotational component during shear deformation. Occasionally, the sheath folds are refolded suggesting progressive deformation during ductile shearing. The shear zone is also intruded by pink granite and grey granite, and is affected by faults of two different orientations: N-S and NE-SW. The faults have developed after the intrusive activity and of these two faults, the NE-SW trend is older than the N-S trend. A number of quartz and quartzo-feldspathic veins, that are parallel to the mylonitic foliation, are seen offset by these faults.

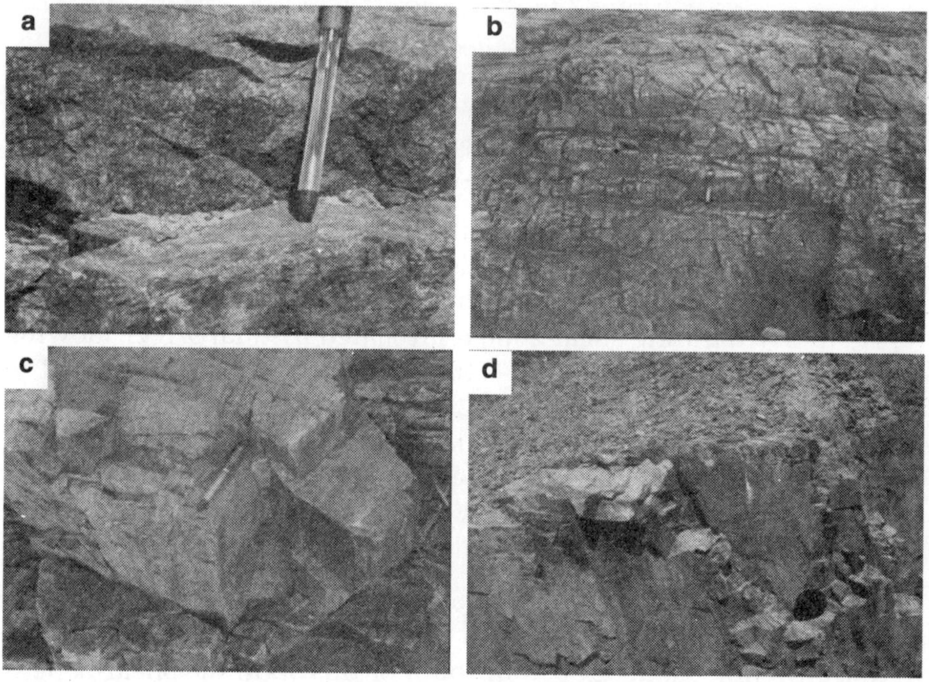

Fig. 4: Field photographs showing the typical occurrence of the vertical shear zones in the Bundelkhand terrain. The rocks are mylonites. (a) Near Shivgarh, (b) At Dhaura, (c) Near Koti, about 6 km SE of Babina, (d) North of Pura, about 5 km NW of Sukumwa Dam.

Near *Sukunwa* Dam, the E-W trending shear zone is noticed at two localities (25^0 12′ 01″: 78^0 30′ 00″ and 25^0 11′ 55″: 78^0 32′ 39″). Here mylonitic gneisses show vertical foliation (Fig. 4 d) with E-W strike. The mylonites of the E-W shear zone show sharp contact with porphyritic pink granite. The undeformed pink granite is an intrusive into the E-W shear zone and is thus post E-W shear. The gneissosity of the surrounding (host) rocks on either side of the shear zone trends NW-SE which is the general regional trend of the gneissosity of the Bundelkhand massif.

At *Jhankri* (25^0 16′ 78″: 79^0 08′ 11″), the granitic and schistose rocks are exposed. The granitic rocks are mostly massive and the foliation is faintly preserved which shows steep dips of about 80^0

towards N. The schistose unit exposes a variety of rocks mainly quartzite, hornblende-chlorite schist, actinolite-chlorite schist, metamorphosed volcano-sedimentary rocks and all these rocks commonly contain pegmatite veins. The rocks of the shear zone show a variety of folds, most of which are refolded hook-shaped folds. These folds occasionally occur in a pack and are involved in shearing with shear planes dipping approximately 80^0 towards N. The shear sense markers indicate top-to-south sense of shear. It appears that the schistose unit, which is relatively older than the granitic units, has been thrust/sheared towards south – a situation also noticed in Sukunwa Dam area.

In the *Saprar* River section near Kuraicha-Mauranipur, augen gneisses are exposed and the gneissosity shows NW-SE strike and dips of 70^0 to 80^0 NE, while north of the river, the gneisses show E-W strike with dips of ca. 80^0 N. Between these two outcrops, the rocks show extremely complicated structures as can be noticed along the river bed (width ca. 20 m) where the rocks show strong evidences of rotation of the marker lines/planes; occurrence of sheath folds is a common features. It is possible that in this area the gneisses show the development of a shear zone with an E-W strike.

At *Roni* (25^0 12′ 70″: 79^0 07′ 07″) granitic gneisses are exposed and contain well developed augens of feldspar and quartzo-feldspathic material. These rocks are affected by a shear zone of ca. 35 m width. The shear sense markers indicate transport towards S.

Discussion and Conclusion

1. Our detailed study indicates that central part of the Bundelkhand craton shows the occurrence of an E-W trending shear zones. The rock types (mostly mylonites and ultramylonites) of these shear zones typically show vertical/subvertical foliation- a fact not reported/highlighted in the literature so far. Report of vertical shear zones from the Bundelkhand craton, as incorporated in this work, appears to be first of its kind in the Indian subcontinent.

2. The shear zone show evidences of their formation at great depths and under highly ductile deformation regime. Some common structures shown by the shear zone rocks include sheath folds, intrafolial folds, highly flattened non-cylindrical folds, S-C fabrics with very low angles between S and C planes, etc.

3. As yet, we do not have any well worked-out model of vertical shear zones from any part of the world. It is therefore too immature to suggest any model at this stage. It is however possible that the vertical shear zones of the Bundelkhand craton may represent the remnant of the root zones of any large scale, crustal shear zones whose upper sedimentary cover may have been eroded away. If it is so, then the vertical shear zones of the Bundelkhand craton may represent culmination of a long period of erosion since the Archaean times.

4. The location of the E-W shear zones almost in the central part of the Bundelkhand craton possibly indicates that the craton is dissected in two tectonic domains. The E-W shear zones of the Bundelkhand massif together constitute a system of very large scale structure that practically dissects the entire Bundelkhand massif.

5. Occurrence of vertical shear zones is very rare in the world. As such, we still do not have any well worked-out model for their formation. It is, therefore, too early to provide a suitable explanation for the occurrence of the Bundelkhand vertical shear zones. As is known, the ductile shear zones rise up in the crust initially vertically because the mass is highly ductile. Under such conditions, ó$_1$ is much stronger such that the difference between ó$_1$ and ó$_3$ is very high and ó$_2$ was close to ó$_1$. This situation promotes vertical rise of the ductile mass. In the context of the Bundelkhand craton, however, it can be said at this stage that the vertical nature of these shear zones could represent the deep seated portion of the associated crustal shear zone whose upper, sedimentary cover possibly may have been eroded away. As such the present vertical nature of the shear zone may mark the culmination of deep erosion since the Archaean times to expose the rather basal part of the shear zones.

6. At most of the places, top-to-south sense of shear is also indicated, *e.g.* Jhankri, Sukunwa Dam site, Dhaura, etc. This situation possibly corresponds to a collision between two blocks or micro-plates during the Archaean times possibly corresponding to some very early, unknown, large scale crustal deformation in the central part of the massif. Singh *et al.* (2007) proposed four tectono-thermal events in the Bundelkhand craton. The 3000 to 3500 Ma radiometric dates (Mondal *et al.*, 2002) point out that

the first phase of metamorphism (M_1 event) was in the Middle Archaean times (Singh and Dwivedi, 2009). Subsequent to this event, the volcano-sedimentrary rocks were deposited and metamorphosed (M_2 event) to the greenschist facies. The intrusives of Bundelkhand granitoids (2500 to 2400 Ma) - that are undeformed and without involving E-W shears indicate that the E-W shear zones were developed after the low grade metamorphism. The E-W shears, thus, imply that there may have been at least one major orogenic movement.

7. These shear zones have transferred a major bulk of sub-crustal/ mantle material on to the upper crust and some of these rocks bear base metals. The E-W shear zones may, thus, hold more potential of showing metallic mineralization in the Bundelkhand terrain than some other lineaments or structural zones. All this also provide an alternative mechanism and additional information in understanding the evolution of the Archaean crust and continental growth (see also Singh *et al.*, 2007; Mondal *et al.*, 2002).
8. The above results also imply that the Bundelkhand craton should not be considered as a part of the prominently NE-SW trending Aravalli craton as is believed by some workers (*e.g.* Naqvi and Rogers, 1987).
9. It is, therefore, possible that in the early stages of crustal evolution, the Archaean crust was thin. The nature of the thin crust may be mainly responsible for the formation and preservation of these vertical shear zones. Subsequently, all these factors gradually thickened the crust and all this changed the crustal components and configuration. As such, the occurrence of vertical shears in Archaean terrains throws significant light in understanding the Archaean tectonics and crustal evolution in a better way.

Acknowledgements

SPS thanks to the Ministry of Mines, Govt. of India, for funds. ARB thanks Prof. N.L. Chhabra, Head, Centre of Advanced Study in Geology, University of Lucknow, for providing working facilities.

References

Arthaund, F. and Matte, P. (1977) Late Paleozoic strike slip faulting in the southern Europe and northern Africa result of aright lateral shear zone between the Applacian and Ural. Bull. Geol. Soc. Amer., v. 88 (9), pp. 1305-1320.

Basu, A.K. 1986. Geology of parts of the Bundelkhand Granite Massif, Central India. Rec. Geol. Surv. Ind., v. 117 (2), pp. 61-124.

Bhattacharya, A.R. (1987) A 'Ductile Thrust' in the Himalaya. Tectonophysics, v. 135, pp. 37-45.

Bhattacharya, A.R. and Weber, K. (2004) Fabric development during shear deformation in the Main Central Thrust zone, NW-Himalaya, India. Tectonophysics, v. 387, pp. 23-48.

Bhattacharya, A.R. (1986) Wavelength-amplitude characteristics of polyphase folds in the Precambrian Bundelkhand Complex, India. Tectonophysics, v. 128, pp. 121-125.

Gansser, A. (1964) Geology of the Himalayas. Interscience Publishers, London, 289 pp.

Brun, J.-P. and Burg, J.-P. (1982) Combined thrusting and wrenching in the Ibero-Armorican Arc: a corner effect during continental collision. Earth and Planetary Science Letters, v. 61 (2), pp. 319–332.

Burg J.-P. (1999) Ductile structures and instabilities: their implication for Variscan tectonics in the Ardennes. Tectonophysics, v. 309 (1-4), pp. 1-25.

Burg, J.P., Bale, P., Brun, J.-P., and Girardeau, J. (1987) Stretching lineations and transport direction in the Ibero-Armorican arc during the Silurodevonian collision. Geodinamica Acta, v. 1, pp. 71–81.

Carreras, J. (2001) Zooming on Northern Cap de Creus shear zones. J. Structural Geol., v. 23, pp. 1457-1486.

Corsini, M., Vauchez, A., and Caby, R. (1996) Ductile duplexing at a bend of a continental-scale strike-slip shear zone: example from NE Brazil. J. Structural Geol., v. 18, pp. 385-394.

Druguet, E., Passchier, C. Carreras, J. and den Brok, V. (1996) Analysis of a complex high-strain zone at Cap de Creus, Spain. Tectonophysics, v. 280, pp. 31-45.

Knipe, R. J. (1989) Deformation mechanisms- recognition from natural tectonites. *J. Structural Geol.*, v. 11, pp. 127-146.

Le Fort, P. (1975) Himalayas, the collided range: present knowledge of the continental arc. Amer. Jour. Sci., v. 275-A, pp. 1-44.

McCaig, A. M. and Knipe, R. J. (1990). Mass-transport mechanisms in deforming rocks; recognition using microstructural and microchemical criteria. Geology, v. 18, pp. 824-827.

McCaig, A. M. (1984) Fluid-rock interaction in some shear zones from the Pyrenees. J. Metamorphic Geology, v. 2, pp. 129-141.

Mondal, M.E.A, Goswami, J.N., Deomurari, M.P. and Sharma, K.K. (2002) Ion microprobe $^{207}Pb/^{206}Pb$ ages of Zircons from the Bundelkhand massif, northern India, implications for crustal evolution of the Bundelkhand-Aravalli Proto-continent. Precambrian Research, v. 117, pp. 85-110.

Naqvi, S.M. and Rogers, J.W. (1987) Precambrian Geology of India. Monographs on Geology and Geophysics, v. 6. Oxford University Press, Oxford, 223 p.

Neves, S.P. Vauchez, A.C.J. and Archanjo J. (1996) Shear zone-controlled magma emplacement or magma-assisted nucleation of shear zones? Insights from northeast Brazil. Tectonophysics, v. 262 (1-4), pp. 349-364

Newman, J. and. Mitra, G. (1993) Lateral variations in mylonite zone thickness as influenced by fluid-rock interactions, Linville Falls Fault, North Carolina. J. Structural Geol. v. 15, pp. 849-863.

Pueyo, E, Liesa C. and Sota, R. (2006) The Varscan-Millares granite (Central Pyrenees): pluton development in a T fracture of a dextral shear zone. Geodynamic Acta, v. 19 (3-4), pp. 197-211.

Roday, P.P., Diwan P. and Singh, S. (1995) A kinematics model of emplacement of quartz reef and subsequence deformation pattern in the central Indian Bundelkhand batholith. Proc. Indian Academy of Sciences (Earth and Planetary Sciences), v. 104 (3), pp. 465-488.

Sharma, K. K. (1998) Geological evolution and crustal growth of the Bundelkhand craton and its relict in the surroundings regions, northern Indian shield. In: *B.S. Paliwal, (ed.),* The Indian Precambrian. Scientific Publishers India, Jodhpur, pp 33-43.

Singh, S.P., Singh, A.N. and Singh, Y. (2007) Lineament fabrics and major shear zones in Bundelkhand and tectonic interpretation from IRS-ID data. Himalayan Geology (Abstract Vol.) 28, pp. 47.

Singh, S.P., and Singh, M. M., Srivastava, G.S., and Basu, A.K. (2007) Crustal evolution in Bundelkhand area, Central India. J. Himalayan Geol., v. 28 (2), pp. 79-101.

Singh, S. P. and Dwivedi, S. B. (2009) Garnet–sillimanite–cordierite–quartz bearing assemblages from early Archean supracrustal rocks of Bundelkhand Massif, Central India. Curr. Sc., v. 97, pp. 103-107.

Singh, S. P., Balaram, V., Satyanarayanan, M., Anjaiah, K. V. and Kharia, Aditya (2010) Platinum group elements in basic and ultrabasic rocks around Madawara, Bundelkhand Massif, Central India. Curr. Sci., v. 99, (3), pp. 375-383.

Westaway, R. and Kusznir, N. (1993) Fault and bed "rotation" during continental extension: block rotation or vertical shear? J. Structural Geol., v. 16, pp. 753-770.

□□□

7

First Report of Spheroidal Rhyolite from Nakora Area of Malani Igneous Suite, Northwestern Peninsular India

G. VALLINAYAGAM AND NARESH KUMAR
Department of Geology, Kurukshetra University, Kurukshetra (Haryana)

Abstract

Occurrence of rhyolite with spheroidal structure is being reported first time in the Neoproterozoic Malani Igneous Suite (MIS) of Northwestern Peninsular India. MIS consists predominantly of felsic volcanic rocks, felsic plutonic rocks and minor amount of mafic rocks. A spheroidal rhyolite that is shaped like a sphere but is not perfectly round, especially an ellipsoid that is generated by concentric shells of different colours with different compositions revolving around one of its axes. The spheroidal rhyolite is exposed in an area of 200 sq.m. radius in the Nakora area. It is pale yellow to light brown colour, fine grained and consists essentially of quartz, alkali feldspar with accessories of altered ferromagnesian (arfvedsonite) and iron oxides (hematite and magnetite). Spheroidal structure is up to 5 m dimension and shows large number (up to 70) of concentric shells in a few cm thickness surrounding a nuclei. Nakora rhyolites display different hues *viz.* dark brown, light pink and pink but dark brown rhyolite is associated with pink colour spheroidal rhyolite where their contact is sharp. Geological setting and the petromineralogical characters of spheroidal rhyolite suggest that at early magmatic processes, the supersaturated liquid crystallized rapidly into relatively differential density silicate liquid phases and growth of crystals lead the formation of spheroidal structure in the studied rhyolite.

Keywords : Spheroidal rhyolite, Neoproterozoic, Malani Igneous Suite, arfvedsonite

Introduction

In igneous rocks, varieties of globular structures *viz.* vesicular, amygdaloidal, orbicular, rapakivi, spherulite, variolite, drusy growth, ocellar, perlitic and spheroidal are realized. They are interpreted as evidence of difference in physical and chemical conditions of the magma at various stages of crystallization processes. Identification and characterization of the minerals crystallization history of the magmatic stages are attached a great importance in the igneous petrology to understand the formation of earth crust. The present paper reports the occurrence of one of such globular structures *i.e.* spheroidal structure associated with the felsic volcanic rocks in the Nakora area of Neoproterozoic Malani Igneous Suite (MIS), Western Rajasthan. The rocks of Nakora belong to MIS. MIS represent the largest felsic magmatism of India and covering an area of approximately 55,000 sq. km in the Trans Aravalli Block (TAB) of Northern Peninsular India. The rocks of MIS are mainly of felsic volcano-plutonic with less amount of mafic lithounits. They are well exposed in Tusham (Haryana), Jhunjhunu, Siwana, Jalor (Rajasthan) and also in Nagar Parkar (Sind-Pakistan), Kirana (Lahore-Pakistan) areas. They are characterized as A-type, peralkaline, peraluminous, anorogenic and owe its origin to hotspot tectonics (Kochhar, 1984, 2000, 2004; Pareek, 1981; Bhushan and Chandrasekaran, 2002; Vallinayagam, 2004).

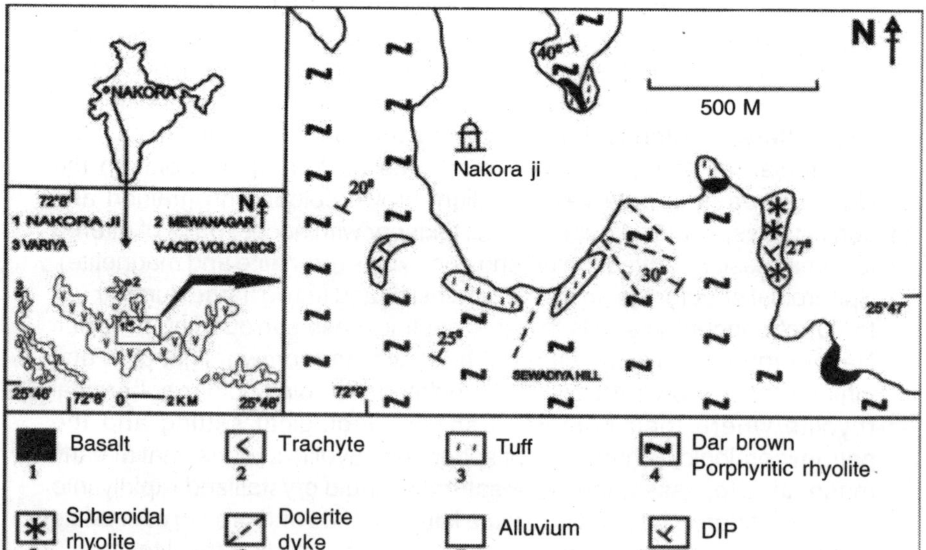

Fig.1: Geological sketch map with Spheroidal Rhyolite outcrop, Nakora area, Rajasthan

Field Features of Spheroidal Rhyolite

During recent geological fieldwork, the spheroidal structure was observed in the felsic volcanic rocks on the Northeastern flank of Sewadiya hill of Nakora area (Fig.1), District Barmer, Rajasthan. Nakora area mainly consists of felsic volcanic rocks (Vallinayagam and Kumar, 2007). These felsic volcanic rocks comprise of rhyolite (with various shades, mainly dark brown and also pink, light brown), pyroclasts and tuff (mainly yellow, also green). The spheroidal rhyolitic rocks are outcropped in the area of having dimension of about 200 sq.m. radius (GPS reading N 25° 47' 21·2", E 72° 9' 61·6") and are associated with sand dunes (Fig. 2A). The spheroidal rhyolites are fine grained, yellow colour and tuffaceous in nature (Fig. 2B, 2C). Spheroidal structures occur as 30cm to 5m dimension and comprise of numerous concentric shells (50–70 shells) of different thickness (3 mm to 7 cm) around a nuclei. They occur mainly in circular/ semicircular forms (Fig. 2B). The shells consist of light and dark colour concentric layers (Fig. 2C). At some places, they also occur in the square forms (Fig. 2D). The light colour layers consist of fine grained quartz, alkali feldspar and dark colour layers mainly consist of hematite and magnetite with minor amount of altered arfvedsonite minerals. Spheroidal rhyolite shows sharp contact with brown rhyolite. The brown rhyolite is porphyritic in nature with orthoclase phenocrysts. Based on the detailed fieldwork, the volcanostratigraphic sequence (older to younger) is established (Vallinayagam and Kumar, 2008) in the present area as follows:

1. Basalt 2. Trachyte 3. Tuff 4. Brown rhyolite 5. Spheroidal rhyolite

The basalt flows underlie the rhyolite flows (Fig. 1). The basalt flows are dark brown and black in colour, fine grained and showing vesicular (Fig. 2E), amygdaloidal structures. The trachytic flows are associated with rhyolite. It is fine grained, light brown, grey, blue in colour and porphyritic as well as non-porphyritic in nature. Xenoliths of basalt (Fig. 2F) and trachyte are seen in the rhyolites which indicate the rhyolite flows are youngest among all volcanic flows.

Petrography and Geochemistry

Rhyolites are fine grained, phyritic and aphyritic types. Aphyritic type rhyolites show flow structures. Rhyolite consists of alkali feldspar (orthoclase and perthite 50-60 vol%), quartz (25-30 vol%), arfvedsonite as ferromagnesian minerals and hematite, magnetite as accessories. Orthoclase phenocryst occurs as tabular form and large size quartz crystals embedded along the flow direction. Perthite is medium grained, anhedral form and vein perthitic texture. The arfvedsonite is coarse grained, dark green colour and pleochroic (X- dark bluish green, Y- pale green, Z-yellow green). Iron oxides are fine to medium grained, irregular shape, brown (hematite) and black (magnetite) colour are scattered in the fine quartzofeldspathic groundmass.

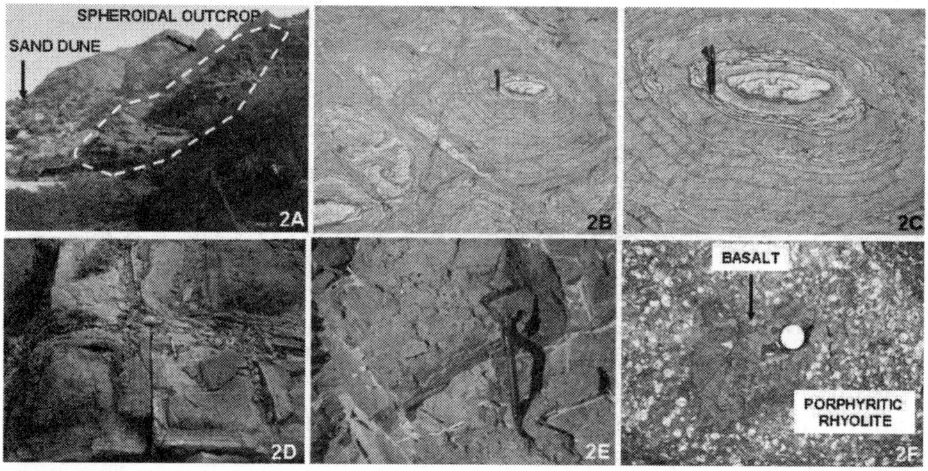

Fig. 2: A. Panoramic view of spheroidal outcrop in the study area. B. Field exposure of Spheroidal Rhyolite shows numerous spheroidal structures in circular/ semicircular form of various dimensions. C. Field photograph shows the enlarged view of one of such spheroidal structure as above. It depicts the alternate arrangement of shells with felsic and mafic minerals layers. D. Top square view of spheroidal rhyolite shows 0.5 m x 0.5 m dimension. E. Basalt shows small scale vesicles (3 mm diameter) adjacent to spheroidal outcrop. F. Xenolith of basalt in the light brown porphyritic rhyolite indicating that basalt is the earliest flow.

The spheroidal rhyolites show numerous concentric shells represented by above similar mineralogical assemblages but they are cryptocrystalline. It exhibits pale to dark brown and whereas light grey layers represent flow texture. The brown layers consist of fine grained mafic (hematite, magnetite and arfvedsonite) minerals. The grey layers consist of medium grained felsic (quartz, orthoclase, perthite) minerals (Fig.3A). The orthoclase crystals are medium grained, lath-prismatic shaped, colourless and show carlsbad twinning. It is randomly fractured and kaolinized (Fig. 3B). The sericitised plagioclase feldspars are cut by fine quartz veins and they show high order interference colour. Quartz is medium grained and occurs as phenocryst as well as in the groundmass. It shows wavy extinction and first order grey interference colour (Fig. 3C,D). Arfvedsonite is fine grained, prismatic, dark green colour and pleochroic. It also occurs in an aggregate form and is wrapped in the flow structure composed of fine quartzofeldspathic minerals. Anhedral hematite and magnetite are associated with arfvedsonite. Trachyte shows cryptocrystalline nature and consists of orthoclase, arfvedsonite and riebeckite minerals with the flow direction (Fig. 3E). The tuffs are micro-crypto-crystalline in nature with flow bands and consist of quartz and alkali feldspar (Fig. 3F). Basalt displays well developed ophitic texture with dominant plagioclase.

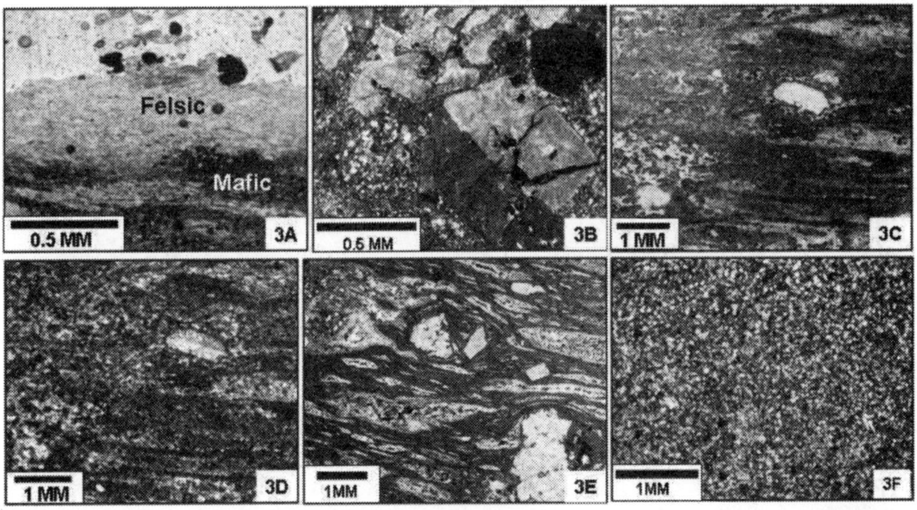

Fig. 3: A. Photomicrograph (PPL) displays the rhythmic arrangement of felsic–mafic minerals in the layers of the shell of spheroidal rhyolite. B Photomicrograph (XPL) shows the prismatic orthoclase showing carlsbad twinning, fracturing and sericitization which is embedded in the ground mass. The groundmass consists of quartz, alkali feldspar, hematite, arfvedsonite and magnetite. C & D. Rhyolite shows orientation of quartz crystals with the flow direction (under PPL and XPL respectively). E. Trachyte shows cryptocystalline nature of quartz, alkali feldspar, hematite and magnetite along the flow direction (PPL). F. Tuffs are micro- to cryptocrystalline in nature with flow bands and consist of quartz and alkali feldspar (XPL).

Based on the field and petromineralogical studies, samples were selected and analyzed for geochemistry. Major elements are analyzed by using wet chemical method after Shapiro and Brannock (1962) using Systronics UV–VIS spectrophotometer - 108 and Atomic Absorption Spectrophotometer (Varian 240 FS –AAS) and Systronics Mediflame Photometer–128 at Department of Geology, Kurukshetra University, Kurukshetra. Standardization for major elements was based on USGS rock standards RGM-1, JG-2 and MRG-1.The analytical precision is found to be in the error level of <5% for major elements.

Table-1 : Major element data (in wt %) of selected Nakora rocks, Barmer District, Rajasthan.

Rock Type Sample no.	Pink rhyolite 170	Dark brown rhyolite 35	Light brown rhyolite 113	Yellow tuffaceous rhyolite 2
SiO_2	68.2	71.9	64.8	65.6
TiO_2	2.34	2.52	2.52	2.94
Al_2O_3	7.6	7.6	11.7	13.8
Fe_2O_3	7.4	9.7	9.2	7.2
MnO	1.9	0.31	0.69	0.81
MgO	1.64	0.98	1.64	2.46
CaO	3.01	1.2	1.6	0.6
Na_2O	1.18	1.04	1.02	1.07
K_2O	6.51	4.48	5.26	4.93
P_2O_5	0.59	0.6	0.66	0.6
H_2O	0.3	0.3	0.2	0.2
TOTAL	100.6	100.6	99.2	100.2
CIPW Norms				
Quartz	33.22	46.37	34.73	37.25
Orthoclase	38.47	26.47	31.08	29.13
Albite	2.85	8.8	8.63	9.05
Anorthite	-	2.03	3.63	0.75
Corundum	-	0.29	3	7.05
Diopside	8.08	-	-	–
Hypersthene	0.34	2.44	4.08	6.13
Acmite	6.29	-	-	–
Illmenite	4.06	0.66	1.48	1.73
Hematite	5.23	9.7	9.2	7.2
Apatite	1.37	1.39	1.53	1.39
Sphene	0.49	-	-	–

Rutile	-	2.77	2.22	2.59
DI	74.54	81.64	74.44	75.43
AI	1.19	0.85	0.62	0.51
Na+K	7.69	5.52	6.28	6
Na/K	0.18	0.23	0.19	0.22
KN/C	2.55	4.6	3.93	1.66
A/CNK	0.54	0.88	1.1	1.75
An/(An+Ab)	-	0.19	0.3	0.08
Density (g/cc)	2.82	2.82	2.86	2.87
Viscosity (pas/sec)	0.66	0.84	0.66	0.69

Generally, Nakora rhyolites show high SiO_2 (64.80 – 71.90 wt %), high Al_2O_3 (7.20 - 13.10 wt %), low total alkalies (5.42 - 7.99 wt %) (Na_2O: 0.94 – 1.18 wt %, K_2O: 4.48 - 6.81 wt %), low total iron (5.40 – 9.70 wt %), low TiO_2 (1.61 - 2.84 wt %), high CaO (0.80 - 3.64 wt %) and high MgO (0.49 – 3.28 wt %). Average values of SiO_2, Al_2O_3, Na_2O, K_2O, Fe_2O_3, TiO_2, CaO, MgO, MnO and P_2O_5 (in wt %) are 68.46, 9.73, 1.05, 5.65, 7.74, 2.40, 1.92, 1.62, 0.73 and 0.55 respectively. The Nakora rhyolites show density (2.76 - 2.86) and viscosity (0.64 - 0.84) and average values are 2.81 and 0.74 respectively.

Two samples (sample no. 170- pink rhyolite and no. 2- yellowish tuffaceous rhyolite) from spheroidal outcrop are compared with other Nakora rhyolite samples (no. 113- light brown rhyolite and no.35- dark brown rhyolite) (Table-1). Pink rhyolite and tuffaceous rhyolite are more enriched in Al_2O_3, Na_2O, K_2O, TiO_2, MnO and MgO whereas it shows low contents of SiO_2 and Fe_2O_3 than the concentrations of other selected Nakora rhyolites. High density and low viscosity is observed in tuffaceous rhyolite than other Nakora rhyolites.

In the normative composition, they show high concentrations of orthoclase, corundum and illmenite with high ratios of AI, Na+K and A/CNK but low contents of quartz, anorthite, hematite with DI, KN/C and An/(An+Ab) ratios.

Discussion

An increasing number of studies have noted that the globular structures occur in the igneous rocks as primary structures. They are formed within the magma before consolidation as a result of the rapid crystallization of a supersaturated liquid. It contains scattered crystal nuclei whose growth is controlled by the mass transfer processes. The concentration of the components across the crystal face, being highest at the edges and lowest near the face centers. Various other studies highlighted the factors in the formation of spheroidal rhyolites viz. impact of changes in vapour pressure during crystallization; supersaturation of magma in the volatile components and bubbles of gaseous phase; increase in the water content and the result of liquid immiscibility above the temperature at which labradorite crystallized (Harker, 1909; Tammann, 1925; Niggli, 1937; Verhoogen, 1951; Philpotts, 1972; Drever, 1960; Bowen, 1928). Nakora spheroidal rhyolites consist of identical mineralogy with those of adjacent rock but without plagioclase feldspar. They are showing a consequent growth from nuclei which is distributed throughout the structure and in turn, is controlled by the liquid immiscibility and subsequent retrograde boiling processes (Phillips, 1973). Spheroidal layers are richer in sodium, aluminium and lower in silica, potassium and magnesium than the groundmass, a trend consistent with their domination by albite rhythms (Gundobin and Paradina, 1987). Nakora spheroidal rhyolites show enrichment of alkali feldspar which point towards possible geochemical differentiation process. The absence of spherical clusters of radiating crystals in Nakora samples, neglect the roll of diffusion and uptake of crystal forming components from the host rhyolite melt or glass and rejection of non-crystal forming components from the crystallizing region (Watkins *et al.*, 2009). Detailed geological field observations and petromineralogical studies of the spheroidal rhyolite of Nakora area suggest that the spheroidal structure is formed due to rapid crystallization of the supersaturated liquid into relatively differential density silicate liquid phases and growth of crystals containing large crystal nuclei at the early stage of magmatic processes. Present observations of shells consist of light and dark colour concentric layers in the spheroidal rhyolites indicate that the differential density silicate liquid phases and growth of dark and light mineral assemblages around the nuclei. The pyrogenetic minerals (olivine, pyroxene, and plagioclase) are absent in the Nakora spheroids which indicates that they are crystallized from a completely anhydrous magma. Hence, Nakora spheroids are not formed under the process of retrograde boiling. The Nakora spheroids are richer in Al_2O_3, Na_2O, K_2O, TiO_2, MnO and MgO and lower in SiO_2 and Fe_2O_3 than the concentrations of other selected Nakora rhyolites which indicates the geochemical differentiation of spheroidal rhyolites. Differential crystal growth in a volcanic environment assign immense significant as it has the possible impact on the magmatic crystallization at an early stage of magma eruption in the earth's crust forming process. Detailed studies on the globular structures in the magmatic rocks will provide the mechanical, chemical characters and related dynamics in the earth system.

Acknowledgements

The authors are thankful to University Grants Commission, New Delhi for the grant in the form of Major Research Project (no: F.31-193/2005 SR, dated 31st March 2006) to GV and Project Fellowship to NK. We sincerely acknowledge the facilities of Department of Science and Technology, New Delhi in the form of FIST Grant and our parent University in the form of University Golden Jubilee Grant. Earlier version of this manuscript was benefited by the expert comments of Professor M. Santosh (Japan). Useful comments of the Journal's referee improved the paper.

References

Bhushan, S.K. and Chandrasekaran, V. (2002) Geology and geochemistry of the magmatic rocks of the Malani Igneous Suite and Tertiary Alkaline Province of Western Rajasthan. Memoirs of Geological Survey of India, v. 126, pp. 1-129.

Bowen, N. L. (1928) The evolution of the igneous rocks. Princeton University Press, 334 p.

Drever, H. I. (1960) Immiscibility in the picritic intrusion at Igdorssuit, West Greenland. 21st International Geologcal Congress, Norden, v. 13, pp. 47-58.

Gundobin, G. M. and Paradina, L. F. (1987) Geochemical differentiation of spheroidal rhyolite from the region North of the Sea of Okhotsk. Doklady Earth sciences, v. 297 (6), pp. 167-171.

Harker, A. (1909) A natural history of igneous rocks. Methuen & Co. London, 384 p.

Kochhar, N. (1984) Malani Igneous Suite: Hot-spot magmatism and cratonization of the Northern part of the Indian shield. J. Geological Society of India, v. 25, pp. 155-161.

Kochhar, N. (2000) Attributes and significance of the A-type Malani magmatism, Northwestern Peninsular Indian shield. In: M. Deb (*ed.*) Crustal Evolution and Metallogeny in the Northwestern India. Narosa Publishers, New Delhi, pp. 158-188.

Kochhar, N. (2004) Geological evolution of the Trans-Aravalli Block (TAB) of the NW Indian Shield: Constraints from the Malani Igneous Suite (MIS) and its Seychelles Connection during Late Proterozoic. Special Publication of Geological Survey of India, v. 84, pp. 247-264.

Niggli, P. (1937) Das Magma und seine produkte unter besonderer Beruchsichtigung des Einflusses der leichfluchtigen Bestandteile. Leipzig, 379 p.

Pareek, H. S. (1981) Petrochemistry and petrogenesis of the Malani Igneous Suite, India. Bulletin of Geological Society of America, v. 92, pp. 206-273.

Phillips, W. J. (1973) Interpretation of crystalline spheroidal structures in igneous rocks. Lithos, v. 6, pp. 235-244.

Philpotts, A. R. (1972) Density, surface tension and viscosity of the immiscible phase in a basic alkaline magma. Lithos, v. 5, pp. 1-18.

Shapiro, L. and Brannock, W.W. (1962) Rapid analysis of silicates, carbonates and phosphate rocks. Bulletin of U.S. Geological Survey, 1144A, 56 p.

Tammann, G. (trans Mehl R F) (1925) States of aggregation. D.van Nostrand, New York, 119 p.

Verhoogen, J. (1951) Mechanics of ash formation. American J. Sciences, v. 249, pp. 729-739.

Vallinayagam, G. (2004) Peralkaline - peraluminous A - type rhyolites, Siwana Ring Complex, Northwestern India: Petrogenetic modelling and tectonic implications. J. Geological Society of India, v. 64, pp. 336–344.

Vallinayagam, G. and Kumar, N. (2007) Volcanic vent in Nakora ring complex of Malani Igneous Suite, Northwestern India. J. Geological Society of India, v. 70 (5), pp. 881-883.

Vallinayagam, G. and Kumar, N. (2008) Flow stratigraphy of Nakora ring complex, Malani Igneous Suite, Rajasthan, NW Peninsular India. Special Publication of Geological Survey of India, v. 91, pp. 127-135.

Watkins, J., Manga, M., Huber, C. and Martin, M. (2009) Diffusion-controlled spherulite growth in obsidian inferred from H2O concentration profiles. Contribution to Mineralogy and Petrology, v. 157, pp. 163-172.

□□□

8

Chemistry of Tourmalines from the Gangotri Granite, Garhwal Higher Himalaya

T. N. JOWHAR
Wadia Institute of Himalayan Geology, 33 GMS Road, Dehra Dun 248 001

Abstract

The Gangotri granite (23 ± 0.2 Ma) is one of the largest bodies of the High Himalayan Leucogranite belt (HHL) located in the Garhwal Himalaya. The Gangotri granite is situated structurally above the kyanite and sillimanite gneisses of the Vaikrita Group, which in turn, overlies the north-dipping Main Central Thrust Zone of inverted metamorphic isograds. Compared to other High Himalayan leucogranites, it is particularly enriched in tourmaline. This study focuses especially on mineral chemistry of tourmalines from the Gangotri granite. The Gangotri granite is composed of quartz+ K-feldspar+ plagioclase+ tourmaline+ muscovite ± biotite ± garnet ± beryl, with apatite as the most abundant accessory mineral. K-feldspar is microcline with microperthite. Tourmaline contains inclusions of plagioclase, apatite and monazite. All the analysed tourmalines belong to Alkali Group and are Schorl. Aluminum in T sites varies from 0.000 to 0.202. In Y sites, Al varies from 0.186 to 0.491, Mg from 0.037 to 0.684, Fe^{2+} from 1.639 to 2.218, Mn from 0.000 to 0.041 and Ti from 0.049 to 0.171. In X sites Na varies from 0.619 to 0.777 and (Na + Ca + K) from 0.645 to 0.831. These tourmalines are zoned and from core to rim Mg decreases whereas Fe^{2+}, Ti, Mn and Ca increase. There is a negative correlation between Mg and Fe^{2+}. These results show that there were changes in physical conditions with increased activity of boron during crystallization of the leucogranite magma. Application of two-feldspar geothermometer gives temperatures of subsolidus equilibration at about 441-270°C and plagioclase-muscovite gives temperature in the range of 448-339°C.

Keywords : Gangotri granite, Vaikrita Group, Tourmaline, Schorl, Geothermometer

Introduction

The importance of tourmaline for petrologic and metallogenetic studies is well established (*e.g.* Grew and Anovitz, 1996; Henry and Dutrow, 1996; London *et al.*, 1996). Tourmaline is stable over a wide range of pressures and temperatures and has a variable composition and is able to exchange components and volatile species with coexisting minerals and fluids as a result of changes in external conditions. Tourmaline is, therefore useful to monitor the physical and chemical environments in which it was developed (Manning, 1982; London and Manning, 1995; Keller *et al.*, 1999). This study focuses especially

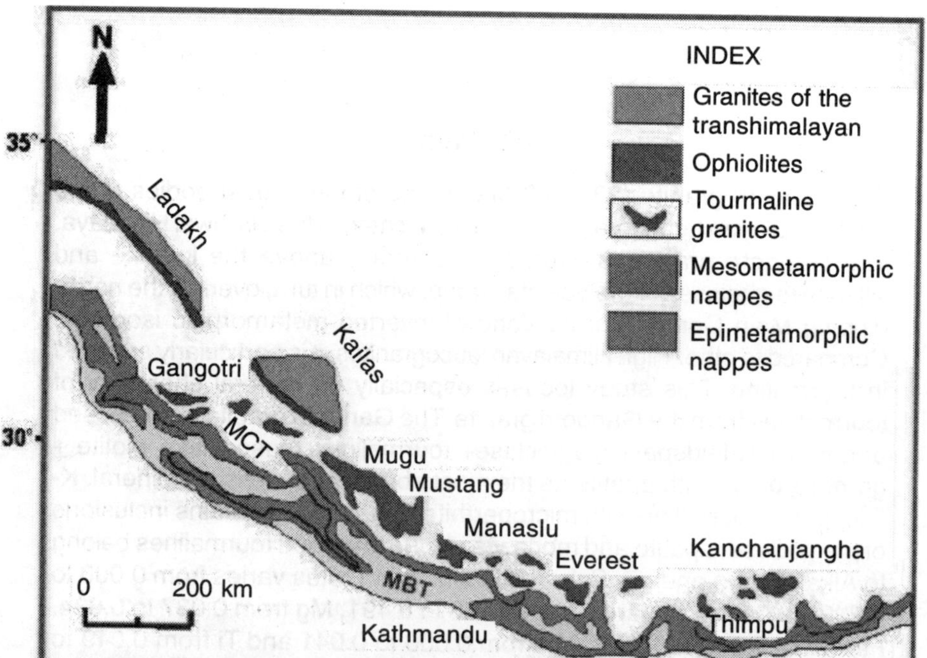

Fig. 1: Map showing the distribution of Higher Himalayan Leucogranites (Tourmaline Granites) in the Himalaya.

on mineral chemistry of tourmalines from the Gangotri granite (Fig.1) and associated pegmatites, which represent one of the Higher Himalayan Leucogranites. The Higher Himalayan Leucogranite Belt (HHL) is a result of the collision-related felsic magmatism with a strong peraluminous character having muscovite ± biotite and tourmaline. These leucogranites are of great interest as they help us to understand evolution of the continental crust.

Geological Setting

The Gangotri granite (Figs. 1 and 2) (23 ± 0.2 Ma, Searle *et al.*, 1999) is one of the largest bodies of the Higher Himalayan Leucogranite (HHL) belt located in the Garhwal Himalaya (Heim and Gansser, 1939; Gansser, 1964; Le Fort, 1975; Yin, 2006). It is exposed along the upper reaches of the Bhagirathi River around the Gangotri glacier region, including the peaks of Thalay Sagar (6904m), Bhagirathi (6856m), Meru (6672m), Shivling (6543m) and Bhigupanth (6044m). The granite was first described by Heim and Gansser (1939) near the village of Badrinath in the upper Alaknanda valley (Jowhar, 1994; Jowhar and Verma, 1995). Later, Auden (1949) described this granite as composed of tourmaline+muscovite+ biotite + garnet from the upper Bhagirathi valley, and is commonly termed as the Gangotri granite (GG). The Gangotri granite is commonly emplaced as lenses, dykes or as small plutons, which are 1.5-2 km thick and 4-5 km long in contrast to a single Manaslu pluton (Scaillet *et al.*,1990, 1995; Searle *et al.*,1993, 1999).The lenses intrude either the metamorphosed base of the Tethyan Sedimentary Zone, here called the Harsil Formation (Pant, 1986) or in a large body of two-mica porphyritic granite, which has been named as Bhaironghati granite (BG) by Pant (1986). Table-1 gives the lithotectonic setting of the Himalayan metamorphic belt in Garhwal Himalaya. Stern *et al.* (1989) noted that the petrographic and geochemical characteristics of BG are very similar to that of the Cambro-Ordovician felsic magmatism defined by Le Fort *et al.* (1986) in the entire Himalayan belt. The leucogranite is a viscous near-minimum melt, emplaced along foliation parallel laccolith via a dyke network not far from its source region. It was emplaced at mid-crustal depths along the footwall of the Jhala fault, a large-scale low-angle normal fault (part of the STD system), above kyanite and sillimante grade gneisses (Searle *et al.*, 1999).

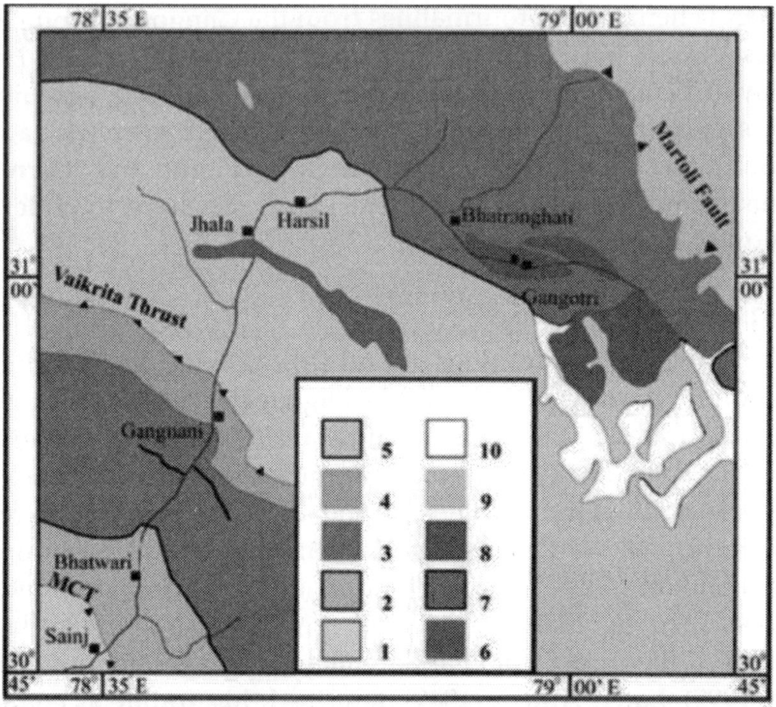

Fig. 2 : Geological map of the Higher Himalayan Crystalline (HHC) Belt along the Bhagirathi Valley, Garhwal. Legend: 1: Lesser Himalayan (LH) Proterozoic sequence; 2: Higher Himalayan Crystallines (HHC), Bhatwari Group-porphyroclastic granite gneiss, garnetiferous mica schist, amphibolites; 3: mylonitized augen gneiss, mica schist, amphibolite, 4: phyllonite, schist, 5: sillimanite/kyanite/staurolite/garneti-ferous schist/gneiss/migmatite, 6: augen gneiss, 7: Bhaironghati granite, 8: Gangotri leucogranite. 9: Tethyan Sedimentary Zone (Martoli Group). 10: Glaciers, debri etc. Abbreviations: MCT - Main Central Thrust, VT - Vaikrita Thrust, MF - Martoli Fault (from Singh *et al.,* 2003).

These two distinct granites (GG and BG) are exposed along the road section from Jangla to Gangotri and beyond. The GG is tourmaline-bearing leucogranite whereas BG is biotite granite/gneiss. At the Jangla bridge, biotite granite is deformed to a medium and coarse-grained granite gneiss at the lower structural levels, close to the contact with the Harsil metamorphics. This granite is strongly foliated to granite gneiss along the contact with the metamorphics. Numerous veins of tourmaline-muscovite leucogranite and garnet-beryl-tourmaline pegmatite intruded the biotite granite and the metamorphics. The temple at Gangotri is built on the tourmaline bearing leucogranite (Jain *et al.,* 2002). The GG intrudes both the BG and the Martoli Formation of the Tethyan Sedimentary Zone. The tourmaline bearing leucogranites cut the foliation of the surrounding older biotite granite gneiss and porphyritic granite (Bhaironghati granite) and metasedimentary host rocks. High Himalayan leucogranites in this area yield Th-Pb monazite ages of 22.4 $\pm$ 0.5 Ma (Gangotri), 21.9 $\pm$ 0.5 Ma (Shivling; Harrison *et al.* 1997), 23.0 $\pm$ 0.2 Ma (Shivling U-Pb age, Searle *et al.*, 1999), whereas 17-22 Ma $^{40}Ar/^{39}Ar$ mica ages were obtained for MCT zone and Higher Himalayan Crystallines (Metcalfe, 1993). Table-2 shows summary of the main tectonic processes and timing constraints for the Garhwal Himalaya and the leucogranite (Prince *et al.*, 1994, Searle *et al.*, 1999).

Table-1 : Lithotectonic setting of the himalayan metamorphic belt, Garhwal Himalaya.

Tectonic unit	Lithostratigraphic unit	Lithology
Tethyan Sedimentary Zone	Martoli Formation	Slate-phyllite-quartzite
	----Martoli Fault----	
Himalayan Metamorphic Belt (HMB)	Higher Himalayan Crystalline Belt (HHC)	Gangotri Granite Bhaironghati biotite granite Megacrystic augen gneiss Sill/Ky/grt/bio/mu schist Gneiss, migmatite (Harsil metamorphics) -------- Vaikrita Thrust----- Phyllonite and chlorite schist Mylonitized megacrystic augen gneiss, mica schist amphibolite-metadolerite
	----Main Central Thrust----	
Lesser Himalayan Sedimentary Zone	Garhwal Group	Orthoquartzite-slate-phyllite-dolomite-volcanics

Sorkhabi *et al.* (1996,1999) documented quantitatively the cooling and denudation history of the Gangotri granites based on fission-track (FT) and $^{40}Ar/^{39}Ar$ ages. Muscovite $^{40}Ar/^{39}Ar$ age of 17.9 $\pm$ 0.1 Ma and a biotite age of 18 $\pm$ 0.1 Ma reflect cooling of the rocks through 300-350 ^{0}C, which is related to an Early Miocene pulse of denudation caused by a basement-cover detachment (the Martoli Normal Fault) above the leucogranites. A total of 15 apatite ages from a vertical profile

(2580-4370m) on the Gangotri granites yielded FT ages in the range of 1.5 ± 0.6 to 2.4 ± 0.5 Ma, indicating that the rock column with a relief of 1800m cooled through 130 ± 10 °C within only one million years during the Late Pliocene. They estimated an average denudation rate of 2 mm/yr for the past 2.4 million years. It is interpreted from these studies that there was one major pulse of tectonic denudation in Early Miocene and another erosional denudation in the Late Pliocene-Quaternary. Searle *et al.* (1999) also reported from the North Ridge of Shivling (from >5000m) K-Ar muscovite ages of 22 ± 1.0 Ma, fission track ages of zircons are 14.2 ± 2.1 and 8.8 ± 1.2 Ma and for apatites are 3.5 ± 0.79 and 2.61 ± 0.23 Ma. They also interpret very rapid cooling of the granite at rates of 200-350° C/Ma between 23-21 Ma, and tectonic unroofing and erosion removed 24-28 km of overburden during this time. Slow steady state cooling at rates of 20-30° C/Ma from 20-1 Ma shows that maximum erosion rates and unroofing of the leucogranite occurred during the early Miocene. This timing coincides with initiation of low-angle, north-dipping normal faulting along the South Tibetan Detachment system.

Table-2 : Summary of the timing of tectonic processes in the Garhwal Himalaya (after Searle *et al.*, 1999)

Time	Process
50-37 Ma	Crustal thickening and prograde metamorphism
37 Ma	Peak metamorphism
37-23 Ma	High-temperature sillimanite-grade metamorphism
23 Ma	Crustal anatexis; crystallization of leucogranite
23-21 Ma	Very rapid cooling (~ 175-350 °C/Ma) of leucogranite; initiation of normal faulting at top of slab along the STD
21-19 Ma	Ductile motion along the MCT shear zone along base of the slab
20-14 Ma	Slow steady cooling at ~ 30°C/Ma
14-1 Ma	Very slow cooling at ~ 16-20°C/Ma
1-0 Ma	Rapid erosion of upper 2-3 km of the slab by glacial erosion

Petrography

The Gangotri Granite (GG) is fine grained (1-2mm) composed of quartz + K-feldspar + plagioclase + tourmaline + muscovite ± biotite ± garnet ± beryl, with apatite as the most abundant accessory mineral (Fig.3). The GG is subdivided into two main types (i) biotite-granite: It is restricted to the boundary of the plutonic lenses, here biotite is the dominant ferromagnesian phase with subordinate tourmaline, and (ii) tourmaline facies: It is tourmaline-rich and the biotite is absent at a macroscopic scale. Tourmaline occurs as black euhedral-subhedral crystals up to 1cm in size, scattered throughout the granite or concentrated along layers by magmatic banding and can reach up to 5 modal%. It contains abundant inclusions of quartz, plagioclase and apatite. Biotite is restricted to the margins of the leucogranite bodies and reaches up to 2-3%. Biotite is enclosed in plagioclase, K-feldspar, muscovite and quartz. Both micas and tourmaline are in textural equilibrium and there is no evidence for late metasomatic activity in the main leucogranite body.

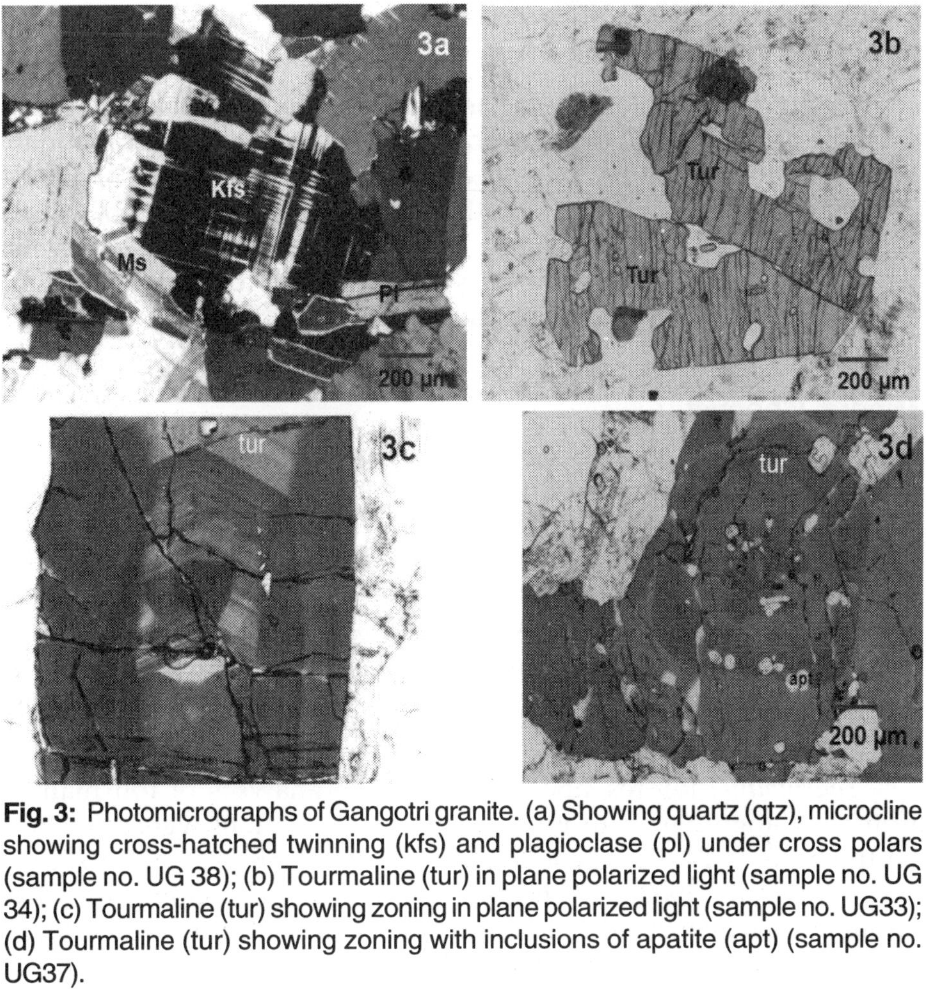

Fig. 3: Photomicrographs of Gangotri granite. (a) Showing quartz (qtz), microcline showing cross-hatched twinning (kfs) and plagioclase (pl) under cross polars (sample no. UG 38); (b) Tourmaline (tur) in plane polarized light (sample no. UG 34); (c) Tourmaline (tur) showing zoning in plane polarized light (sample no. UG33); (d) Tourmaline (tur) showing zoning with inclusions of apatite (apt) (sample no. UG37).

Muscovite is present in both varieties of GG, and its abundance decreases from the biotite to the tourmaline facies (from 13-10%, to 4-5% modal) of GG. Muscovite is mainly enclosed in plagioclase and K-feldspar and is in textural equilibrium with all the other phases. Plagioclase occurs as euhedral-subhedral crystals and contains inclusions of biotite, muscovite, tourmaline and grains of rounded apatite. Optical zonation is extremely rare. K-feldspar is characterized by anhedral habit and contains abundant inclusions of quartz and plagioclase. Exsolution lamellae are very thin. Quartz is homogeneously distributed as interstitial phase. Apatite, zircon and monazite are present in accessory amounts. Apatite is the most abundant accessory mineral and occurs as rounded crystals of 100-200 um in diameter (Fig. 3d). Zircon is present as inclusions in apatite, biotite and muscovite. Monazite occurs as inclusions in apatite and K-feldspar. Opaque minerals are very scarce.

Biotite, muscovite, plagioclase, tourmaline and quartz begin to crystallize early. K-feldspar is xenomorphic in habit and with abundant inclusions and therefore, crystallized late in the sequence. In GG, biotite and plagioclase are magmatic rather than restitic because, the biotite is poor in inclusions when compared to the biotite of paragneisses and the fact that plagioclase show rare optical zonation. Since tourmaline and muscovite define a magmatic layering, this implies they are of magmatic origin.

Mineral Chemistry

The mineral analyses were performed with Cameca SX50 microprobe at Geological Survey of India, Faridabad and Cameca SX100 at Wadia Institute of Himalayan Geology, Dehra Dun. A probe current of 20 nA at an accelerating voltage of 15 KeV and a beam size of 1 microns were used. Standardization was conducted against natural standards using ZAF corrections after Philibert (1963). Representative microprobe analyses of tourmalines, plagioclaes, K-feldspar and muscovites from the Gangotri granite are given in Tables 3, 4, 5 and 6 respectively.

Tourmaline: Representative microprobe analyses of tourmalines from the Gangotri granite are given in Table-3. Cations were calculated on the basis of 24.5 oxygens using a computer program CLASTOUR by Yavuz *et al.* (2002). All the analysed tourmaline belongs to Alkali Group and is Schorl. Aluminum in T sites varies from 0.000 to 0.202. In Y sites, Al varies from 0.186 to 0.491, Mg from 0.037 to 0.684, Fe^{2+}

Table-3

	UG33/1 CORE	UG33/1 RIM	UG33/2 CORE	UG33/2 RIM	UG33/3 CORE	UG33/3 RIM	UG37/1 CORE	UG37/1 RIM	UG37/2 CORE	UG37/2 RIM
SiO2	35.16	34.83	35.14	34.90	35.26	34.46	34.99	34.43	35.07	34.58
TiO2	0.39	0.69	0.69	0.58	0.52	0.76	0.65	1.34	0.68	0.81
Al2O3	32.91	32.33	32.24	31.72	32.97	32.68	32.44	31.56	32.85	32.40
Cr2O3	0.03	0.00	0.08	0.00	0.00	0.07	0.01	0.00	0.00	0.01
FeO	14.97	14.46	14.26	15.61	14.07	14.98	13.41	14.57	13.45	13.56
MnO	0.17	0.05	0.08	0.17	0.15	0.18	0.12	0.19	0.09	0.01
MgO	1.22	1.72	1.52	1.29	1.61	1.38	2.14	1.68	2.29	2.08
CaO	0.11	0.27	0.20	0.25	0.15	0.19	0.20	0.27	0.21	0.20
Na2O	1.93	1.96	2.13	2.10	2.00	2.02	2.17	2.21	2.07	2.06
K2O	0.06	0.05	0.06	0.03	0.09	0.06	0.06	0.09	0.05	0.04
Total	86.94	86.37	86.39	86.63	86.81	86.77	86.20	86.34	86.75	85.74
	CATIONS PER 24.5 OXYGENS								CATIONS PER 24.5 OXYGENS	
Si	5.915	5.895	5.938	5.930	5.916	5.829	5.905	5.858	5.876	5.869
Al (T)	0.085	0.105	0.062	0.070	0.084	0.171	0.095	0.142	0.124	0.131
SUM T	6.000	6.000	6.000	6.000	6.000	6.000	6.000	6.000	6.000	6.000
Al (Z)	6.000	6.000	6.000	6.000	6.000	6.000	6.000	6.000	6.000	6.000
Mg	0.306	0.434	0.383	0.327	0.403	0.348	0.539	0.426	0.572	0.526
Fe2+	2.106	2.046	2.015	2.218	1.974	2.119	1.894	2.073	1.884	1.924
Mn	0.024	0.007	0.011	0.024	0.021	0.026	0.018	0.027	0.013	0.001
Al (Y)	0.440	0.343	0.359	0.282	0.434	0.343	0.356	0.186	0.362	0.350
Ti	0.049	0.088	0.088	0.074	0.066	0.097	0.082	0.171	0.086	0.103
Cr3+	0.004	0.000	0.011	0.000	0.000	0.009	0.002	0.000	0.000	0.001
SUM Y	2.930	2.918	2.867	2.925	2.898	2.941	2.890	2.883	2.916	2.907
Ca	0.020	0.049	0.036	0.046	0.027	0.034	0.036	0.049	0.038	0.036
Na	0.629	0.643	0.698	0.692	0.651	0.662	0.709	0.729	0.672	0.678
K	0.013	0.011	0.013	0.007	0.019	0.013	0.012	0.020	0.011	0.009
SUM X	0.662	0.703	0.747	0.744	0.697	0.710	0.757	0.798	0.721	0.723
CATIONS	15.592	15.621	15.614	15.669	15.595	15.651	15.647	15.681	15.637	15.630

	UG38/1 CORE	UG38/1 RIM	UG38/2 CORE	UG38/2 RIM	UG41/1 CORE	UG41/2 CORE	UG41/2 RIM	UG42/2 CORE	UG42/2 RIM	UG42/1 RIM
SiO2	34.99	34.32	36.58	35.76	35.71	35.21	34.39	34.55	34.04	34.48
TiO2	0.79	0.82	0.75	0.81	0.65	0.90	1.14	0.54	0.72	0.86
Al2O3	31.76	31.42	31.74	31.52	32.37	32.01	32.08	32.67	31.91	31.77
Cr2O3	0.00	0.00	0.00	0.00	0.00	0.03	0.00	0.01	0.05	0.08
FeO	12.46	14.04	13.13	13.92	12.26	13.19	14.34	13.51	14.82	14.80
MnO	0.09	0.20	0.17	0.18	0.06	0.13	0.13	0.18	0.22	0.16
MgO	2.52	1.60	2.08	1.44	2.74	2.32	1.29	1.80	1.30	1.38
CaO	0.19	0.23	0.22	0.20	0.19	0.19	0.38	0.19	0.35	0.37
Na2O	2.20	2.17	2.15	2.09	2.13	2.22	2.07	2.04	2.21	2.02
K2O	0.06	0.07	0.06	0.06	0.06	0.03	0.10	0.04	0.11	0.06
Total	85.03	84.85	86.86	85.95	86.17	86.23	85.92	85.53	85.73	85.98
						CATIONS PER 24.5 OXYGENS				
Si	5.957	5.919	6.000	6.000	5.980	5.934	5.863	5.878	5.843	5.888
Al (T)	0.044	0.081	0.000	0.000	0.020	0.066	0.137	0.122	0.157	0.112
SUM T	6.000	6.000	6.000	6.000	6.000	6.000	6.000	6.000	6.000	6.000
Al (Z)	6.000	6.000	6.000	6.000	6.000	6.000	6.000	6.000	6.000	6.000
Mg	0.640	0.410	0.516	0.363	0.684	0.583	0.328	0.456	0.333	0.351
Fe2+	1.773	2.025	1.831	1.972	1.717	1.859	2.044	1.922	2.127	2.113
Mn	0.013	0.029	0.024	0.026	0.009	0.019	0.019	0.026	0.032	0.023
Al (Y)	0.328	0.303	0.236	0.292	0.368	0.292	0.309	0.428	0.298	0.282
Ti	0.101	0.106	0.094	0.103	0.082	0.114	0.146	0.069	0.093	0.110
Cr3+	0.000	0.000	0.000	0.000	0.000	0.004	0.000	0.001	0.007	0.011
SUM Y	2.855	2.872	2.700	2.754	2.858	2.870	2.846	2.902	2.889	2.891
Ca	0.034	0.043	0.039	0.036	0.034	0.034	0.069	0.035	0.064	0.068
Na	0.726	0.726	0.695	0.687	0.691	0.725	0.684	0.673	0.735	0.669
K	0.012	0.016	0.012	0.012	0.013	0.006	0.022	0.009	0.024	0.013
SUM X	0.772	0.784	0.746	0.734	0.738	0.766	0.775	0.716	0.824	0.750
CATIONS	15.627	15.656	15.446	15.488	15.596	15.636	15.621	15.618	15.713	15.641

Table-4

	UG33/1 CORE	UG33/1 RIM	UG33/2 CORE	UG33/2 RIM	UG37/1 CORE	UG37/1 RIM	UG37/2	UG38/1 CORE	UG38/1 RIM	UG38/1A CORE	UG38/1A RIM	UG38/2 CORE	UG38/2 RIM	UG38/3 CORE	UG41/1 CORE	UG41/1 RIM	UG41/2 CORE	UG41/2 RIM	UG42/2	UG42/2	UG42/2
SiO2	68.11	68.04	67.12	68.98	69.27	68.38	68.82	68.12	67.62	67.09	68.89	69.34	66.89	67.65	67.59	68.55	67.55	67.84	66.93	66.72	67.21
TiO2	0.00	0.04	0.04	0.00	0.00	0.00	0.00	0.02	0.02	0.00	0.02	0.02	0.01	0.01	0.00	0.03	0.00	0.03	0.00	0.04	0.00
Al2O3	19.62	19.98	20.06	19.95	20.37	20.30	20.28	19.78	20.08	20.33	19.84	19.93	20.40	20.02	19.58	19.82	19.94	19.44	19.72	20.08	19.66
Cr2O3	0.02	0.07	0.06	0.03	0.00	0.01	0.07	0.00	0.00	0.00	0.00	0.00	0.00	0.00	0.00	0.00	0.00	0.00	0.00	0.05	0.00
FeO	0.03	0.00	0.07	0.02	0.00	0.01	0.00	0.03	0.08	0.09	0.02	0.05	0.09	0.03	0.01	0.19	0.00	0.00	0.14	0.00	0.00
MnO	0.00	0.00	0.04	0.05	0.04	0.03	0.07	0.02	0.03	0.05	0.03	0.01	0.01	0.04	0.06	0.05	0.00	0.00	0.04	0.09	0.01
MgO	0.00	0.01	0.00	0.01	0.00	0.01	0.03	0.00	0.01	0.00	0.01	0.00	0.00	0.01	0.00	0.02	0.00	0.01	0.01	0.00	0.00
CaO	0.52	0.86	1.00	0.47	0.64	0.73	0.65	0.44	0.50	0.82	0.46	0.44	1.09	0.54	0.47	0.31	0.63	0.36	0.61	1.01	0.77
Na2O	11.38	11.19	8.88	11.39	9.88	11.06	10.60	11.30	11.39	11.20	11.31	11.52	11.04	11.50	11.17	11.43	11.07	11.51	11.09	10.92	11.26
K2O	0.12	0.14	0.10	0.07	0.14	0.18	0.08	0.10	0.14	0.12	0.09	0.06	0.13	0.13	0.08	0.04	0.13	0.11	0.06	0.10	0.10
Total	99.82	100.31	97.36	100.97	100.34	100.70	100.60	99.78	99.85	99.70	100.67	101.37	99.66	99.90	99.06	100.54	99.54	99.32	98.63	99.08	99.03
	Cations per 8 Oxygens							Cations per 8 Oxygens							Cations per 8 Oxygens						
Si	2.984	2.968	2.989	2.984	2.997	2.968	2.981	2.982	2.964	2.948	2.988	2.988	2.942	2.965	2.984	2.983	2.972	2.987	2.969	2.951	2.971
Al	1.013	1.027	1.053	1.017	1.039	1.038	1.035	1.020	1.038	1.053	1.014	1.012	1.057	1.034	1.019	1.016	1.034	1.009	1.031	1.047	1.024
Na	0.966	0.946	0.767	0.955	0.829	0.931	0.890	0.959	0.968	0.954	0.951	0.962	0.941	0.977	0.956	0.964	0.944	0.982	0.954	0.936	0.965
Ca	0.024	0.040	0.048	0.022	0.030	0.034	0.030	0.021	0.024	0.039	0.021	0.020	0.051	0.025	0.022	0.014	0.030	0.017	0.029	0.048	0.036
K	0.007	0.008	0.006	0.004	0.008	0.010	0.004	0.006	0.008	0.007	0.005	0.003	0.007	0.007	0.005	0.002	0.007	0.006	0.003	0.006	0.006
Ti	0.000	0.001	0.001	0.000	0.000	0.000	0.000	0.001	0.001	0.000	0.001	0.001	0.000	0.000	0.000	0.001	0.000	0.001	0.000	0.001	0.000
Fe2+	0.001	0.000	0.003	0.001	0.000	0.000	0.000	0.001	0.003	0.003	0.001	0.002	0.003	0.001	0.000	0.007	0.000	0.000	0.005	0.000	0.000
Mg	0.000	0.001	0.000	0.001	0.000	0.001	0.002	0.000	0.001	0.000	0.001	0.000	0.000	0.001	0.000	0.001	0.000	0.001	0.001	0.000	0.000
Mn	0.000	0.000	0.002	0.002	0.001	0.001	0.003	0.001	0.001	0.002	0.001	0.000	0.000	0.002	0.002	0.002	0.000	0.000	0.002	0.003	0.000
Cr3+	0.001	0.002	0.002	0.001	0.000	0.000	0.002	0.000	0.000	0.000	0.000	0.000	0.000	0.000	0.000	0.000	0.000	0.000	0.000	0.002	0.000
Total	4.996	4.993	4.871	4.987	4.904	4.983	4.947	4.989	5.005	5.006	4.983	4.988	5.001	5.010	4.988	4.990	4.987	5.003	4.994	4.994	5.002
XAB	0.969	0.952	0.934	0.973	0.956	0.955	0.963	0.973	0.968	0.954	0.973	0.977	0.942	0.968	0.973	0.984	0.962	0.977	0.968	0.945	0.958
XAN	0.024	0.040	0.058	0.022	0.035	0.035	0.032	0.021	0.024	0.039	0.021	0.020	0.051	0.025	0.022	0.014	0.031	0.017	0.029	0.048	0.036
XOR	0.007	0.008	0.007	0.004	0.009	0.010	0.004	0.006	0.008	0.007	0.005	0.003	0.007	0.007	0.005	0.002	0.007	0.006	0.003	0.006	0.006

Table-5

	UG33/1	UG33/2	UG33/3	UG37/1 CORE	UG37/1 RIM	UG37/1	UG37/2	UG38/1 CORE	UG38/1 RIM	UG38/2 CORE	UG38/2 RIM	UG38/3 CORE	UG38/3 RIM	UG41/1 CORE	UG41/2 CORE	UG41/2 RIM	UG42/1 CORE	UG42/1 RIM	UG42/2	UG42/2	UG42/2	UG42/3	UG42/3	UG42/3
SiO2	64.68	65.33	65.25	66.12	64.87	65.19	65.73	65.83	65.35	65.11	66.13	64.83	66.59	65.08	63.66	63.55	64.87	64.69	64.21	63.96	61.78	62.54	63.59	64.05
TiO2	0.00	0.00	0.03	0.02	0.07	0.00	0.00	0.01	0.02	0.00	0.00	0.01	0.00	0.00	0.00	0.00	0.00	0.00	0.00	0.00	0.00	0.00	0.00	0.01
Al2O3	18.60	18.69	18.56	18.82	18.21	18.91	18.86	18.30	18.18	18.69	18.38	18.45	18.53	18.24	17.89	18.12	18.05	18.21	18.02	18.27	17.28	17.82	18.07	18.48
Cr2O3	0.02	0.00	0.10	0.00	0.05	0.00	0.00	0.00	0.00	0.00	0.00	0.00	0.00	0.04	0.00	0.09	0.05	0.01	0.00	0.01	3.67	0.08	0.00	0.00
FeO	0.04	0.00	0.05	0.01	0.00	0.01	0.02	0.06	0.09	0.06	0.10	0.03	0.01	0.06	0.00	0.00	0.01	0.16	0.02	0.06	0.05	0.00	0.01	0.00
MnO	0.04	0.01	0.05	0.03	0.00	0.02	0.06	0.02	0.00	0.03	0.02	0.01	0.00	0.00	0.04	0.00	0.00	0.00	0.04	0.00	0.00	0.00	0.00	0.00
MgO	0.01	0.00	0.00	0.00	0.01	0.01	0.00	0.00	0.00	0.00	0.01	0.00	0.01	0.02	0.00	0.00	0.02	0.00	0.01	0.01	0.06	0.01	0.00	0.01
CaO	0.00	0.00	0.00	0.00	0.00	0.00	0.00	0.00	0.02	0.01	0.00	0.00	0.00	0.00	0.00	0.00	0.00	0.00	0.00	0.03	0.00	0.01	0.00	0.00
Na2O	0.51	0.71	0.63	0.78	0.56	0.27	0.36	0.51	0.45	0.88	0.52	0.53	0.67	0.47	0.78	0.38	0.70	0.36	0.45	0.41	0.38	0.55	0.55	0.79
K2O	15.74	15.64	15.40	15.37	15.84	15.84	15.44	15.78	16.04	15.10	15.74	15.78	15.30	16.55	15.96	16.58	16.32	16.46	16.21	16.07	16.12	15.54	16.56	16.07
Total	99.65	100.38	100.08	101.14	99.59	100.24	100.47	100.50	100.13	99.86	100.88	99.63	100.11	100.52	98.38	98.75	100.08	99.90	98.97	98.85	99.50	96.57	99.08	99.42
	Cations per 8 Oxygens							Cations per 8 Oxygens							Cations per 8 Oxygens								Cations per 8 Oxygens	
Si	2.993	2.997	3.000	3.003	3.004	2.994	3.004	3.016	3.012	2.997	3.016	3.000	3.021	3.000	2.997	2.987	3.003	2.999	3.002	2.993	2.920	2.993	2.988	2.982
Al	1.014	1.011	1.006	1.007	0.994	1.023	1.016	0.988	0.988	1.014	0.988	1.006	0.991	0.991	0.993	1.004	0.985	0.995	0.993	1.008	0.963	1.005	1.001	1.014
Na	0.046	0.063	0.056	0.069	0.050	0.024	0.032	0.045	0.040	0.078	0.046	0.048	0.059	0.042	0.071	0.035	0.063	0.032	0.041	0.037	0.035	0.051	0.050	0.071
Ca	0.000	0.000	0.000	0.000	0.000	0.000	0.000	0.000	0.001	0.000	0.000	0.000	0.000	0.000	0.000	0.000	0.000	0.000	0.000	0.002	0.000	0.001	0.000	0.000
K	0.929	0.915	0.903	0.891	0.936	0.928	0.900	0.922	0.943	0.887	0.916	0.932	0.885	0.973	0.959	0.994	0.964	0.973	0.967	0.959	0.972	0.949	0.993	0.954
Ti	0.000	0.000	0.001	0.001	0.002	0.000	0.000	0.001	0.001	0.000	0.000	0.000	0.000	0.000	0.000	0.000	0.000	0.000	0.000	0.000	0.000	0.000	0.000	0.000
Fe2+	0.002	0.000	0.002	0.000	0.000	0.000	0.001	0.002	0.004	0.003	0.004	0.001	0.001	0.002	0.000	0.000	0.000	0.006	0.001	0.002	0.002	0.000	0.000	0.000
Mg	0.001	0.000	0.000	0.000	0.001	0.001	0.000	0.000	0.000	0.000	0.001	0.000	0.001	0.001	0.000	0.000	0.001	0.000	0.001	0.001	0.004	0.001	0.000	0.001
Mn	0.002	0.000	0.002	0.001	0.000	0.001	0.002	0.001	0.000	0.001	0.001	0.000	0.000	0.000	0.002	0.000	0.000	0.000	0.002	0.000	0.000	0.000	0.000	0.000
Cr3+	0.001	0.000	0.004	0.000	0.002	0.000	0.000	0.000	0.000	0.000	0.000	0.000	0.000	0.001	0.000	0.003	0.002	0.000	0.000	0.000	0.137	0.003	0.000	0.000
Total	4.988	4.986	4.974	4.972	4.989	4.971	4.955	4.974	4.986	4.978	4.971	4.986	4.956	5.010	5.022	5.023	5.018	5.005	5.007	5.002	5.033	5.003	5.032	5.022
XOR	0.953	0.936	0.942	0.928	0.949	0.975	0.966	0.953	0.958	0.919	0.952	0.951	0.938	0.959	0.931	0.966	0.939	0.968	0.959	0.961	0.965	0.948	0.952	0.931
XAB	0.047	0.064	0.058	0.072	0.051	0.025	0.034	0.047	0.041	0.081	0.048	0.049	0.063	0.041	0.069	0.034	0.061	0.032	0.041	0.037	0.035	0.051	0.048	0.069
XAN	0.000	0.000	0.000	0.000	0.000	0.000	0.000	0.000	0.001	0.000	0.000	0.000	0.000	0.000	0.000	0.000	0.000	0.000	0.000	0.002	0.000	0.001	0.000	0.000

Table-6

	UG33/2 CORE	UG33/2 RIM	UG37/1	UG37/2	UG38/3 CORE	UG38/3 RIM	UG41/2 CORE	UG41/2 RIM	UG42/2	UG42/2	UG42/3	UG42/3	UG42/3
SiO2	47.61	46.67	47.66	48.76	48.27	47.69	45.42	46.17	45.24	45.22	45.92	45.87	45.39
TiO2	0.23	0.31	0.71	0.17	0.49	0.43	0.53	0.41	0.40	0.32	0.44	0.33	0.37
Al2O3	34.58	33.96	34.89	34.90	32.95	32.60	33.59	32.56	33.17	33.13	32.41	31.37	33.27
Cr2O3	0.00	0.01	0.04	0.04	0.00	0.00	0.01	0.00	0.00	0.00	0.00	0.03	0.00
FeO	3.32	3.45	2.48	2.77	3.51	3.72	3.07	3.80	3.30	3.59	3.73	3.94	3.23
MnO	0.17	0.07	0.15	0.15	0.17	0.18	0.09	0.15	0.18	0.07	0.08	0.18	0.13
MgO	0.41	0.38	0.59	1.51	0.34	0.38	0.37	0.53	0.41	0.41	0.53	0.53	0.40
CaO	0.00	0.02	0.00	0.02	0.00	0.03	0.00	0.04	0.00	0.00	0.02	0.00	0.00
Na2O	0.34	0.33	0.36	0.26	0.24	0.35	0.31	0.14	0.33	0.44	0.26	0.28	0.43
K2O	9.72	9.45	9.97	9.55	10.93	10.58	10.12	9.82	10.43	10.36	9.96	9.73	10.11
Total	96.38	94.65	96.85	98.13	96.89	95.94	93.62	93.64	93.46	93.60	93.36	92.30	93.46
	Cations per 22 Oxygens								Cations per 22 Oxygens				
Al(Total)	5.377	5.378	5.388	5.305	5.137	5.136	5.410	5.242	5.368	5.362	5.239	5.134	5.377
Si	6.281	6.271	6.245	6.289	6.386	6.375	6.207	6.307	6.212	6.210	6.298	6.369	6.225
Al(IV)	1.719	1.729	1.755	1.711	1.614	1.625	1.793	1.693	1.788	1.790	1.702	1.631	1.775
Total	8.000	8.000	8.000	8.000	8.000	8.000	8.000	8.000	8.000	8.000	8.000	8.000	8.000
Al(VI)	3.658	3.649	3.633	3.594	3.523	3.511	3.617	3.549	3.580	3.572	3.537	3.503	3.602
Ti	0.023	0.031	0.070	0.016	0.049	0.043	0.054	0.042	0.041	0.033	0.045	0.034	0.038
Fe(2+)	0.366	0.388	0.272	0.299	0.389	0.416	0.351	0.434	0.379	0.412	0.428	0.457	0.370
Mn	0.019	0.008	0.017	0.016	0.019	0.020	0.010	0.017	0.021	0.008	0.009	0.021	0.015
Mg	0.081	0.076	0.115	0.290	0.067	0.075	0.075	0.108	0.084	0.084	0.108	0.110	0.082
Cr	0.000	0.001	0.004	0.004	0.000	0.000	0.001	0.000	0.000	0.000	0.000	0.003	0.000
Total	4.147	4.153	4.111	4.219	4.046	4.064	4.108	4.150	4.105	4.109	4.127	4.128	4.107
Ca	0.000	0.003	0.000	0.003	0.000	0.004	0.000	0.006	0.000	0.000	0.003	0.000	0.000
Na	0.087	0.086	0.091	0.065	0.062	0.090	0.082	0.037	0.088	0.117	0.069	0.075	0.114
K	1.636	1.620	1.666	1.571	1.845	1.804	1.764	1.711	1.827	1.815	1.743	1.723	1.769
Total	1.723	1.709	1.757	1.639	1.906	1.897	1.846	1.754	1.915	1.932	1.815	1.798	1.883
Sum Cations	13.870	13.862	13.868	13.858	13.952	13.961	13.954	13.904	14.020	14.041	13.942	13.926	13.990

from 1.639 to 2.218, Mn from 0.000 to 0.041 and Ti from 0.049 to 0.171. In X sites Na varies from 0.619 to 0.777 and (Na + Ca + K) from 0.645 to 0.831. These tourmalines are zoned (Figs. 3c and 3d) and from core to rim Mg decreases, Fe^{2+}, Ti, Mn and Ca increases. There is a negative correlation between Mg and Fe^{2+}.

Plagioclases: Plagioclases are rich in albite component, X_{AB} varies from 0.934 to 0.984 and X_{AN} varies from 0.014 to 0.058.

K-feldspar: In K-feldspar (microcline microperthite, Fig. 3a) X_{OR} varies from 0.904 to 0.975 and X_{AB} varies from 0.025 to 0.096. Both plagioclase and K-feldspar from Gangotri granite can be treated as binary solid solutions.

Muscovites: In muscovites Al(IV) varies from 1.561 to 1.793, Al(VI) from 3.503 to 3.658, Fe^{2+} from 0.272 to 0.457, Mg from 0.065 to 0.290 and Na from 0.037 to 0.117.

Geothermometry

The geothermometric calculations were made using two feldspar thermometry in order to estimate the temperature of chemical re-equilibration of feldspars as the feldspars grew and re-equilibrated at different stages of teconothermal history of the Gangotri granite. The methods of Stormer (1975), Whitney and Stormer (1977), Powell and Powell (1977) and Perchuk *et al.* (1991) were followed (Table-7). The plagioclase-alkali feldspar geothermometer formulated by Stormer (1975) does not take into account the effect of calcium in the alkali feldspar. Powell and Powell (1977) calibration takes into account the Ca content of the alkali feldspars. Since there is very little X_{AN} in both plagioclase and K-feldspar from Gangotri granite, they can be treated as binary solid solutions and it is important to note that same temperature estimates are obtained by using Stormer (1975) and Powell and Powell (1977) geothermometric calibrations (Table-7). The temperatures calculated by two-feldspar thermometry gives temperature of subsolidus equilibration and it varies from 441 to 270^0 C. Plagioclase-muscovite geothermometer of Green and Usdansky (1986) was also utilized, and it gives temperature estimates in the range of 448-339^0 C (Table-7).

Conclusion

Mineralogical studies on tourmalines from Gangotri granite reveals that they belong to Alkali Group and are Schorl. These tourmalines are zoned and from core to rim, Mg decreases, Fe^{2+}, Ti, Mn and Ca

Table-7

Two Feldspar	UG33/1 CORE	UG33/1 RIM	UG33/2 CORE	UG33/2 RIM	UG37/1 CORE	UG37/1 RIM	UG37/2	UG41/1 CORE	UG41/1 RIM	UG41/2 CORE	UG41/2 RIM	UG42/2
1. Whitney & Stormer 1977	398	397	432	427	441	404	366	382	373	436	363	375
2. Stormer 1975	350	349	385	379	394	356	316	333	324	388	314	326
Average of 1 and 2	374	373	408	403	418	380	341	357	349	412	339	351
3. Powell & Powell 1977	350	349	385	380	394	356	317	334	325	389	315	326
4. Perchuk et al. 1991	311	310	352	344	362	318	273	292	281	355	270	284
Plagioclase-Muscovite												
Green & Usdansky 1986			425	366	438	448				404	339	391

increases. There is a negative correlation between Mg and Fe^{2+}. These results show that there were changes in physical conditions during crystallization of the leucogranite magma. Further thermodynamic modelling of zoned tourmalines is in progress, which will establish the variation in intensive parameters during crystallization of leucogranite magma. Application of two-feldspar geothermometer gives temperature of subsolidus equilibration of 441-270^0 C and plagioclase-muscovite gives temperature in the range of 448-339^0 C.

Acknowledgements

I wish to express my thanks to Dr. N. C. Pant, Mr. A. Kundu and Ms. Anubhooti Saxena for their help in EPMA analysis and Prof. Santosh Kumar, Kumaon University, Nainital, for valuable suggestions. The facilities and encouragement provided by Dr. A. K. Dubey, Director, Wadia Institute of Himalayan Geology, Dehra Dun, to carry out this research work are thankfully acknowledged.

References

Auden, J. B. (1949) Tehri Garhwal and British Garhwal. Records of the Geological Survey of India, v.76, pp.74-78.

Gansser, A. (1964) Geology of the Himalayas. 289 p., Interscience, New York.

Green, N. L. and Usdansky, S. I. (1986) Towards a practical plagioclase-muscovite thermometer. American Mineralogist, v. 71, pp.1109-117.

Grew, E. S. and Anovitz, L. M. (1996) Boron: mineralogy, petrology and geochemistry. Mineralogical Society of America, Reviews in Mineralogy v.33.

Harrison, T. M., Grove, M. and Lovera, O. M. (1997) New insights into the origin of two contrasting Himalayan granite belts. Geology, v.25, pp.899-902.

Heim, A. and Gansser, A. (1939) The Central Himalayas: Geological observations of the Swiss Expedition of 1936. Mem. Soc. Helv. Sci. Nat., v.73, pp.1-245.

Henry, D. J. and Dutrow, B. L. (1996) Metamorphic tourmaline and its petrologic applications. In: *E. S. Grew and L. M. Anovitz (eds.)* Boron: mineralogy, petrology and geochemistry. Mineralogical Society of America, Reviews in Mineralogy, v.33, pp.503-557.

Jain, A. K., Singh, S. and Manickavasagam, R. M. (2002) Himalayan Collision Tectonics. Gondwana Research Group. Memoir No. 7, 114p.

Jowhar, T. N. (1994) Crystal parameters of K-feldspar and geothermometry of the Badrinath Crystalline Complex, Himalaya, India. Neues Jahrbuch fur Mineralogie Abhandlungen, v.166, pp.325-342.

Jowhar, T. N. and Verma, P. K. (1995) Alkali feldspars from the Badrinath Crystalline complex and their bearing on the Himalayan Metamorphism. The Indian Mineralogist, v.29, pp.1-12.

Keller, P., Robles, E. R., Perez, A. P. and Fontan, F. (1999) Chemistry, paragenesis and significance of tourmaline in pegmatites of the Southern Tin belt, Central Namibia. Chemical Geology, v.158, pp.203-225.

Le Fort, P. (1975) Himalayas: the collided range, present knowledge of the continental arc. American J. Science, v.275, pp.1-44.

Le Fort, P., Debon, F., Pecher, A., Sonet, J. and Vidal, P. (1986) The 500 Ma magmatic event in alpine southern Asia, a thermal episode at Gondwana scale. Sciences de la Terre, v.47, pp.191-209.

London, D. and Manning, D. A. C. (1995). Chemical variation and significance of tourmaline from Southwest England. Economic Geology, v.90, pp.495-519.

London, D., Morgan VI, G. B., and Wolf, M. B. (1996) Boron in granitic rocks and their contact aureoles. In: *E. S. Grew and L. M. Anovitz (eds.)* Boron: mineralogy, petrology and geochemistry. Mineralogical Society of America Reviews in Mineralogy, v.33, pp. 299-330.

Manning, D. A. C. (1982) Chemical and morphological variation in tourmalines from the Hub Kapong batholith of peninsular Thailand. Mineralogical Magazine, v.45, pp.139-147.

Metcalfe, R. P. (1993) Pressure, temperature and time constraints on metamorphism across the Main Central Thrust zone and High Himalayan Slab in the Garhwal Himalaya. In: *P.J. Treloar, and M. P. Searle (eds.)* Himalayan Tectonics. Geological Society of London Special Publication, v.74, pp. 485-509.

Pant, R. (1986) Petrochemistry and petrogenesis of the Gangotri granite and associated granitoids, Garhwal Himalaya. Ph.D. Thesis, University of Roorkee, Roorkee, India.

Perchuk, L. L., Podlesskii, K. K. and Aranovich, L. Y. (1991) Thermodynamics of some framework silicates and their equilibria: application to geothermobarometry. In: *L.L. Perchuk (ed.)* Progress in Metamorphic and Magmatic Petrology, A Memorial volume in Honour of D. S. Korzhinskiy. Cambridge University Press, Cambridge, pp.131-164.

Philibert, J. (1963) X-ray optics and X-ray microanalysis. 329p., Academic Press, New York.

Powell, M. and Powell, R. (1977) Plagioclase-alkali feldspar geothermometry revisited. Mineralogical Magazine, v. 41, 253-256.

Prince, C. I., Vance, D. and Harris, N., (1994). Timing of metamorphism in the Garhwal Himalaya. Mineralogical Magazine, L62a, pp.1210-1211.

Scaillet, B., France-Lanord, C. and Le Fort, P., (1990). Badrinath-Gangotri plutons (Garhwal, India): petrological and geochemical evidence for fractionation processes in a high Himalayan leucogranite. Journal of Volcanology and Geothermal Research, v.44, pp.163-188.

Scaillet, B., Pecher, A., Rochette, P. and Champenois, M. (1995) The Gangotri granite (GarhwaHimalaya): laccolithic emplacement in an extending collisional belt. J. Geophysical Research, v.100, pp.585-607.

Searle, M. P., Metcalfe, R. P., Rex, A. J. and Norry, M. J. (1993) Field relations, petrogenesis and emplacement of the Bhagirathi leucogranite, Garhwal Himalaya. In: *P. J. Treloar and M. P. Searle (eds.)* Himalayan Tectonics. Geological Society of London Special Publication, v.74, pp. 429-444.

Searle, M. P., Noble, S. R., Hurford, A. J. and Rex, D. C. (1999) Age of crustal melting, emplacement and exhumation history of the Shivling leucogranite, Garhwal Himalaya. Geological Magazine, v.136, pp.513-525.

Singh, S., Mukherjee, P. K., Jain, A. K., Khanna, P. P., Saini, N. K. and Kumar, Rajeev (2003) Source characterization and possible emplacement mechanism of collision- related Gangotri leucogranite along Bhagirathi valley, NW-Himalaya. In: *S. Singh (ed.)* Granitoids of the Himalayan Collision belt. J. Virtual Explorer, v.11, pp. 60-73.

Sorkhabi, R. B., Stump, E., Foland, K. A. and Jain, A. K. (1996) Fission-track and $^{40}Ar/^{39}Ar$ evidence for episodic denudation of the Gangotri granites in the Garhwal Higher Himalaya, India. Tectonophysics, v.260, pp.187-199.

Sorkhabi, R. B., Stump, E., Foland,K.A. and Jain, A. K. (1999) Tectonic and cooling history of the Garhwal Higher Himalaya (Bhagirathi Valley): constraints from thermochronological data. In: *A. K. Jain and R. M. Manickavasagam (eds.)* Geodynamics of the NW Himalaya. Gondwana Research Group Memoir, v.6, pp. 217-235.

Stern, C. R., Kligfield, R., Schelling, D., Virdi, N. S., Futa, K., Peterman, Z. E. and Amini, H. (1989) The Bhagirathi leucogranite of the High Himalaya (Garhwal, India): age, petrogenesis and tectonic implications. Geological Society of America, SpecialPublications, v.232, pp.33-45.

Stormer, J. C., Jr. (1975) A practical two-feldspar geothermometer. American Mineralogist, v. 60, pp. 667-674.

Whitney, J. A. and Stormer, J. C., Jr. (1977) The distribution of $NaAlSi_3O_8$ between coexistingmicrocline and plagioclase and its effect on geothermometric calculations. American Mineralogist, v. 62, pp.687-691.

Yavuz, F., Gultekin, A. H. and Karakaya, M. C. (2002) CLASTOUR: a computer program for classification of the minerals of the tourmaline group. Computers and Geosciences, v.28, pp.1017-1036.

Yin, A. (2006) Cenozoic tectonic evolution of the Himalayan orogen as constrained by along-strike variation of structural geometry, exhumation history, and foreland sedimentation. Earth Science Reviews, v.76, pp.1-131.

□□□

9

Study of Porosity Loss Due to Compaction in the Cretaceous Upper Bima Sandstone, Upper Benue Trough, N.E. Nigeria

N.K. SAMAILA[1] AND G.P. SINGH[2*]

[1]Geology Programme, Abubakar Tafawa Balewa University, Bauchi, P.M.B. 0248 Bauchi, Nigeria.

[2]Department of Geology, Gombe State University, P.M.B. 127, Gombe, Gombe State, Nigeria.

Abstract

Present paper entails the study carried out on the fluvial Cretaceous Upper Bima Sandstone of the Upper Benue Trough. Compositional and textural changes due to diagenesis have altered the reservoir properties of the formation. Modal analyses reveal that these sandstones comprise of 71.21% quartz, 18.66% feldspars and 10.05% rock fragments. The average compactional porosity loss (COPL), cementational porosity loss (CEPL), and compaction index (ICOMPACT) in the Upper Bima Sandstone have been computed as 13.77%, 3.05% and 0.82 respectively. The sum of these values subtracted from the original porosity indicates that the initial porosity of the formation has been reduced to 18.18%. A cross plot of COPL versus CEPL shows that compaction is responsible for reducing the initial porosity of the sediments.

Keywords : Bima Sandstone, Benue Trough, diagenesis, porosity loss, compection

Introduction

The Benue Trough is a NE-SW trending sedimentary basin extending for over 1000 km from the Niger Delta to the Chad Basin

(Wilson and Guiraud, 1992). It consists of up to 5000m of Cretaceous sediments and is contiguous with the rift basins of the Niger, Chad and Sudan Republics (Akande, 2004). Benkhelil (1989) gave detail description of tectonic setting, sedimentary fill and tectonic phases of the Benue Trough. Accounts on the geology of the Benue Trough have been given by Carter *et al.* (1963), Cratchley and Jones (1965) amongst many other authors, indicating that the trough is flanked on either side by crystalline rocks of the Nigerian Basement Complex composed mainly of Precambrian to Early Paleozoic granites and gneisses.

The Bima Sandstone, an entirely continental formation throughout the Upper Benue Trough (Fig. 1), is the basal part of the sedimentary succession (Table- 1) in the basin. The sandstone has been interpreted to be of alluvial fan to braided river origin (Benkhelil, 1989; Guiraud, 1990). It lies unconformably on the Precambrian Basement Complex rocks. The Bima Sandstone is a sequence that shows an upward maturation reflecting change from periods of tectonic activity to stable condition (Guiraud, 1993). Lithostratigraphic terms (the lower Bima

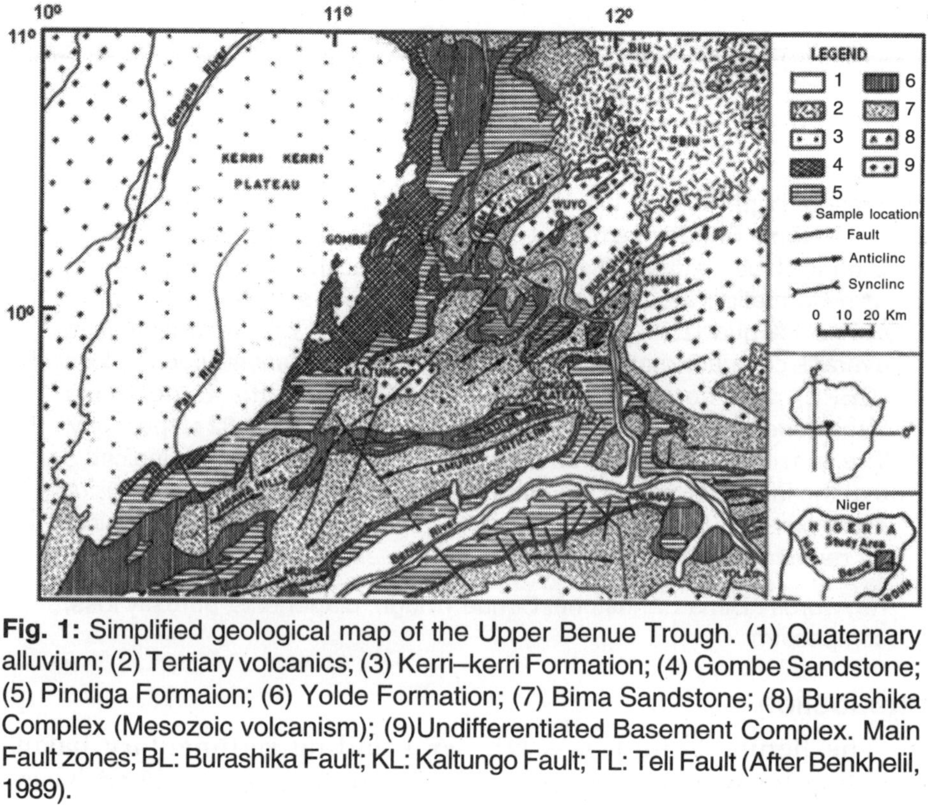

Fig. 1: Simplified geological map of the Upper Benue Trough. (1) Quaternary alluvium; (2) Tertiary volcanics; (3) Kerri–kerri Formation; (4) Gombe Sandstone; (5) Pindiga Formaion; (6) Yolde Formation; (7) Bima Sandstone; (8) Burashika Complex (Mesozoic volcanism); (9)Undifferentiated Basement Complex. Main Fault zones; BL: Burashika Fault; KL: Kaltungo Fault; TL: Teli Fault (After Benkhelil, 1989).

member, middle Bima member, and upper Bima member) were given to the formation to informally denote major units of sediments formed in direct relationship with tectonic events that took place during the Early Cretaceous times (Guiraud, 1990). The upper Bima member, referred to in this paper as the Upper Bima Sandstone, consists of sandy deposits containing ubiquitous cross bedding and diversity of soft-sediment sedimentary structures (Samaila *et al.*, 2005, 2006; Samaila, 2007).

Table-1 : Stratigraphic succession of the Upper Benue Trough (From Samaila *et al.*, 2008)

Age	Paleo-Environment	Gongola Basin	Yola Basin	Lamurde-Lau Basin
Quaternary	Conntinental	Biu Basalts	Longuda Basalts	
Pliocene				
Miocene				
Oligocene		Kerri-Kerri Fm		
Eocene				
Paleocene				
Maastrichtian	Continental/ Transitional	Gombe Sandstone		
Campanian				
Santonian			Lamja Sandstone	Lamja Sandstone
Coniacian			Numanha Fm	Numanha Fm
	Marine	Pindiga Formation		
Turonian			Sekuleye Fm	Sekuleye Fm
Cenomanian			Jessu Fm	Jessu Fm
			Dukul Fm	Dukul Fm
			Yolde Formation	
Upper Albian			Bima Sandstone (B_3)	
Late Aptian			Bima Sandstone (B_2)	
Early Aptian	Continental		Bima Sandstone (B_1)	
Late Jurassic?				
Pre-Cambrian			Basement Complex	

Study of porosity and permeability of rocks is of prime importance in relation to the search for oil, gas and underground water since the pore system is the channel for movement as well as the storage of fluids. Permeability of a rock, which is the property of allowing the passage of fluids through its pore spaces, is related to grain size; it increases with increase in pore-throats size (Beard and Weyl, 1973). Scherer (1987) explained further that porosity and permeability generally decreases in poorly sorted sands.

Indurated samples of the Upper Bima Sandstone were cut into thin sections and studied to determine composition of the samples by carefully counting individual grains, analyzing their shapes and inter-granular relationships. Percentages of framework, non-framework and non-granular components of the grains in thin sections were calculated. Parameters used for the estimation of diagenetic processes (compaction and cementation) that affected porosity in the formation were determined from the modal analysis data. The data generated from each thin section include the volumetric proportion of minerals in the rock (modal composition), macro-porosity, grain sizes, sorting, roundness and sphericity. This paper is aimed at showing the effects of diagenesis on reservoir potential of the Upper Bima Sandstone.

Petrography of the Upper Bima Sandstone

Modal study of 17 thin sections of the Upper Bima Sandstone shows quartz, feldspars and rock fragments constituting the main components of the Upper Bima Sandstone. Raw values from the modal analysis showed that 25-50% volume of the samples is quartz, 5-25% feldspar, 0.5-15% rock fragments, 1-3% mica, 0.8-10% heavy minerals, 0.2-5% opaque minerals, 2-20% cements and 10-23% micro-porosity. The framework components were recalculated and tabulated (Table-2). Results show an average framework composition of 71.21% quartz, 18.66% feldspars and 10.05% rock fragments (Table-2). Overgrowth of quartz was not observed in thin sections of the Upper Bima Sandstone (Samaila, 2007).

Some granular but non-framework components (void fillers including cement and matrix) (Tables-3) as well as non-granular components (including grain size, sorting, rounding and sphericity) (Table-4) were also determined from the thin section study. Roundness, which is a function of grain composition, grain size, and type of transport process and distance of transport, shows that the representative samples of the Upper Bima Sandstone are generally angular to sub-angular grains with a few that are sub-rounded (Fig. 2).

Table-2 : Modal analysis data of the framework components of some representative samples of the Upper Bima Sandstone.

S No.	Sample No.	Quartz		Feldspars			Rock Fragments		Q/F/L
		Modal value	Modal value (%)	Potassium Feldspar	Plagioclase Feldspar	Modal value (%)	Modal Value	Modal Value (%)	
1	SB10	30.00	78.95	8.00	0.00	13.16	0.00	0.00	79/13/0
2	SB11	35.00	58.33	15.00	0.00	25.00	10.00	16.67	58/25/17
3	SB9	50.00	70.42	12.00	0.00	16.90	9.00	12.68	70/17/13
4	BSB8	25.00	62.50	10.00	0.00	25.00	5.00	12.5	63/25/13
5	BST10	45.00	81.82	0.00	0.00	0.00	10.00	18.18	82/0/18
6	BST14	40.00	100.00	0.00	0.00	0.00	0.00	0.00	100/0/0
7	SB2	30.00	46.15	20.00	0.00	30.77	15.00	23.08	46/31/23
8	SB14	45.00	57.69	25.00	5.00	38.46	3.00	3.85	58/38/4
9	BSS3	54.00	77.70	0.00	15.00	21.58	0.50	0.72	78/21/1
10	BSS5	45.50	60.26	25.00	0.00	33.11	5.00	6.62	60/33/7
11	SB 1	50.40	73.68	10.00	00.00	14.71	8.00	11.76	74/15/12
12	BSB 4	54.00	90.00	0.00	0.00	0.00	6.00	10.00	90/0/10
13	BSS4	50.00	66.67	5.00	0.00	6.67	20.00	26.67	67/7/27
14	SB7	32.00	47.76	25.00	0.00	37.31	10.00	14.93	48/37/15
15	BST2	45.00	90.00	0.00	0.00	0.00	5.00	10.00	90/0/10
16	BSB9	58.00	72.50	15.00	0.00	18.75	7.00	8.75	73/19/8
17	SB4	45.00	69.23	20.00	0.00	30.77	0.00	0.00	69/31/0
	Average		**71.28**			**18.66**		**10.05**	

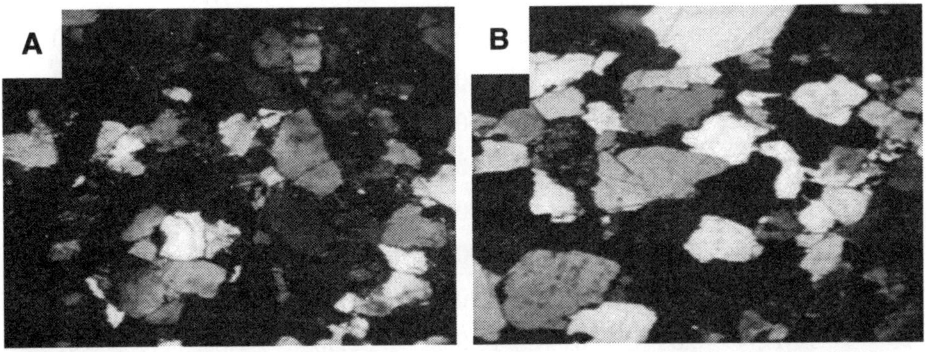

Fig.2 : Thin section photomicrographs of some representative samples of the Upper Bima Sandstone (A and B) under cross polarizer, showing highly fractured medium- to coarse-grained, poorly sorted arkose (x60).

Some parameters from Tables-3 & 4 were substituted into the three equations developed by Ehrenberg (1989) and Lundergard (1992) to determine the porosity loss due to compaction and cementation using the initial porosity (P_i) of the formation assumed as 35%. Total optical porosity or macro-porosity (P_o), and volume-percent pore filling cement (C) attributed to compaction and cementation, was calculated (Table- 4). The sum of P_o and C is equal to intergranular volume (IGV). Cement was not distinguished into primary and secondary because of equipment limitation, however, iron oxide cement, undifferentiated authigenic and detrital clays were detected from the petrographic study (Fig. 2). The compactional porosity loss (COPL), cementational porosity loss (CEPL), and compaction index (ICOMPACT) were clearly shown in Table-5.

Table-3 : Estimated percentages of non-framework components from modal analysis of some representative samples of the Upper Bima Sandstone.

S. No.	Sample No.	Mica	Heavy minerals	Opaque Minerals	Cements	Optical Porosity%	Detrital clay (D)
1	SB10	0.00	0.00	10.00	22.00	11.00	1.00
2	SB11	0.00	00.00	5.00	15.00	12.00	2.00
3	SB9	0.00.	0.00	0.00	12.00	8.00	1.00
4	BSB8	0.00	0.00	0.00	18.00	10.00	1.00
5	BST10	0.00	0.00	5.00	18.00	8.00	3.00
6	BST14	0.00	0.00	5.00	30.00	10.00	5.00
7	SB2	0.00	0.00	10.00	20.00	11.00	10.00
8	SB14	0.00	5.00	5.00	10.00	8.00	2.00
9	BSS3	0.00	0.00	5.00	16.00	12.00	0.00
10	BSS5	0.00	5.00	7.00	25.00	13.00	12.00
11	SB 1	0.00	0.00	0.00	15.00	5.00	0.00
12	BSB 4	0.00	0.00	0.00	18.00	5.00	0.00
13	BSS4	0.00	0.00	5.00	22.00	10.00	0.00
14	SB7	0.00	0.00	0.00	15.00	6.00	0.00
15	BST2	0.00	5.00	10.00	35.00	9.00	6.00
16	BSB9	0.00	0.00	0.00	18.00	10.00	0.00
17	SB4	0.00	0.00	0.00	8.00	9.00	0.00

Table-4 : Parameters used in determining porosity reduction in the Upper Bima Sandstone from the initial porosity of 35% at the time of deposition

S.No.	Sample No.	*Grain size	*Sorting	**Rounding	**Sphericity
1	SB10	0.21	1.20	0.30	0.50
2	SB11	0.84	2.00	0.30	0.30
3	SB9	0.84	2.80	0.30	0.30
4	BSB8	0.42	1.40	0.30	0.80
5	BST10	0.42	1.40	0.30	0.80
6	BST14	0.84	2.80	0.30	0.30
7	SB2	0.84	1.50	0.10	0.90
8	SB14	0.84	2.30	0.40	0.90
9	BSS3	0.84	1.40	0.30	0.60
10	BSS5	0.84	1.10	0.50	0.30
11	SB 1	0.84	2.70	0.10	0.70
12	BSB 4	0.84	1.30	0.30	0.50
13	BSS4	0.59	1.20	0.70	0.90
14	SB7	0.15	2.70	0.50	0.50
15	BST2	0.17	2.00	0.50	0.50
16	BSB9	0.84	1.30	0.30	0.60
17	SB4	0.84	1.50	0.50	0.50

Discussion and Conclusion

The Upper Bima Sandstone has shown varying degrees of diagenetic effects that contributed in reducing the reservoir properties of the formation. The sediments were subjected to severe diagenetic processes as a result; most of the feldspars, rock fragments and other detrital grains were partly dissolved or removed (Fig. 2). From both petrological and petrographical studies, the Upper Bima Sandstone is texturally and mineralogically immature (Samaila, 2007). Although development of secondary minerals (kaolinite, goethite, hematite, gibbsite, etc.) in the formation is capable of decreasing the porosity and permeability through pore-space occupancy (Samaila, 2007), this study is focused on the effect of compaction and cementation on diagenetic processes that significantly altered the original grain matrix ratio in the Upper Bima Sandstone.

Table-5 : Modal analysis of non-granular components of some representative samples of the Upper Bima Sandstone (based on**Pettijohn, 1975, *Beard and Weyl, 1953).

S. No.	Sample No.	P_i	IGV	IGV*100	P_i*IGV	100-IGV	CEM	COPL	CEPL	ICOMPACT
1	SB10	35.00	20.00	2000.00	700.00	80.00	0.00	18.75	0.00	1.00
2	SB11	35.00	22.00	2200.00	770.00	78.00	0.00	16.67	0.00	1.00
3	SB9	35.00	24.00	2400.00	840.00	76.00	0.00	14.47	0.00	1.00
4	BSB8	35.00	21.00	2100.00	735.00	79.00	0.00	17.72	0.00	1.00
5	BST10	35.00	28.00	2800.00	980.00	72.00	0.00	9.72	0.00	1.00
6	BST14	35.00	25.00	2500.00	875.00	75.00	0.00	13.33	0.00	1.00
7	SB2	35.00	22.00	2200.00	770.00	78.00	0.00	16.67	0.00	1.00
8	SB14	35.00	30.00	3000.00	1050.00	70.00	0.00	7.14	0.00	1.00
9	BSS3	35.00	23.00	2300.00	805.00	77.00	10.00	15.58	8.44	0.65
10	BSS5	35.00	27.00	2700.00	945.00	73.00	0.00	10.96	0.00	1.00
11	SB 1	35.00	20.50	2050.00	717.50	79.50	15.00	18.24	12.26	0.60
12	BSB 4	35.00	23.00	2300.00	805.00	77.00	0.00	15.58	0.00	1.00
13	BSS4	35.00	20.00	2000.00	700.00	80.00	10.00	18.75	8.13	0.70
14	SB7	35.00	28.00	2800.00	980.00	72.00	8.00	9.72	7.22	0.57
15	BST2	35.00	25.00	2500.00	875.00	75.00	10.00	13.33	8.67	0.61
16	BSB9	35.00	33.00	3300.00	1155.00	67.00	10.00	2.99	9.70	0.24
17	SB4	35.00	23.00	2300.00	805.00	77.00	0.00	15.58	0.00	1.00
	Average	35.00	24.45	2445.00	855.75	77.00	3.53	**13.77**	**3.05**	**0.82**

Detailed diagenetic analyses have not been carried out to ascertain whether different phases of dissolution, diffusion and precipitation of silica cements existed during the evolution of the Upper Bima Sandstone. However, the drastic decrease in grain size and sorting observed in some thin sections around the extensional fault zones in the formation is an indication that great amount of pressure solution was promoted (Samaila *et al.*, 2008).

Overgrowth of quartz, generally related to silica derivation from pressure solution, and usually signified by line of very small clay or dust particles or ferruginous coatings was not observed in thin sections of the Upper Bima Sandstone (Samaila, 2007). Sippel (1968) observed that light microscopy alone does not reveal all of quartz overgrowth; it takes the use of cathodoluniscence to show all of it. The undetected/ and or absence of overgrowth of quartz in the thin sections of the Upper Bima Sandstone is an indication that pressure solution due to compaction is not the most important source of silica in the entire formation.

Compaction induced fracturing in form of microcracks on brittle quartz grains is fairly common in the Upper Bima Sandstone (Fig. 2). These fractured grains are mostly associated with outcrops that are adjacent to the extensional fault zones in the formation. Healing of the microcracks by silica was observed in the field, characterized by several sets of faults and microfaults of more resistant deformed sandstone (Fig. 3). The phenomenon is associated; with cataclasis along breccia-gouge zone induced by tectonic pressure solution (Samaila, *et al.*, 2008).

Fig.3 : Microfaults (deformation bands) affecting the Upper Bima Sandstone, appearing as series of light colored, more resistant strands with slip surfaces on the order of few millimeters to a few centimeters.

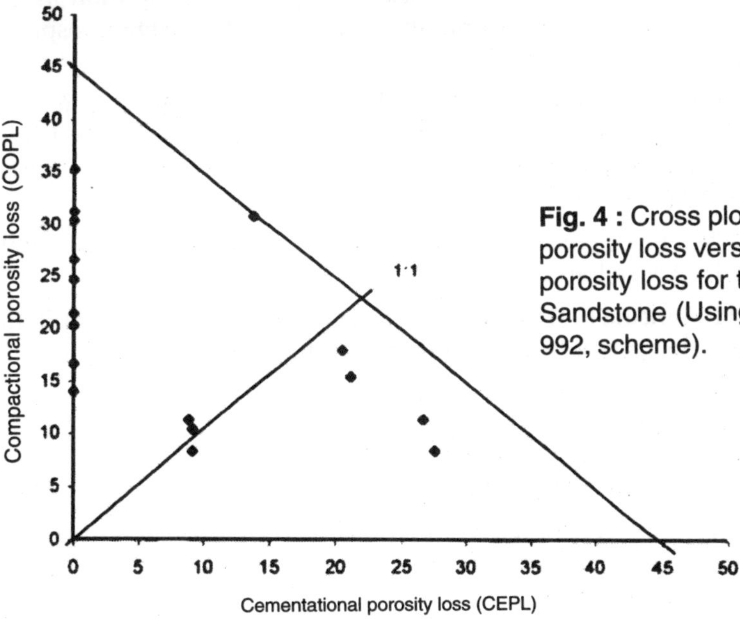

Fig. 4 : Cross plot of compaction porosity loss versus cementation porosity loss for the Upper Bima Sandstone (Using Lundergard's, 992, scheme).

The disaggregated samples of the Upper Bima Sandstone used for sieve analyses are generally poorly sorted, therefore its initial porosity was assumed to be 35% (Samaila, 2007), although Manus and Cogan (1974) showed that initial porosity of freshly deposited sands is 40-45%. The average values of COPL, CEPL and ICOMPACT in the Upper Bima Sandstone have been computed as 13.77%, 3.05% and 0.82 respectively (Table-5). The sum of these values was subtracted from the original porosity. The result showed a reduction of the initial porosity of the Upper Bima Sandstone to 18.18%. This value is in the range of values for some giant and super giant oil and gas fields found in siliclastic reservoir rocks of the world (14-32%) (Morse, 1994). It is clear from the COPL-CEPL diagram, that apart from the isolated cases around the major faults, where cementation played bigger role in reducing the initial porosity of the sediments, compaction was generally responsible for the porosity reduction (Fig. 4). The result agrees with work done on other sandstone formations (Ehrenberg, 1989; Houseknecht, 1987; Lundergard, 1992). The diagonal line in Fig. 4 indicates a divide between porosity loss due to compaction or cementation (Lundergard, 1992). Compaction index equals 1.0 when all porosity loss is by compaction but equals 0.0 when porosity loss is by cementation.

References

Akande, S.O. (2004) Cretaceous source rocks and thermal maturation in the Nigerian sedimentary basins: Implication for offshore petroleum prospects. NAPE news, v.3 (5) Sept-Oct. 2003.

Akande, S.O. and Erdtmann, B.D. (1998) Burial metamorphism (thermal maturation) in Cretaceous sediments of the Southern Benue Trough and Anambra Basin, Nigeria. American Association of Petroleum Geologists, Bulletin, v. 82 (6), pp. 1191-1206.

Benkhelil, J. (1989) The origin and evolution of the Cretaceous Benue Trough (Nigeria). J. African Earth Sciences, v. 8, pp. 251-282.

Benkhelil, J. and Robineau, B. (1983) Le fosse de la Benoue est-il un rift? Bulletin des Centres de Recherches Exploration- production Elf- Aquitaine, v. 7, pp. 315-321.

Carter, J.D., Baber,W., Tait, E.A., and Jones, G.P. (1963) The Geology of Parts of Adamawa, Bauchi, and Borno Provinces in North Eastern Nigeria. Geological Survey of Nigeria Bulletin. v. 30, 99p.

Cratchley, C.R. and Jones, G.P. (1965) An interpretaion of the geology and gravity anomalies of the Benue valley, Nigeria. Overseas Geological Survey and Geophysics Paper, v. 1, 24p.

Ehrenberg, S.N.(1987) Assessing the relative importance of compaction processes and cementation to reduction of porosity in sandstone: discussion; compaction and porosity evolution of Pliocene sandstone Venture Basin, California: discussion. American Association of Petroleum Geologits, Bulletin, v. 73, pp.1274-1276.

Guiraud, M. (1990) Tectono-sedimentary framework of the Early Cretacous Continental Bima Formation (Upper Benue Trough, N.E. Nigeria). Journal of African Earth Sciences (Special Publication), v. 10 (1/2), pp. 341-353.

Guiraud, M. (1993) Late Jurassic rifting - Early Cretaceous rifting and Late Cretaceous transpressional inversion in the Upper Benue Basin (N.E. Nigeria). Bulletin Centres Recherches Exploration et Production Elf Aquitane, v. 17, pp. 371-383.

Houseknecht, D.W. (1987) Assessing the relative importance of compaction processes and cementation to reduction of porosity in sandstones. American Association of Petroleum Geologists, Bulletin, v. 71, pp. 633-642.

Manus, R.W. and Coogan, A.H. (1974) Bulk volume reduction and pressure-solution derived cement. J. Sedimentary Petrology, v. 44, pp. 466-471.

Morse, D.G. (1994) Siliclastic reservoir rocks. In: *L.B. Magoon, and W.G. Dow (eds.)* The Petroleum system- from source rock to trap. American Association of Petroleum Geologists Memoir, v. 60, pp. 121-139.

Samaila, N.K. (2007) Reservoir Potentials of the Upper Bima Sandstone in the Yola and Lau-Lamurde Basins, Upper Benue Trough, Northeastern Nigeria. Unpublished Ph.D. thesis, 201p.

Samaila, N.K. Dike, E.F.C. and Obaje, N.G. (2008) Microstructures in the Cretaceous Bima Sandstone, Upper Benue Trough, N.E. Nigeria: Implication for hydrocarbon migration. Journal of African Earth Sciences v. 50 (1), pp. 6-15.

Samaila, N.K. Abubakar, M.B., Dike, E.F.C. and Obaje, N.G. (2006) Description of soft-sediment deformation structures in the Cretaceous Bima Sandstone from the Yola Arm, Upper Benue Trough, Northeastern Nigeria. J. African Earth Sciences, v. 44, pp. 66-74.

Samaila, N.K. Braide, S.P., Dike E.F.C. and Suh, C.E. (2005) Study of soft-sediment structures in the Cretaceous Bima Sandstone of the Yola Arm, Upper Benue Trough, northeastern Nigeria. J. Mining and Geology, v. 41(1), pp. 81-86.

Sippel, R.F. (1968) Sandstone petrology, evidence from luminescence petrography. J. Sedimentary Petrology, v. 38, 530-554.

Wilson, M. and Guiraud, R. (1992) Magmatism and rifting in Western and Central Africa, from Late Jurassic to Recent times. Tectonophysics, v. 213, pp. 203-225.

❑❑❑

10

Use of Bulk Sample and Hot Acids in Stream Sediment Survey for Gold

K. BHEEMALINGESWARA
Department of Earth Science, College of Natural and Computational Sciences
Mekelle University, P. O. Box 3066, Mekelle, Ethiopia

Abstract

Fifteen bulk stream sediment samples (minus 0.59mm size) from an area known for gold mineralization in Ramagiri Gold Fields, India are tested for gold. In addition, different size fractions (-0.590 + 0.259; -0.259 + 0.177 and -0.177) of the stream sediment are also analysed for gold. The data when compared suggests that the bulk samples data not only tally well with the previous stream sediment survey data but also provide additional anomalies for gold. The method suggested in this paper is bulk leach extractable gold (BLEG) using hot aqua regia and methyl isobutyl ketone solvents (MIBK-AR). The bulk leach extractable gold data of the new method seems do not suffer from iron oxide coatings, nugget effects, invisible gold etc as compared to the data of different size fractions (coarse/ fine size fractions) of stream sediment sample. Hence it provides near total values for gold present in the sample. The proposed method differs from the widely used BLEG in terms of hot aqua regia treatment in place of cold cyanide solution. The procedure suggested in this paper is simple to use and it provides an alternative to the commonly used coarse/finer fractions or BLEG (using cyanide) methods in stream sediment surveys for Au particularly in areas where oxide/hydroxide coatings are conspicuous.

Keywords : Iron oxide/hydroxide coating, Bulk sample, Bulk leach gold, Stream sediment survey.

Introduction

Geochemical surveys these days have become an integral part of mineral exploration particularly in the search for different metallic deposits of Cu, Pb, Zn, Au etc. Sophistication in technology, which resulted in the development of advanced instrumentation techniques in recent years, has further helped to improve and modify the geochemical survey methods to a great extent. Of course, the need to look for subsurface or blind deposits also has provided impetus to this development.

Depending on the nature of the geochemical survey- regional or local and type of deposit under consideration, the sampling scheme is planned. Stream sediment, soil or rock sampling either one or in combination is a common practice. In the case of stream sediment and soil samples, it is important to decide whether a particular size fraction of the sample or the bulk sample is used for analysis. A brief account on the advantages and limitations of the commonly used geochemical survey methods for gold, like panning in the field, stream sediment or soil sampling using coarse/ finer fractions or bulk (for bulk leach extractable gold (BLEG)) is provided here so as to understand the scope of the proposed method in the paper.

Panning method is in practice from time immemorial. The nuggets are recovered by panning under water. This field method on the basis of number of visible gold nuggets present in the panned concentrates helps to obtain information related to the source, resource size, type and extent of the deposit. Reliability of the data of panning method has been improved significantly by shifting from field to laboratory and adopting a systematic and scientific approach. In the laboratory, heavy mineral fractions are separated from the bulk sample using heavy liquids and run in Isodynamic separator to separate magnetic and non-magnetic components. Gold bearing non-magnetic components are analysed for gold after pre-concentration using organic solvents like isobutyl methyl ketone (MIBK). Further developments in the panning methodology have helped to modify this method with regard to secondary iron oxide/hydroxide coating over gold grains which was hampering identification of visible gold and separation from magnetic fractions in Isodynamic separation (Bheemalingeswara 1995a, 1995b). Furthermore, secondary iron oxide-hydroxide coating being common in the secondary environment, there were many efforts by different workers to understand the possible relationship between the coatings and metal ions including gold in the secondary environment

(Schooven *et al.*, 1992; Enzweiter and Joekes, 1991; Crone *et al.*, 1984; Bheemalingeswara, 1996; Chao *et al.*, 1976). Limitations in this method are the nugget effects, intensity of the secondary iron oxide coatings and sample processing problems.

Finer fractions of stream sediments or soil samples are used in place of coarse to avoid processing problems and nugget effects and there-by improving reliability of the data (Bheemalingeswara,1997; Xie and Wang, 1991; Fletcher and Muda, 2003; Cohen *et al.*, 2005). Hot acid treatment is preferred to leach total gold present in the finer fraction of the sample *i.e.* 100% recovery. Limitations in this method are: 1) finer fractions generally will be <1% in the bulk sample, hence sieving of large quantities of bulk is required; 2) sieving should be done in the field itself; 3) concentrations will generally be in ppb ranges and 4) in the case of intense coating proper separation of the fraction may be difficult.

Another important method recently came into practice is bulk leach extractable gold (BLEG). Significant change in this method is use of the bulk sample instead of a particular size fraction and leaching of gold (irrespective of size) with cyanide solution. It does not involve any sample processing. Data becomes more reliable and reproducible. This method has helped to locate many gold deposits including low grade in different parts of the globe (Beeson, 1995). Limitations or disadvantages in this method are:

a) The dilute cyanide solution react only the top layers of the gold grains not the complete grain.
b) Leaching of gold in the presence of other metals like iron etc. may not be effective.
c) It may not be effective in the case of invisible gold (Spry and Thieben, 2000) or trapped gold (Liang *et al.*, 1980) and
d) Cyanide being highly poisonous, it has to be handled with care.

Keeping in view the above mentioned shortcomings, a new method is proposed in this paper. That is bulk leach extractable gold using hot aqua regia acids. It is denoted as BLEG-AR to differentiate it from BLEG (using cyanide). It is tested on 15 stream sediment samples collected from an area known for gold mineralization near Ramagiri, Anantapur district, Andhra Pradesh, India. The study area (about 15km^2) is covered earlier by a regional stream sediment survey (about

100km^2) covering the Ramagiri Gold Fields for gold by the author using minus 0.177 mm size fraction.

Geologically, the study area forms part of Ramagiri schist belt, locally and Ramagiri-Penakacherla schist belt, regionally. It is one of the greenstone belts in Precambrian Dharwar Craton. The belt forms a Y-shaped outcrop near Ramagiri village and surrounded on all sides by Peninsular Gneiss. The quartz bodies in the form of veinlets, pods, reefs etc., occur along foliation planes in the area. Three types of quartz veins observed in the area are related to different generations and vary in color from grey to white. The gray fine-grained dull variety is generally considered to be gold bearing (Ghosh *et al.*, 1970).

Present study is confined to the western limb of the schist belt and the western part of the Kottapalli Block of Ramagiri Gold Fields. It is about 1-2 km north of Ramagiri town (Fig.1). The study area is dominated by granites, breccia, quartz-chlorite schist (with sericite) and Banded Iron Formation (BIF). The iron formation varies in thickness (about 100m) may be due to repeated folding.

Previous Survey

A stream sediment survey and a follow-up lithogeochemical survey were carried out in the Ramagiri Gold Fields covering both eastern and western limbs of the Ramagiri schist belt from Jibutal, in the south, to Chennabhavi (beyond Kottapalle, Fig.1) in the north. The stream sediment survey results (minus 0.177mm size fractions, hot aqua regia and MIBK treated) confirmed the presence of an ore body in both eastern and western limbs. The old workings at Jibutal, Chennabhavi, Kottapalle and working mine at Yappamana (all lie in the eastern limb) (not shown in the figure) correlate well with the anomalies recorded in the survey. Interestingly, the survey has also noted anomalies for gold in the western limb (passing Ramagiri village, Fig.1) which were not known earlier. The follow-up lithogeochemical survey further confirmed presence of gold mineralization in association with breccia, quartz-chlorite schist (Bheemalingeswara *et al.*, 1996) (Fig.1). Ore petrographic investigation indicated presence of pyrite, chalcopyrite, pyrrhotite, sphalerite and minor arsenopyrite in the auriferous zone in the Kottapalle Block of the Ramagiri ore zone (Deb and Bheemalingeswara, 2005). Among sulfides, pyrite has been the dominating sulfide mineral.

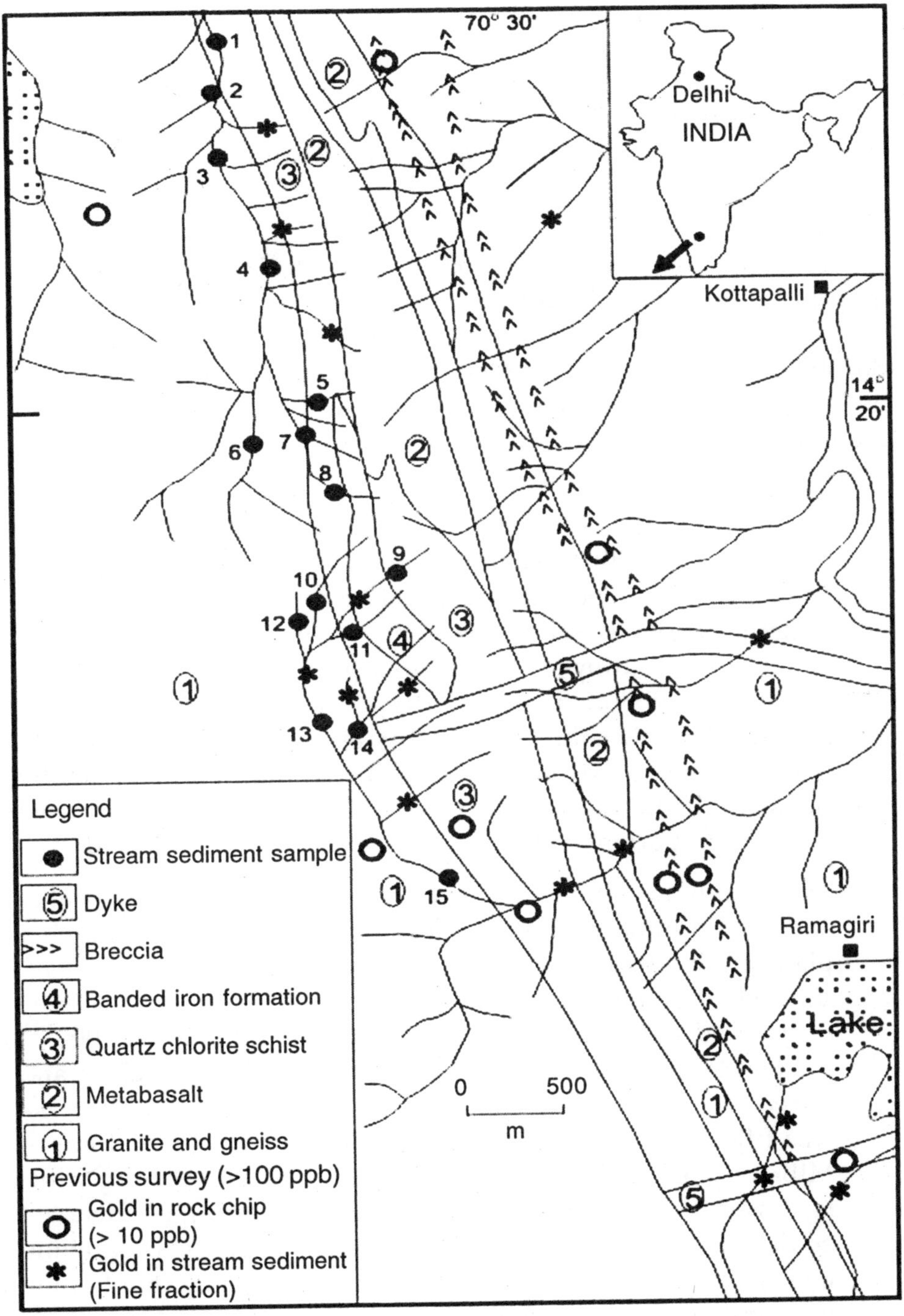

Fig. 1 : Geological and stream sediment sample location map, Ramagiri Gold Fields, Anantapur district, Andhra Pradesh, India (after Ghosh *et al.,* 1970).

Selection of the area for the present study to test BLEG-AR method is quite appropriate because a) it is known for gold mineralization; b) iron hydroxide coating is quite conspicuous (as indicated in the previous geochemical survey); c) banded iron formation (BIF) is quite prominent; and d) pyrite is prominent among sulfide minerals.

Sampling and Analysis

About 5-6 kilograms of –30 (-0.59mm) mesh size stream sediment samples (n=15) were collected after removing the top layer from each site (Fig.1). About 2 kg of the bulk sample was further sieved to obtain –30+50 (-0.59+0.297mm) and -80 (0.177mm) mesh size fractions. Later, the bulk (-0.59 mm), coarse (-0.59+0.297mm) and finer (0.177mm) fractions were further processed and analysed. Following is the procedure adopted to obtain the solutions.

I. The bulk samples of -0.59 mm size were treated initially with 1+1 HCl, later with concentrated aqua regia and iso- butyl methyl ketone (MIBK). The purpose of this scheme is to eliminate the mineral phases like carbonates, remove iron hydroxide coating and liberate coating- bounded gold grains using 1+1 HCl and dissolve the liberated and other types of gold using aqua regia and transfer gold solution into MIBK organic solvent so that gold is transferred to MIBK. Since, all the available gold is finally transferred into the organic solvent, the volume of MIBK used becomes important rather than the volume of the hot acids.

   The sample weighing 2 kg, was transferred into a glass beaker and treated initially with 250-500 ml of 1+1 HCl and kept on hot plate till the acid volume was reduced approximately to half. 150-250 ml of aqua regia was added and heated till the volume of the acid reduced to minimum. After cooling, about 200-300 ml double distilled water (DDW) was added to the beakers and mixed well and filtered into a 500ml volumetric flask. The flask was filled with DDW close to the mark after adding 10ml of MIBK organic solvent and mixed well. The top gold bearing MIBK layer was then transferred into a 100ml volumetric flask, denoted as BULK, for analysis. The organic layer which was showing intense red color for some of the samples, possibly due to high iron content, was repeatedly washed with dilute HCl and DDW and removed iron before analysis

II. The –30+50 (-0.59+0.297mm) mesh size fraction, weighing about 1kg was treated with bromoform to separate heavy mineral components (HMC) and run in Iso-dynamic separator, at standard settings (Bheemalingeswara, 1995a) to separate magnetic (M) and non-magnetic (NM) components. 100 gm of NM components was treated with aqua regia and MIBK. The solution thus prepared was denoted as NM.

III. The finer fractions (-0.177mm) weighing about 50gm were treated with aqua-regia and followed by MIBK and the solutions were denoted as FINE.

IV. All the solutions were analysed using Atomic Absorption Spectrophotometer (AAS) (Perkin-Elmer, 2380) in the Department of Geology, Delhi University. One gram of analytical grade chloroauric acid (E-MERCK) was dissolved using dilute hydrochloric acid and made the volume to 1 litre. This solution was used as standard/reference for gold analysis. From the stock solution of gold, different standards, 1, 2, 3, 4 and 5ppm were prepared and together with blank used to set AAS and suggested standard analytical conditions. Both standard and stream sediment solutions were treated with MIBK. In the case of gold standard (reference) one ml of 10 ppm gold was added to 100 ml MIBK to prepare 0.1 ppm gold standard. Likewise gold standards with different strengths were prepared and used to construct the working curve against which the unknown sample solutions were run. Since the standards fed were in ppm, the values thus obtained for samples were also in ppm. Known standards (as unknown) and stream sediment samples were re-run many times to check the reliability and reproducibility of the data. The precision of the data has been within the acceptable limits (± 10%). Data generated for fifteen stream sediment samples in the present study and data recorded in the previous survey are given in Table-1.

Discussion

Gold data generated for BULK (-0.59mm), NM (–0.59+0.297mm) and FINE (-0.177mm) indicate presence of gold anomalies. They not only tally well with the earlier data (Bheemalingeswara *et al.*, 1996) and anomalies (Fig.1) but also show additional new anomalies compared to the previous survey (Fig. 2). Gold concentrations (maximum) vary significantly in bulk and size fractions *i.e.* 0.45 ppm in BULK, 0.20 ppm in NM and 90 ppb in FINE in the present survey. The values of FINE are comparable with the data of the finer fractions (100 ppb) of the stream sediments in the previous survey. In the same area, maximum value recorded for gold in rock samples is 2.47 ppm (Bheemalingeswara *et al.,* 1996).

Table-1 : Gold concentrations (ppm) in BULK, NM and FINE fractions stream sediment samples, Ramagiri Gold Fields (previous data is only for comparison).

S. No.	Bulk sample (BULK) (-0.59mm)	Magnetic component (-0.59+ 0.297mm)	Non-magnetic component (NM) (-0.59+0.297mm)	Fine fraction (FINE) (- 0.177mm)	Previous survey data (- 0.177mm)
1	0.08	<0.01	<0.01	0.06	No data
2	0.16	0.09	0.03	<0.01	0.06
3	0.04	0.02	<0.01	0.02	<0.01
4	0.08	<0.01	0.03	0.06	0.02
5	0.04	<0.01	<0.01	0.06	0.04
6	0.45	0.11	0.20	< 0.01	0.03
7	0.12	0.08	0.06	0.05	0.04
8	0.14	0.03	0.08	< 0.01	<0.01
9	0.05	<0.01	0.02	< 0.01	<0.01
10	0.08	<0.01	0.04	< 0.01	<0.01
11	0.22	0.06	0.11	0.07	0.06
12	0.19	0.05	0.03	0.11	0.10
13	0.05	0.02	0.02	0.05	<0.01
14	0.08	0.05	<0.01	0.04	0.05
15	0.17	0.06	0.08	0.04	0.04

Distribution of gold in the three different fractions/components is quite conspicuous. Bulk samples (BULK) show higher values for gold compared to both coarse (non-magnetic component) and finer fractions. Gold data obtained for finer fractions in both present and previous surveys, which tally well, amply indicate the reliability of the method. Gold values in the bulk sample and different fractions vary in the order of BULK>NM>FINE. Reproducibility of the data though a problem particularly in the case of coarse fractions, present data (with limitations) suggests that the values vary within 10-15%. Among the three fractions, both BULK and FINE seems to be more reliable compared to NM, where the values fluctuate significantly due to nugget effects.

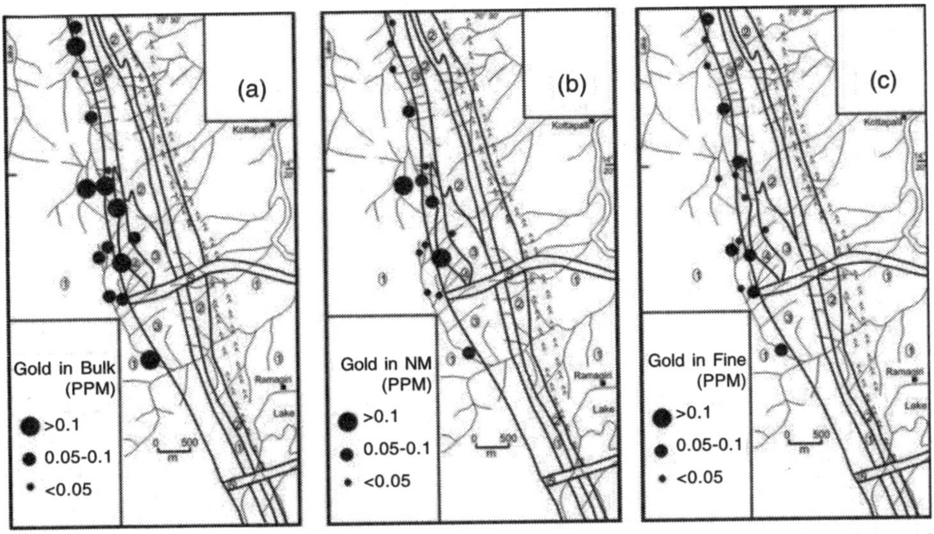

Fig. 2 : Gold concentrations (ppm) in a. BULK; b. NM and c. FINE fractions of stream sediment samples, Ramagiri Gold Fields.

Iron oxide/hydroxide coating is quite significant on the grains in the study area. Its intensity is quantified approximately as 20%. Though it is difficult to say whether the coating is uniform or partial or selective, its effect or influence on the sample seems to be about 20%. The effect of these secondary iron oxide/hydroxide coatings will be in the form of 1) accumulation of small grains of magnetite, hematite, goethite etc, (Liang Pan *et al.*, 1980); 2) fine to coarse grained particles of gold to get attached to these grains; and 3) growth of gold nuggets. The coatings also act like adhesive and influence different grains particularly small/fine grains including gold to stick together and attain bigger sizes. Higher the intensity of coating bigger the grain sizes and reduction in finer sizes. This affects proper recovery of different size grains during sieving and Iso-dynamic separation and thus causing errors in the data (Bheemalingeswara, 1995b). In addition, higher the intensity of coating over the coarse detrital grains, higher will be its ability to adsorb small/fine size detrital mineral grains (ferromagnetic or paramagnetic) on to its surface including gold. This is noticed when the samples were treated with dilute HCl to remove the coatings. Such fine magnetic or paramagnetic grains actually influence the separation process not the coating itself while running in Isodynamic separation. Once coating is removed, the grains become liberated and behave accordingly and achieve better separation.

Another problem that may still persist is the mineral inclusions like magnetite, ilmenite etc. in gold nugget or as nucleus for the growth of gold (nugget) (Chapman *et al.*, 2000; Bheemalingeswara, 2004). This kind of association is more common with dark colored minerals compared to the light colored minerals like quartz. Further, gold can also stretch into thin plates and fold around different mineral grains (Vasconcelos and Kyle, 1991) including magnetite, ilmenite etc. Dilute acid treatment in this case does not help. The survey programs, which intend to use coarse (particulate gold) or finer fractions particularly in areas where coating is conspicuous, need to take into account the above mentioned problems to achieve better results. Use of bulk leach extractable gold (BLEG) method (Beeson, 1995) and finer fractions (Xie and Wang, 1991) in place of traditional NM method though could overcome the problems like nugget effects, the question of total recovery of gold in the bulk or a particular fraction of interest still remained unanswered particularly where gold is invisible or secondary coatings are conspicuous.

In routine surveys, sample volumes in the case of coarse (-0.59+0.297mm) and fine (-0.177mm) will be generally about one kg and 200 gm respectively collected in the field after sieving the bulk sample. The proportion of these fractions is generally expected to be about 50 and 10% respectively in the bulk sample. This of course depends on the nature of the geological condition locally. This is an average figure obtained using the data from various case studies conducted by the author, predominantly the areas experiencing semi-arid conditions (central, west and south India). In all the cases, the sample volume used in the analysis was about 50 or 100gm. Since the volume of bulk sample used in the present analysis is about 2kg or more, the volumes of these fractions will also become more and the final solution prepared for analysis is 10ml. It means that gold available in BULK (out of 2kg) which includes all the fractions will be much higher compared to NM (100gm) and FINE (50mg) data. It is also important to note here that the BULK values are free from the secondary iron oxides and nugget effects and the recovery of gold is almost total and the reliability and reproducibility of the data is also better.

Bulk sampling and cyanide treatment (BLEG), though is a proven method and widely used to recover total gold in the sample, the question of its ability to bring gold into solution in the case of coatings, invisible gold etc., has remained unanswered. In the case of new hot aqua regia method suggested here, the problems related coatings, size of the grains, inclusions, nuggets, and invisible gold etc do not affect the gold recovery. Gold, irrespective of its type is brought into solution and the data reflects the total gold present in the sample.

Present method (MIBK-AR) of using bulk sample and treating with aqua regia though tested on only 15 samples, it has the potential to be an alternative to the existing methods in terms of total recovery of gold and reliability of the data. It will also take care of the problems related to the size fraction, processing, nugget effect, invisible nature, recovery and secondary iron oxide/hydroxide coatings etc. It may be best suited for orientation surveys where it can provide information to set parameters required for detailed survey.

Conclusion

- BULK data shows anomalous values for gold and tally well with the previous survey data. Better contrast for gold is achieved for bulk (-0.59mm) compared to finer fractions.
- Data suggests that the selection of the size fraction in areas where iron oxide/hydroxide coating is conspicuous needs caution.
- Since BLEG-AR method does not suffer from nugget effects, coating, invisible, nucleus and mineral inclusion effects, the data reflects near total concentration for gold present in the sample irrespective of its forms.
- BLEG-AR method seems to provide better results than BLEG (using cyanide) at least in some cases.
- The method may also be tested using other sizes like –10 or –20 mesh.

Acknowledgements

This work was carried out by the author in the Department of Geology, Delhi University, Delhi while working as post-doctoral fellow. Thanks are due to Prof. M. Deb for his cooperation and encouragement to do this work in Ore Geology Laboratory.

References

Beeson, R. (1995) A drainage sediment geochemical orientation study at Boddington, Western Australia. J. Geochemical Exploration, v.54, pp. 63-71.

Bheemalingeswara, K. (1995a) Possible effects of iron oxide coating in the recovery of particulate gold from stream sediments. J. Geochemical Exploration, v. 52, pp. 373-380.

Bheemalingeswara, K. (1995b) Geochemical analysis of bulk and non-magnetic component of acid-treated and untreated heavy mineral concentrates: Implication on gold exploration. Indian Minerals, v. 49, no.3, pp.197-200.

Bheemalingeswara, K. (1996) Geochemical analysis of iron oxide/hydroxide coated non-magnetic component of the bulk sample: possible utility of the coated component in a stream sediment survey for gold. Current Science, v.71, no.5, pp. 410-411.

Bheemalingeswara, K. (1997) Stream sediment survey for gold in the vicinity of Paroli, Bhilwara district, Rajasthan. Indian J. Geochemistry, v.12, pp. 11-16.

Bheemalingeswara, K. (2004) Orientation survey results from an area known for gold mineralization, Bhiwapur, Nagpur district, Maharashtra. Indian Journal of Applied Geochemistry, v.6, pp.61-67.

Bheemalingeswara, K., Deb, M. and Dutta, A. (1996) New potential ore zones in the Ramagiri Gold Field: Results of stream sediment and a follow-up lithogeochemical survey. *In: S.M. Naqvi (ed.),* Gold resources in India, Workshop volume, Hyderabad, India, pp. 89-94.

Chao, T.T. and Theobald, P.K. (1979) The significance of iron and manganese oxides in geochemical exploration. Economic Geology, v.71, pp.1560-1569.

Chapman, R.J., Leake, R.C. and Moles, N.R. (2000) The use of microchemical analysis of alluvial gold grains in mineral exploration: experiences in Britain and Ireland. J. Geochemical Exploration, v.71, pp. 241–268.

Cohen, D.R., Dunlop, A.C. and Rose, T. (2005) Contrasting dispersion patterns for gold in stream sediments at Timbarra, NSW, Australia. J. Geochemical Exploration, v.85, pp. 1-46.

Crone, W., Larson, L.T., Carpenter, R.H., Chao, T.T. and Sanzolone, R. F. (1984) A comparison of iron oxide-rich joint coatings and rock chips as geochemical sampling media in exploration for disseminated gold deposits. J. Geochemical Exploration, v. 20, pp. 161-178.

Deb, M. and Bheemalingeswara, K. (2005) Stable isotope geochemistry of the gold-sulfide mineralized zone of the Kottapalle block of the Ramagiri greenstone belt, Dharwar craton, South India. *In: J. Mao, and F.P. Bierlein, (eds.),* Mineral Deposit Research: Meeting the Global Challenge, Proceedings of the Eighth Biennial SGA Meeting, Beijing, China, Springer Verlag, pp.743-746.

Enzweiler, J. and Joekes, I. (1991) Adsorption of colloidal gold on colloidal iron oxides. J. Geochemical Exploration, v.40, pp.133-142.

Fletcher, W. K. and Joanes Muda. (2003) Dispersion of gold in stream sediments in the Sungai Kuli region, Sabah, Malaysia. Geochemistry: Exploration, Environment, Analysis, v. 3, pp. 51-56.

Ghosh, D. B., Sastry, B. B. K., Rao, A. J. and Rahim, A. A. (1970) Ore environment and ore genesis in Ramagiri Gold Field, Andhra Pradesh, India. Economic Geology, v.65, pp. 801-814.

Lange, I.M. and Gignoux, T. (1999) Distribution, characteristics and genesis of high fineness gold placers, Ninemile Valley, Central-Western Montana. Economic Geology, v.94, pp. 375-386.

Liang Pan K., Overstreet, W.C., Robbinson, K., Hubert, K.E. and Crenshaw, G.L. (1980) Equivalent uranium and selected minor elements in magnetic concentrates from the Candle Quadrangle, Soloman Quadrangle and elsewhere in Alaska. United States Geological Survey Professional Papers, v.1135, pp.1-108.

Schoonen, M.A.A., Fisher, N.S. and Wente, M. (1992) Gold sorption onto pyrite and goethite :A radiotracer study. Geochimica et Cosmochimica Acta, v.56, pp. 1801-1814.

Simon, G., Kesler, S.E. and Chryssoulis, S. (1999) Geochemistry and textures of gold-bearing arsenopyrite, Twin Creeks, Nevada: Implications for deposition of gold in Carlin type deposits. Economic Geology, v. 94, pp. 404-422.

Spry, P.G. and Thieben, S.E. (2000) The distribution and recovery of gold in the Golden Sunlight gold-silver telluride deposit, Montana, USA. Mineralogical Magazine, v. 64, no.1, pp. 31-42.

Vasconcelos, P. and Kyle, J.R. (1991) Supergene geochemistry and crystal morphology of gold in a semi-arid weathering environment: application to gold exploration. J. Geochemical Exploration, v. 40, pp. 115-132.

Xie,, Xuejing and Wang Xueqiu (1991) Geochemical exploration for gold: a new approach to an old problem. J. Geochemical Exploration, v. 40, pp. 25-48.

□□□

11

Distribution of Mangroves along the Red Sea Coast of the Arabian Peninsula : Part-I : the Northern Coast of Western Saudi Arabia

ARUN KUMAR, M. ASIF KHAN AND ABDUL MUQTADIR
Center for Petroleum and Minerals, Research Institute
King Fahd University of Petroleum and Minerals, Dhahran 31261, Saudi Arabia

Abstract

Historical Landsat Enhanced Thematic Mapper (ETM) data of the years 2000 and 2001 were used to numerically illustrate the distribution of mangroves that occur along the northern Saudi Arabian Red Sea coast from Sharm Zubair in the north to Al-Lith in the south. The distribution of mangroves is patchy and they do not occur as a continuous forest and are generally restricted to the low energy quieter environments like bays, narrow channels, and inland faces of offshore islands. However, they also occur in the intertidal environments. At certain locations they occupy very small area that could indicate depletion of mangroves. Due to rapid urbanization and industrialization of this region, traditional land use is changing and adversely impacting sustenance and growth of mangroves. There should be an inclusive policy that promotes economic development and protects mangroves also; this can be achieved as shown by Khan and Kumar (2009) on certain locations along the Arabian Gulf coast.

Keywords : Red Sea, Remote Sensing, Mangroves, Infrastructure Development, Image Registration, Normalized Difference Vegetation Index (NDVI)

Introduction

Mangroves are trees and shrubs that grow in brackish to saline coastal environments in the tropics normally between latitudes 25° N and 25° S. They belong to the plant family Rhizophoraceae and along the Red Sea coast of Saudi Arabia *Avicennia marina* is the dominant species but *Rhizophora mucronata* too occurs in patches. They occur in inter-tidal zones, estuaries and offshore islands. Mangroves protect coastal areas from erosion, storm surges, hurricanes and tsunamis their massive root system is efficient at dissipating wave energy. The unique ecosystem found in the intricate mesh of mangrove roots offers a quiet marine habitat for diverse groups of organisms.

Mangrove habitats support marine and avian life in the Red Sea. These environments have been under stress because of overgrazing, cutting, pollution, physical modification of the coastal habitat, and land use change for quite some time. There is a need for their conservation. Mangroves occur along both east and west coasts of the Arabian Peninsula and early research on the mangrove ecology and distribution along the Red Sea coast of Saudi Arabia were by published by Vesey-FitzGerald (1955, 1957); Zahran (1974, 1980) and EESAL, 1983. Price *et al.* (1987) surveyed the mangal (mangroves and other associated communities of flora and fauna) ecology on several locations along the Saudi Arabian coast of the Red Sea. A detailed environmental assessment study along the Red Sea coast of Saudi Arabia by Price *et al.* (1998) found that mangroves were significantly more abundant at the coastal sites than offshore and on the southern latitudes than in the north.

A comprehensive report of PERSGA (The Regional Organization for the Conservation of the Environment of the Red Sea and Gulf of Aden) provides the current status of mangroves in the Red Sea and Gulf of Aden (Khalil, 2004). This report discusses various threats to mangroves due to camel grazing, cutting, pollution due to construction, industrialization and changes in land use and shrimp farming. It also recommends guidelines for conservation, rehabilitation and management of the mangroves in the region.

We are studying distribution of mangroves all along the Red Sea coast of the Arabian Peninsula. This paper is first part of our study that covers the geographical distribution of mangroves along the northern Red Sea coast of Saudi Arabia. A similar study was recently published by Khan and Kumar (2009) along the Arabian Gulf coast of Saudi Arabia and the United Arab Emirates. The distribution of mangroves along the RED sea coast is not continuous, instead occurs in patches. The main objective of this paper is to numerically show the geographic distribution of mangroves in the region and create a point of reference about their distribution for the year 2000-2001. This paper reports on patchy distribution of mangroves on twenty locations along the northern Saudi Arabian Red Sea coast (Fig. 1) using Lnadsat ETM images of the years 2000 and 2001 (Fig. 2).

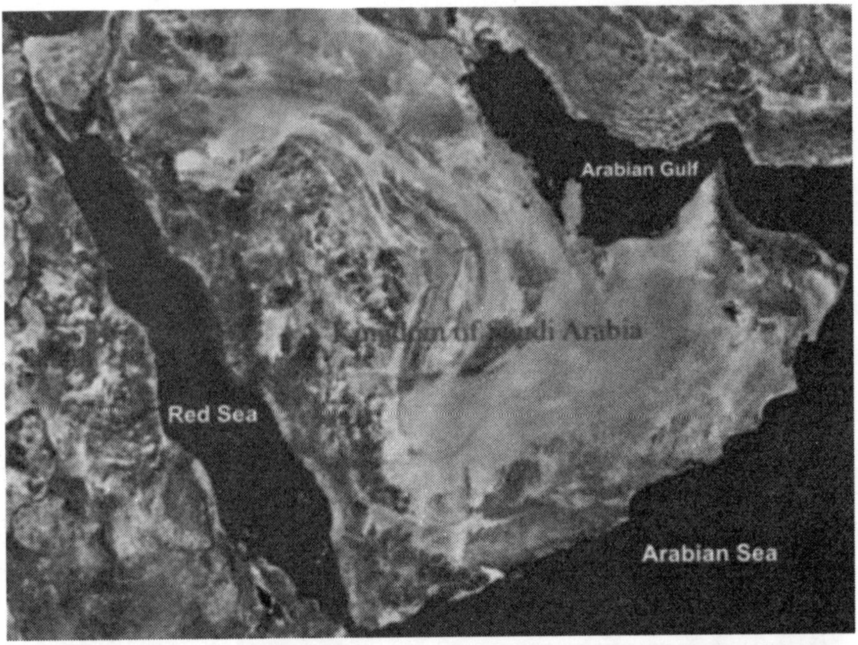

Fig. 1 : Study area along the northern Red Sea coast of Saudi Arabia.

Red Sea

Red Sea is a long and narrow body of water separating northeast Africa on the west with the Arabian Peninsula to the east (Fig. 1). It is connected with the Indian Ocean at its southern end through the Bab el Mandeb strait and Gulf of Aden and is the world's northernmost tropical sea. Its northern end is bifurcated into two narrow elongated bodies of water on both sides of the Sinai Peninsula. On the west side is the Gulf of Suez and on the east side is the Gulf of Aqaba. Some basic statistics about the Red Sea is given in Table-1.

Fig. 2 : Locations of 20 study sites along the northern Red Sea coast of Saudi Arabia.

Surface water temperatures are constant year round at 21–25 °C. The salinity is higher than the world average ~ 40 (ppt). This is due to (a) high rate of evaporation and very little precipitation (b) lack of significant rivers or streams draining into the sea, and (c) limited connection with the Indian Ocean, which has lower water salinity. The climate of the Red Sea is the result of two distinct monsoon seasons; a northeasterly monsoon and a southwesterly monsoon. The rainfall over the Red Sea and its coasts is extremely low averaging 0.06 m/year; the rain is mostly in the form of showers of short spells often associated with thunderstorms and occasionally with dust storms. In summer NW winds drive surface water south, whereas in winter the flow is reversed resulting in the inflow of water from the Gulf of Aden into the Red Sea. The net value of the latter predominates, resulting in an overall drift to the northern end of the Red Sea. Thus the prevailing north and northeastern winds influence the movement of water in the coastal inlets to the adjacent sabkhas, especially during storms. Winter mean sea level is 0.5 m higher than in summer. Some information used in this paper is taken from Wikipedia.

Table-1 : Some basic statistics about the Red Sea

Length: ~2,250 km (79% of the eastern Red Sea with numerous coastal inlets)
Maximum Width: ~ 306–355 km (across Massawa in Eritrea)
Minimum Width: ~ 26–29 km (across Bab el Mandeb Strait in Yemen)
Average Width: ~ 280 km
Average Depth: ~ 490 m
Maximum Depth: ~2,211 m
Surface Area: 438-450 x 10^2 km^2
Volume: 215–251 x 10^3 km^3
Approximately 40% of the Red Sea is <100 m and about 25% is < 50 m deep
About 15% of the Red Sea is over 1,000 m deep that forms the deep axial trough
Shelf breaks are marked by coral reefs
Continental slope has an irregular profile (series of steps down to ~500 m)
Centre of Red Sea has a narrow trough (~ 1,000 m; exceeding 2,500 m)

Materials and methods

The study area is located along the northern Saudi Arabian Red Sea coast from Sharm Zubair in the north to Al-Lith in the south between Latitudes 27° 26' 00"and 20° 09' 00" (Fig. 2). Historical Landsat Enhanced Thematic Mapper (ETM) data of the years 2000 and 2001 were used (Table-2). The ETM data has seven spectral bands (Table-3). The spatial resolution of ETM data is 28.5 meters.

Table-2 : Satellite image data used in the study

Location	Sensor	Date
site-1	ETM	05-10-2001
site-2	ETM	05-10-2001
site-3	ETM	05-10-2001
site-4	ETM	10-07-2000
site-5	ETM	10-07-2000
site-6	ETM	10-07-2000
site-7	ETM	10-07-2000
site8	ETM	23-10-2000
site-9	ETM	23-10-2000
site-10	ETM	23-10-2000
site-11	ETM	23-10-2000
site-12	ETM	12-07-2000
site-13	ETM	12-07-2000
site-14	ETM	12-07-2000
site-15	ETM	01-11-2000
site-16	ETM	01-11-2000
site-17	ETM	01-11-2000
site-18	ETM	01-11-2000
site-19	ETM	01-11-2000
site-20	ETM	03-06-2000

Table-3 : Landsat spectral bands

Landsat 7 (ETM+ sensor)	Wavelength (micrometers)
Band 1	0.45 - 0.515
Band 2	0.525 - 0.605
Band 3	0.63 - 0.69
Band 4	0.75 - 0.90
Band 5	1.55 - 1.75
Band 6	10.40 - 12.5
Band 7	2.09 - 2.35

Image processing

Satellite images and geographic information systems are useful tools to detect and compute aerial distribution of mangroves (Long and Skewes, 1996; Riaza *et al.*, 1998; Verstraete *et al.*, 2008). The first step in the processing of remotely sensed images is the geometric correction (Richards, 1986). In this study map-to-image ground control points were used to geometrically rectify all of the Landsat ETM data to UTM, WGS84 coordinates. All the images were radiometrically corrected using dark-pixel subtraction (Crane, 1971), also known as histogram minimum method (Chavez *et al.*, 1977).

The second step in the processing of the satellite images is to exclude irrelevant land and water areas that can be readily identified as non mangroves habitats. This was accomplished by creating region masks by visual interpretation of color composites of Landsat ETM bands 4, 2 and 1, and high resolution Google Earth images (Manson *et al.*, 2001). The coastal vegetation and algal mats were also excluded from the ETM images by creating region masks using ETM color composite of bands 4, 5 and 1, where vegetation species can be differentiated by their color and tone variations (Coleman *et al.*, 1990). The vegetation associated with the mangroves such as halophytes which are mingled with mangroves and also has similar spectral signature may not have been excluded.

The most widely used technique for mapping vegetation with remotely sensed images is the Normalized Difference Vegetation Index (NDVI). Studies using NDVI and other vegetation indices are varied and include those involved with local, regional, and global mapping of vegetation (Townsend and Justice, 1986; Choudhury and Tucker, 1987; Jackson and Huete, 1991; Justice *et al.*, 1991; Tucker *et al.*, 1991; Richardson and Evert, 1992). Recently, Helldén and Tottrup (2008) used time series of annually integrated and standardized annual NDVI anomalies to study trends in regional desertification and suggested that this.methodology is a robust and reliable way to assess and monitor vegetation trends and related desertification on a regional-global scale.

The Normalized Difference Vegetation Index (NDVI) is defined as:

$$NDVI = (RIR-R) / (RIR+R)$$

Where RIR = Reflective Infrared band (MSS bands 6 and 7; TM band 4) and R = Red band (MSS band 5; TM band 3)

The masked Landsat ETM images were used to create NDVI images. NDVI for Landsat ETM images was calculated using bands 3 (0.63-0.69 mm) and 4(0.76-0.90 mm).

Results and Discussion

In Landsat images the red band records the absorption of red wavelength by chlorophyll; lower values indicate higher chlorophyll. Reflected Infrared (RIR) band records the reflection of IR wavelengths by the cell structures of leaves; higher values indicate more vigorous growth (Richardson and Evert, 1992). Values of NDVI images ranged from -1.0 to 1.0. Higher values indicate higher concentrations of vegetation, which in this study are mangrove stands. Lower values indicate non-vegetated regions like water bodies and bare soil. Thresholds of NDVI greater than 0.1, were applied to the NDVI images to classify the mangrove stands in the images. The threshold criteria, was determined from the false color composites of two visible and one near infrared bands. In this color composite vegetation appears in hues of red and can be easily differentiated from bare land, mud flats and water bodies along the coastal regions. The familiarity with the area and known locations of mangroves were also helpful in determining the threshold boundary.

Fig. 3 through 22 show images of geographical distribution and area of mangrove forests at sites 1 through 20 respectively (Table-4). Their distribution is patchy and the areas covered by various mangrove stands vary considerably. They do not occur as a continuous forest and mostly their distribution is restricted to the quieter environments, for example, bays, narrow channels, and inland face of offshore islands. However, not all mangroves occupy quieter environments they inhabit higher energy intertidal environments also. At places they occupy very small area, for example, locations, 1, 3, 6 and 8 which could be a sign of depletion due to changes in land use and pollution. The Red Sea coastal regions of Saudi Arabia are witnessing rapid changes in urbanization and industrialization altering the traditional land use. These factors and pollution are adversely impacting the mangrove cover in the region.

The result of this study indicates that a variety of human activities in the region are adversely impacting the mangrove forests. It is advised that due caution is required to protect the mangroves while planning for any economic and developmental activity. This can be achieved as shown by Khan and Kumar (2009) in their study of the mangroves of the Arabian Gulf coast where with in a span of thirty years' of 'developmental' activities mangroves were depleted but in few cases mangroves grew more along with all the development al activities. Thus there is a need for clear policy to protect mangroves along with any developmental activity that takes place in the region.

Table-4 : Area of mangrove forests (hectares) at the twenty locations along with their latitude and environmental setting

Site	Latitude (range) (Deg. Min. Sec.)	Mangrove area (Hectares)	Environment
1	27 26 03	4.3	Bay
2	26 56 34 to 26 54 11	21.5	Channel between mainland and nearshore island
3	26 38 07	0.6	Bay
4	26 21 06 to 25 55 40	36.1	Intertidal
5	25 54 41 to 25 22 00	661.9	Landward side of the nearshore island
6	25 18 24	2.8	Intertidal
7	24 48 02	10.1	Bay
8	24 11 57	8.5	Bay
9	24 00 27 to 23 57 16	121.0	Bays and intertidal
10	23 57 04 to 23 54 26	125.2	Intertidal
11	23 21 16 to 23 19 40	22.9	Bays
12	23 09 19 to 23 07 38	24.9	Bays and intertidal
13	22 58 22 to 22 50 39	136.7	Island with in the bay
14	22 45 40 to 22 45 07	13.6	Bay
15	22 23 52 to 22 12 50	64.7	Bays and intertidal
16	22 09 42 to 21 59 55	150.5	Bays and intertidal
17	21 23 36 to 21 22 06	11.4	Landward side of the nearshore island
18	20 49 44 to 20 44 27	166.7	Bay
19	20 32 59 to 20 32 22	23.9	Bay
20	20 17 59 to 20 09 32	418.0	Island with in the bay

Fig. 3 : Satellite image of mangroves shown at site 1.

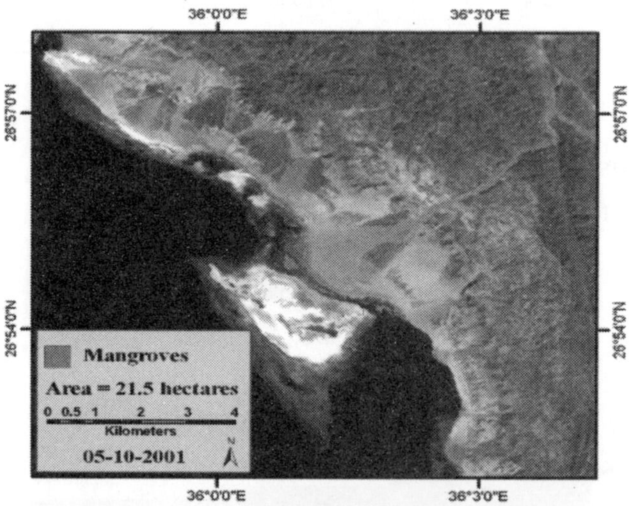

Fig. 4 : Satellite image of mangroves shown at site 2.

Fig. 5 : Satellite image of mangroves shown at site 3.

Fig. 6 : Satellite image of mangroves shown at site 4.

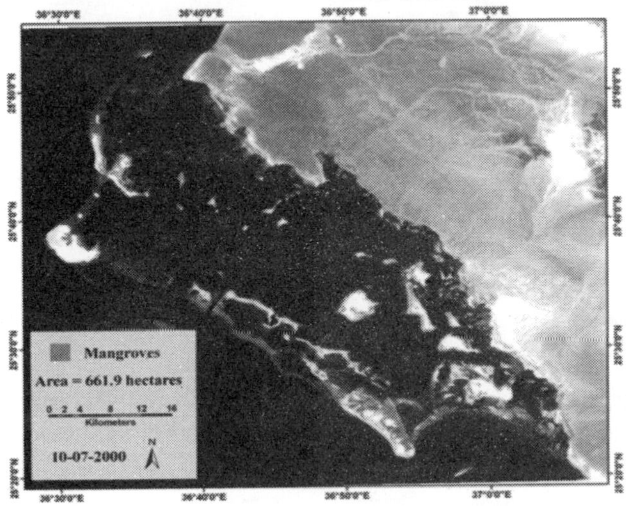

Fig. 7 : Satellite image of mangroves shown at site 5.

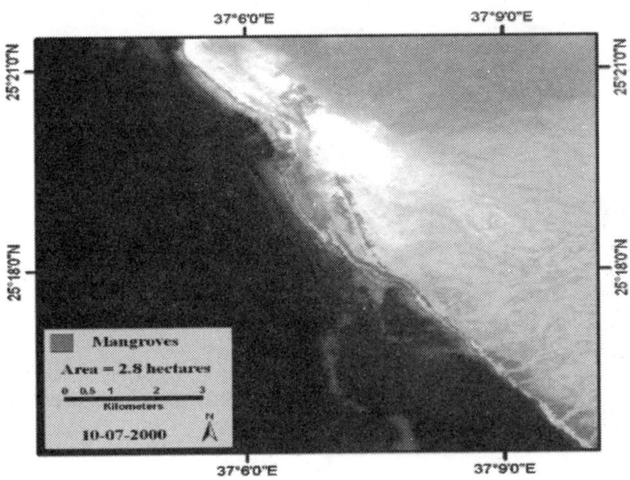

Fig. 8 : Satellite image of mangroves shown at site 6.

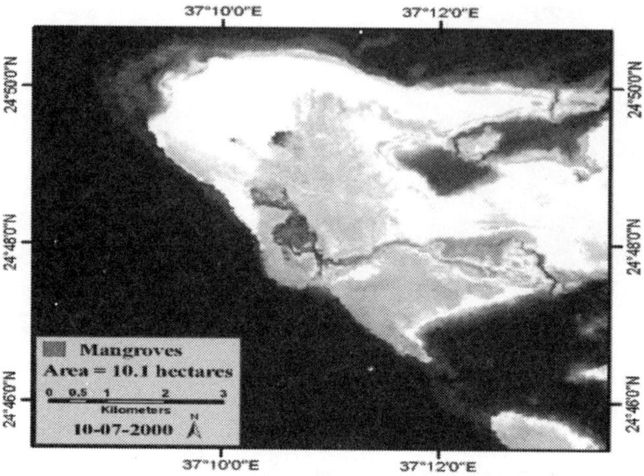

Fig. 9 : Satellite image of mangroves shown at site 7.

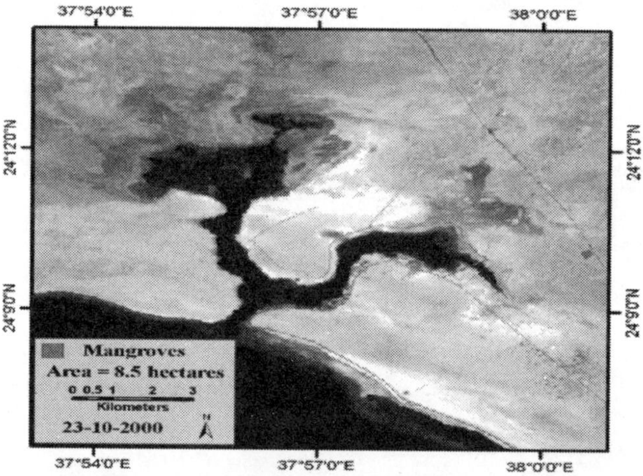

Fig. 10 : Satellite image of mangroves shown at site 8.

Fig. 11 : Satellite image of mangroves shown at site 9.

Fig. 12 : Satellite image of mangroves shown at site 10.

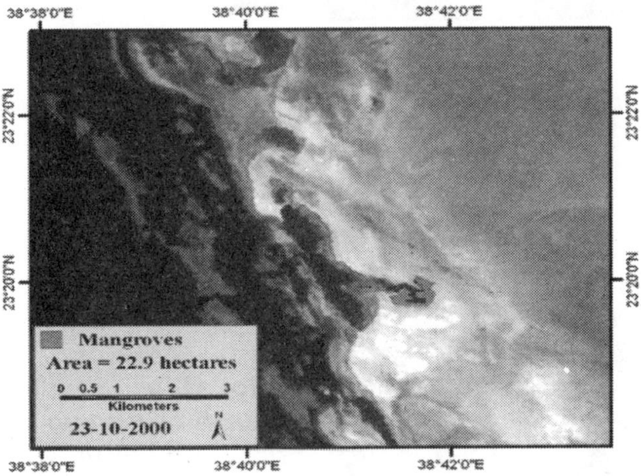

Fig. 13 : Satellite image of mangroves shown at site 11.

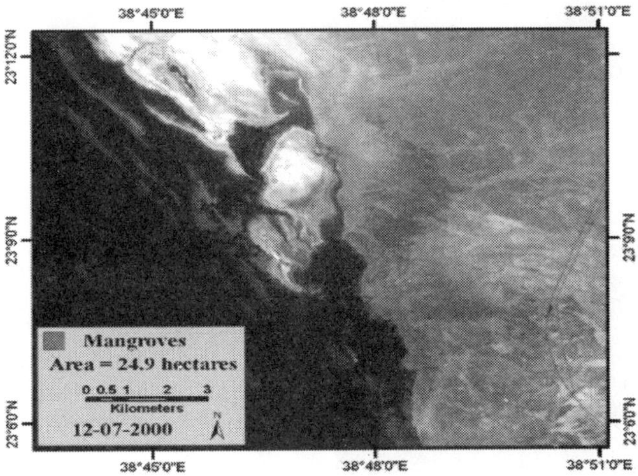

Fig. 14 : Satellite image of mangroves shown at site 12.

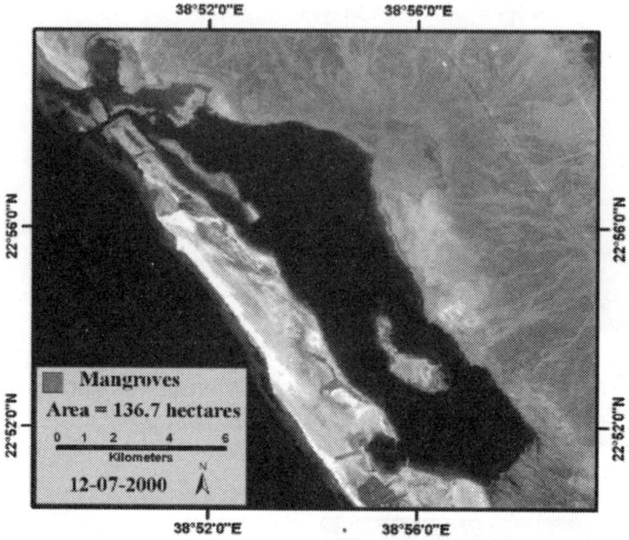

Fig. 15 : Satellite image of mangroves shown at site 13.

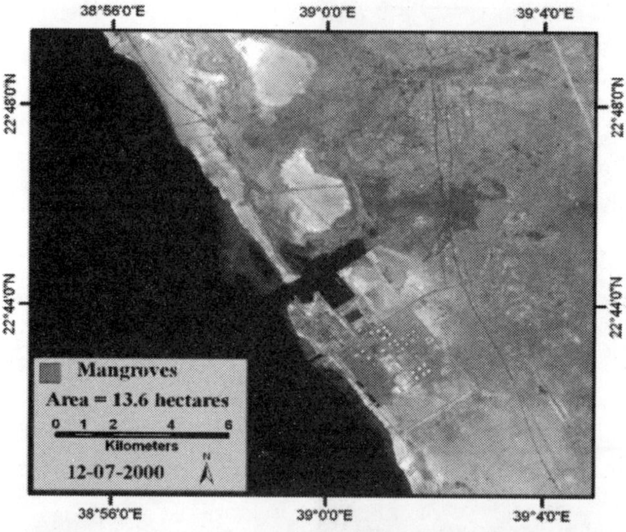

Fig. 16 a : Satellite image of mangroves shown at site 14.

Fig. 16 b : Study site 14. A recent view of mangrove stands close to the Red Sea coastline near Petro-Rabigh in Rbigh.

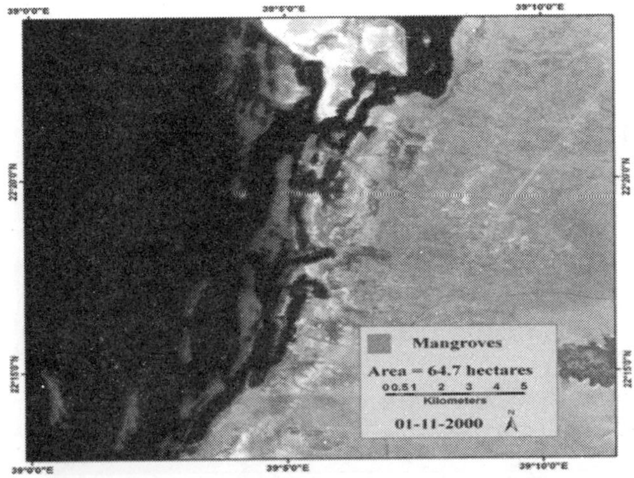

Fig. 17 : Satellite image of mangroves shown at site 15.

Fig. 18 : Satellite image of mangroves shown at site 16.

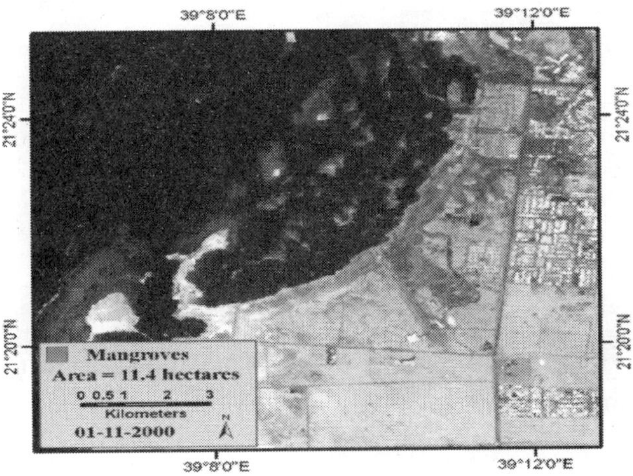

Fig. 19 : Satellite image of mangroves shown at site 17.

Fig. 20 : Satellite image of mangroves shown at site 18.

Fig. 21 : Satellite image of mangroves shown at site 19.

Fig. 22 : Satellite image of mangroves shown at site 20.

Conclusion

1. The distribution of mangroves along the northern regions of the Red Sea coast of Saudi Arabia is patchy and they do not occur as a continuous forest.
2. Mostly their distribution is restricted to the quieter environments, for example, bays, narrow channels, and inland face of offshore islands.
3. Mangroves also inhabit intertidal environments.
4. At places they occupy very small area, for example, locations, 1, 3, 6 and 8. This could be a sign of depletion due to changes in land use and pollution.
5. The Red Sea coastal regions of Saudi Arabia are witnessing rapid changes in urbanization and industrialization leading to alteration in the traditional land use. This is adversely impacting the mangrove cover in the region.
6. It is advised that due caution is required to protect the mangroves while planning for any economic and developmental activity. This can be achieved as shown by Khan and Kumar (2009).

Acknowledgements

We thank King Fahd University of Petroleum and Minerals (KFUPM), Dhahran, Saudi Arabia for permission to publish this paper.

References

Chavez, P. S. Jr., Berlin, G. L. and Mitchell, W. B. (1977) Computer enhancement techniques of Landsat MSS images for land use/land cover assessments. Remote Sensing Earth Resources, University of Tennessee, v. 6, 259p.

Crane, R. B. (1971) Preprocessing techniques to reduce atmospheric and sensor variability in multispectral scanner data. Proceedings of the 7th International Symposium on Remote Sensing of Environment (Ann Arbor: University of Michigan), p. 1345.

Choudhury, B. J. and Tucker, C. J. (1987) Satellite observed seasonal and inter-annual variations of vegetation over the Kalahari, the Great Victoria Desert, and the Great Sandy Desert: 1979-84. Remote Sensing of Environment, v. 23, pp. 233-241.

Coleman, T. L., Gudapati, L. andDerrington, J. (1990) Monitoring forest plantations using Landsat Thematic Mapper Data. Remote Sensing of Environment., v. 33, pp. 211-221.

EESAL (1983) Overview and status report of the Mangroves at Madinat Yanbu al-Sinaiyah (Red Sea Saudi Arabia). Ecology and Environment of Saudi Arabia Co. Ltd. (EESAL).

Helldén, U. and Tottrup, C. (2008) Regional desertification: a global synthesis. Global and Planetary Change, v. 64, pp. 169-176.

Jackson, R. D. and Huete, A. R. (1991) Interpreting vegetation indices. Preventive Veterinary Medicine, v. 11, pp. 165-200.

Justice, C. O., Townsend, J. G. R. and Kalb, V. I. (1991) Representation of vegetation by continental data sets derived from NOAA-AVHRR data. International Journal of Remote Sensing, v. 12(5), pp. 999-1021.

Khalil, A. S. M. (2004) Status of mangroves in the Red Sea and Gulf of Aden. The Regional Organization for the Conservation of the Environment of the Red Sea and Gulf of Aden (PERSGA), Draft 09-05-2004, PERSGA Technical Series No. 11.

Khan, A. M. and Kumar, A. (2009) Impact of 'urban development' on mangrove forests along the west coast of the Arabian Gulf. e-journal Earth Science India, v. 2(III), pp. 159-173.

Long, B. G. and Skewes, T. D. (1996) A technique for mapping mangroves with Landsat TM satellite data and Geographic Information System. Estuarine Coastal and Shelf Science, v. 43(3), pp. 373-381.

Manson, F. J., Loneragan, McLeod, I. M. and Kenyon, R. A. (2001) Assessing techniques for estimating the extent of mangroves: topographic maps, aerial photographs and Landsat images. Marine and Freshwater Research, v. 52(5), pp. 787-792.

Price, A. R. G., Medley, P. A. H., McDowall, R. J., Dawson Shepherd, A. R., Hogarth, P. J. and Ormond, R. F. G. (1987) Aspects of mangal ecology along the Red Sea coast of Saudi Arabia. Journal of Natural History, v. 21, pp. 449-464.

Price, A. R. G., Jobbins, G., Dawson Shepherd, A. R. and Ormond, R. F. G. (1998) An integrated environmental assessment of the Red Sea coast of Saudi Arabia. Environmental Conservation, v. 25(1), pp. 65-76.

Riaza, A., Martinez-Torres M. L., Ramon-Lluch, R., Alonso, J. and Heras, P. (1998) Evolution of equatorial vegetation communities mapped using Thematic Mapper images through a geographical information system (Guinea, Equatorial Africa). Inter. J. Remote Sensing, v. 19(1), pp. 43-54.

Richardson, A.J. and Evert, J. H. (1992) Using spectral vegetation indices to estimate rangeland productivity. Geocarto International, v. 1, pp. 63-69.

Richards, J. A. (1986) Remote Sensing, Digital Image Processing and Analysis – An Introduction. New York: Springer-Verlag.

Townsend, J. G. R. and Justice, C. O. (1986) Analysis of the dynamics of African vegetation using the normalized difference vegetation index. Inter. Jour. Remote Sensing, v. 7(11), pp. 1435-1445.

Tucker, C J., Newcomb, W. W., Los, S. O. and Price, S. D. (1991) Mean and inter-year variance of growing-season normalized difference vegetation index for the Sahel 1981-1989. Inter. J. Remote Sensing, v. 12(6), pp. 1133-1135.

Verstraete, M. M., Brink, A. B., Scholes, R. J., Beniston, M. and Smith, M. S. (2008) Climate change and desertification: where do we stand, where should we go? Global and Planetary Change, v. 64, pp. 105-110.

Vesey-Fitzgerald, D. F. (1955) Vegetation of the Red Sea coast south of Jedda. Saudi Arabia. J. Ecology, v. 43, pp. 477-498.

Vesey-Fitzgerald, D. F. (1957) Vegetation of the Red Sea coast north of Jedda. Saudi Arabia. J. Ecology, v. 45, pp. 547-562.

Zarhan, M. A. (1974) Biogeography of mangrove vegetation along the Red Sea coasts. Proceedings of the International Symposium on Biology and Management of Mangroves, Honolulu, Hawaii. pp. 43-51.

Zarhan, M. A. (1980) Mangroves and shoreline development in the Arabian Peninsula. Proceedings of the Regional Symposium on Prospects of Development and Environmental Protection in the Arab Gulf Countries. Qatar, University of Qatar. pp. 51-61.

□□□

12

Distribution of Mangroves along the Red Sea Coast of the Arabian Peninsula : Part-II : The Southern Coast of Western Saudi Arabia

M. ASIF KHAN, ARUN KUMAR*AND ABDUL MUQTADIR
Center for Petroleum and Minerals, Research Institute
King Fahd University of Petroleum and Minerals, Dhahran 31261, Saudi Arabia

Abstract

Historical Landsat Enhanced Thematic Mapper (ETM) data of the years 1999, 2000 and 2001 were used to numerically illustrate the distribution of mangroves along the southern Saudi Arabian Red Sea coast from Al-Lith (Latitude 20° 08' 22", and Longitude 40° 16' 17.23") in the north to the Saudi Arabia -Yemen border (Latitude16° 41' 16", and Longitude 42° 46' 27.18") in the south. A similar study covering the northern part of the Saudi Arabian Red Sea coast from Sharm Zubaïr in the north to Al-Lith in the south has already been published as first part of this ongoing study. The distribution of mangroves in both northern and southern Red Sea coast of Saudi Arabia is patchy and they do not occur as a continuous forest and are generally restricted to the low energy quieter environments. Mangroves occupy similar environments in both regions and are under similar stresses caused by various anthropogenic activities and camel grazing. However, mangrove growth in south is relatively denser in comparison to north due mainly to the tropical climate, more wadis, higher rainfall and nutrients and less saline waters.

Introduction

Saifullah (1994) presents an overview of various mangrove studies along the Red Sea coast of Saudi Arabia and concluded that information on the subject is inadequate. Early publications merely reported on mangrove occurrences (Vessey-Fitzgerald, 1955, 1957; Migahid, 1978) and their brief ecological studies (Zahran *et al.*, 1983; Frey *et al.* 1984). However, detailed ecological studies of mangrove ecosystem along the Saudi Arabian Red Sea coast were published by Mandura *et al.* (1987, 1988) Khafaji *et al.* (1988, 1991) and Saifullah *et al.* (1989). However, there are no published accounts of geographical distribution, numerical data about area covered and environments of mangrove habitats along the Saudi Arabian Red Sea coast.

Present study provides numerical information about area covered by mangroves at several sites and the environments in which they grow along the Red Sea coast of the Arabian Peninsula. Historical Landsat Enhanced Thematic Mapper (ETM) data of the years 1999, 2000 and 2001 was used in this study. First part of this study covered twenty locations covering a length of 930 km of the northern Red Sea coast of Saudi Arabia from Sharm Zubair in the north to Al-Lith in the south (Kumar *et al.* 2009, Fig. 1&2). We extend that work to the southern Red Sea coast of Saudi Arabia (Fig.1) covering a length of 500 km and it forms second part on this ongoing study. This paper reports similar study on fifteen more sites (Fig.2; sites 21 through 35) and two more locations on offshore islands (sites 36 and 37). Detailed introductory remarks and information about Red Sea is given in Kumar *et al.* (2010).

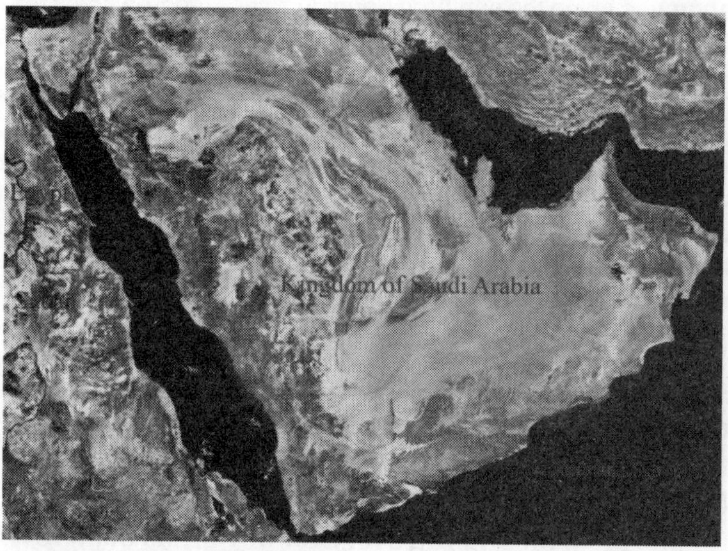

Fig. 1 : Study area along the southern Red Sea coast of Saudi Arabia.

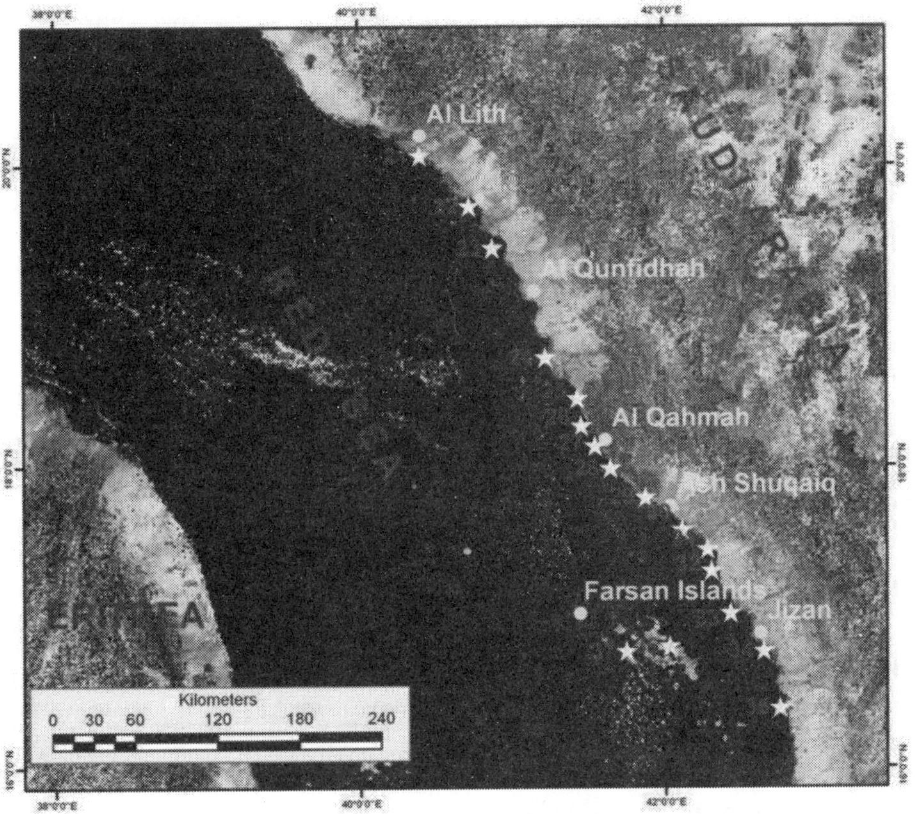

Fig. 2 : Locations of 17 study sites along the southern Red Sea coast of Saudi Arabia.

Red Sea Coastline of the Arabian Peninsula

Red Sea coastline of the Arabian Peninsula is dry without much vegetation except for few species of halophytes and mangroves. Geomorphologically coastal regions vary from rugged coast due mainly to marine terraces to coastal sabkhas, alluvial plains and wadis of up to thirty km width occasionally having small hills of Neogene sediments (Fig. 3). There are a large number of dry river beds, alluvial fans and estuaries locally known as wadis that dot all along this long coastline. The mountains to the east and the coastal plains have large number fluvial channels that transport water and sediment during rains either into the lagoons or in the open sea (Morley, 1975). Due to Quaternary sea-level changes diverse types coastal environments have developed that range from open shelf basins to fringing, barrier or patch reefs to various littoral environments like lagoons (Jado and Zölt, 1984). A large number of coral reef islands of various sizes and shapes are present some near shore and others far offshore in the Red Sea, the most notable being the Farasan Islands which are located around forty km away from the coastline. Several lagoons and bays both small and large meander through this long coastline and at times become saline lakes because they get disconnected from the sea due either to local tectonics or Quaternary sea-level changes.

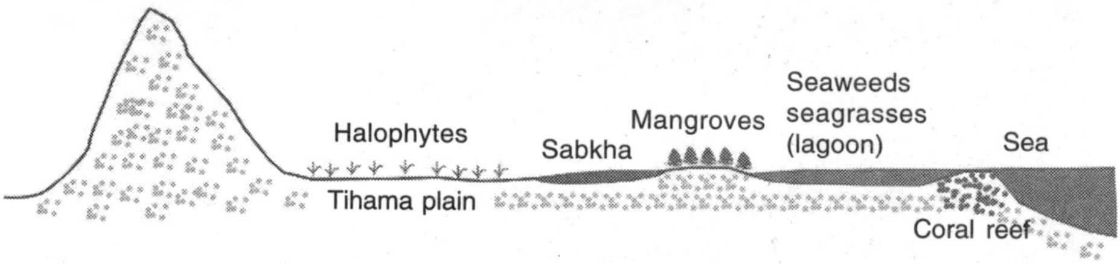

Fig. 3: A generalized geomorphological cross-section of the Red Sea coast of Saudi Arabia showing various environments and plant communities (modified after Saifullah, 1994).

Climatically this long coast line covers both tropical and subtropical zones. Monsoons originating in the Indian Ocean cause rain fall in southwestern Saudi Arabia and the area around Al Lith approximately form the northern limit of influence of monsoon in this region. The yearly average rainfall along the foot hills is 100 mm and approximately 300 mm on mountains 1000 m above sea level (Jado and Zölt (1984). Thus rainfall and groundwater enrichment increases significantly from north to south. This demarcation is very interesting and impacts environmental variables and mangrove growth (Table- 1). Surface temperatures in the Red Sea increases southward in response to latitude but salinity of surface water increases northwards indicating intrusion of low salinity water of Gulf of Aden into the Red Sea (Edwards, 1987). The tidal amplitude along the Red Sea coast is very low, which is ~50 cm in the northern and southern coast that gradually decreases towards the center, close to zero near Jeddah (Edwards, 1987).

Materials and Methods

The study area is located along the southern Red Sea coast of Saudi Arabia from Al-Lith in the north to Saudi Arabia – Yemen border in the south between Latitudes 20° 08' 22" and 16° 41' 6" (Fig. 1 & 2). Historical Landsat Enhanced Thematic Mapper (ETM) data of the years 1999, 2000 and 2001 were used (Table-2). The ETM data has seven spectral bands (Table-3). The spatial resolution of ETM data is 28.5 meters. Research methodology and details about image processing techniques used in this paper is given in Kumar *et al.* (2010).

Table-1 : Environmental differences between mangroves of northern and southern regions of the Red Sea coast of Saudi Arabia (after Saifullah, 1994).

	North	South
1.	Sub-tropical and tropical	Tropical only
2.	Rocky substrate	Muddy substrate
3.	Fewer Wadis	Many Wadis
4.	Less rainfall	More rainfall
5.	Low nutrients	High nutrients
6.	More saline	Less saline
7.	One species of mangroves	Two species of mangroves
8.	Poor growth of mangroves	Dense growth of mangroves
9.	Dwarf forest	Fringe forest
10.	Flowering and fruiting October-April	Flowering and fruiting March-August
11.	Litter fall 2.16 gm m day	3gm m day
12.	*Bostrychia tenella* absent	Present
13.	*Mudskippers absent*	Present

Table-2 : Satellite image data used in the study.

Location	Sensor	Image Acquisition Date
site-21	ETM	03-06-2000
site-22	ETM	03-06-2000
site-23	ETM	28-04-2001
site-24	ETM	28-04-2001
site-25	ETM	28-04-2001
site-26	ETM	28-04-2001
site-27	ETM	28-04-2001
site28	ETM	28-04-2001
site-29	ETM	13-08-1999
site-30	ETM	05-04-2001
site-31	ETM	05-04-2001
site-32	ETM	05-04-2001
site-33	ETM	05-04-2001
site-34	ETM	05-04-2001
site-35	ETM	09-09-2000
site-36	ETM	05-04-2001
site-37	ETM	05-04-2001

Table-3 : Landsat spectral bands.

Landsat 7 (ETM+ sensor)	Wavelength (micrometers)
Band 1	0.45 - 0.515
Band 2	0.525 - 0.605
Band 3	0.63 - 0.69
Band 4	0.75 - 0.90
Band 5	1.55 - 1.75
Band 6	10.40 - 12.5
Band 7	2.09 - 2.35

Results and Discussion

Fig. 4 through 20 show images of geographical distribution as well as areas of mangroves at sites 21 through 37 respectively (Table-4). Like distribution in the northern coast mangroves along the southern coast also occur in small patches and the areas covered by them vary considerably. They do not occur as a continuous forest and mostly their distribution is restricted to the quieter environments, for example, bays, lagoons, narrow channels, and inland face of offshore islands. However, not all mangroves occupy quieter environments they also inhabit higher energy intertidal environments and seaward face of the near shore islands. The environmental and anthropogenic stress is evident here also as was observed in the northern coast.

Saifullah (1994) lists several environmental differences in the mangroves of the northern and the southern Saudi Arabian Red Sea coast (Table-1). We observe denser growth of mangroves along the southern coast in comparison to the northern coast and thus support the observation of Saifullah (1994). This dense growth in south is due mainly to the tropical climate, more wadis, higher rainfall and nutrients and less saline waters along the Red Sea coast. The small size, scattered and sparse distribution of individual trees in the north is due to narrow tidal zone, oligotrophic waters and high salinity of waters along the Red Sea coast (Saifullah, 1994). The mangrove species *Avicennia marina* occurs all along the Saudi Arabian coast and on the islands, but *Rhizophora mucronata* occurs only in the Farasan Islands (Mandura *et al.*, 1987). The third and final part of this study will include a similar study along the Red Sea coast of Yemen.

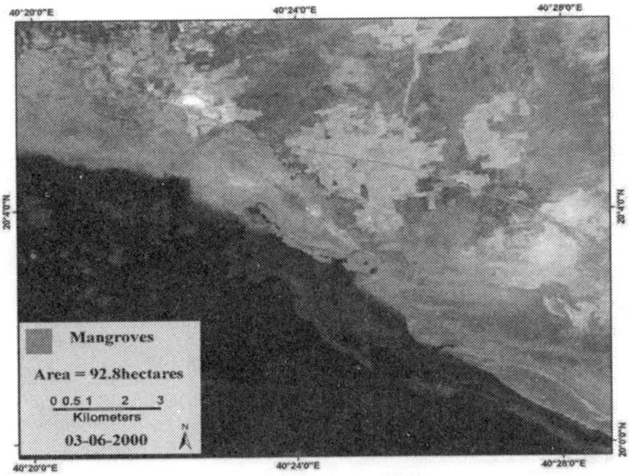

Fig. 4 : Satellite image of mangroves shown at site 21.

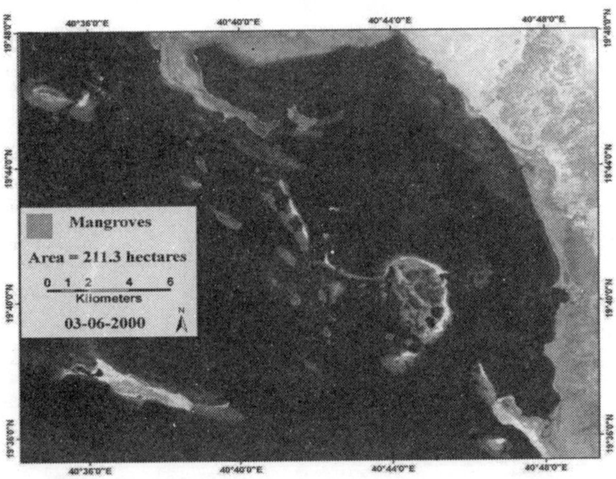

Fig. 5 : Satellite image of mangroves shown at site 22.

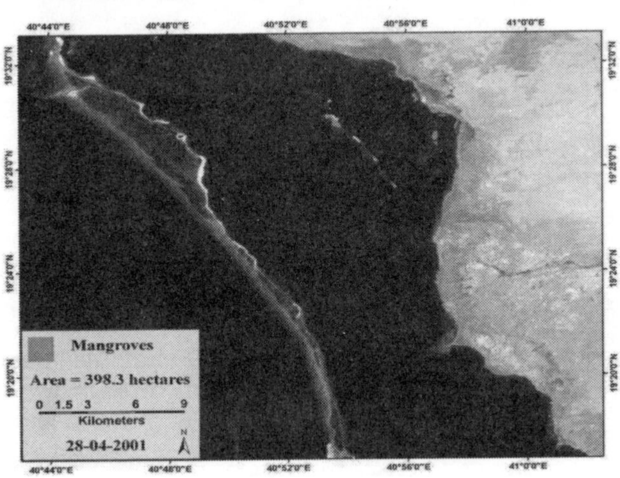

Fig. 6 : Satellite image of mangroves shown at site 23.

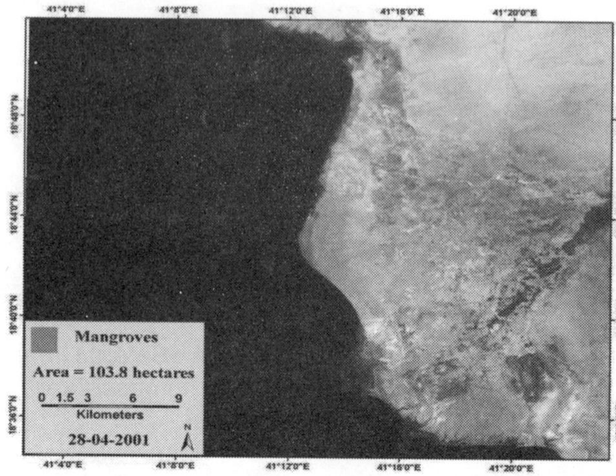

Fig. 7 : Satellite image of mangroves shown at site 24.

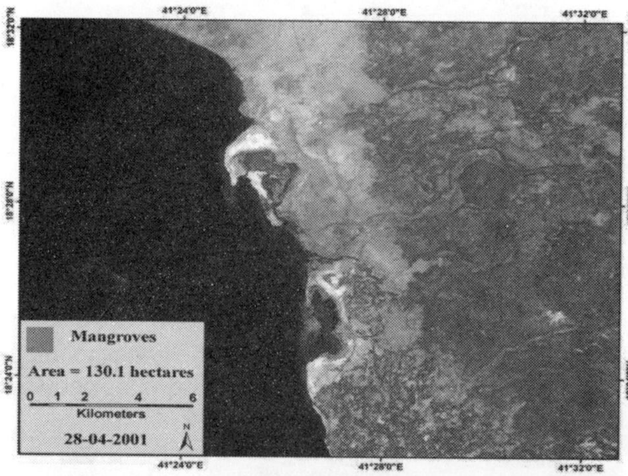

Fig. 8 : Satellite image of mangroves shown at site 25.

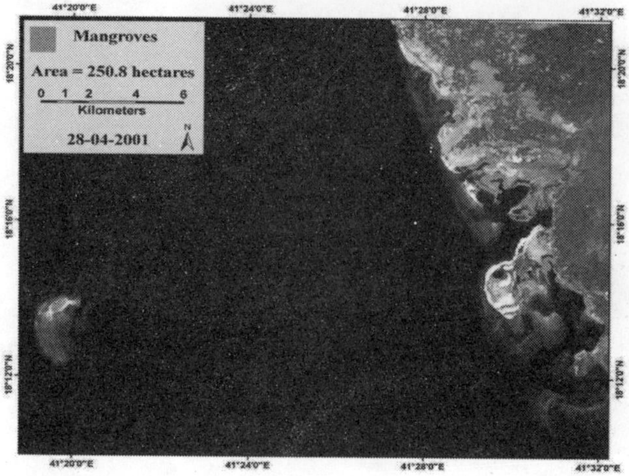

Fig. 9 : Satellite image of mangroves shown at site 26.

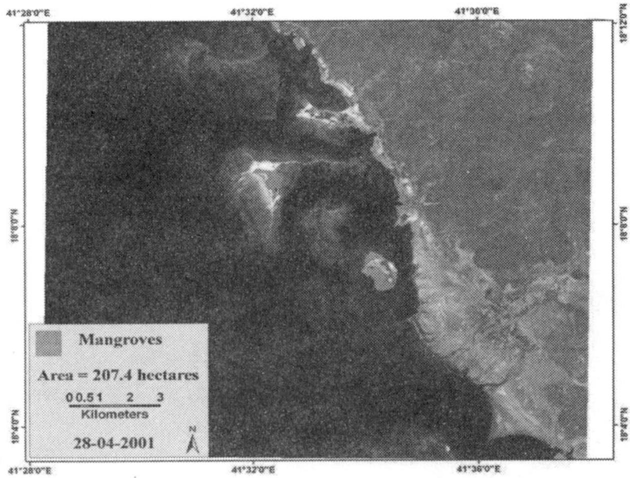

Fig. 10 : Satellite image of mangroves shown at site 27.

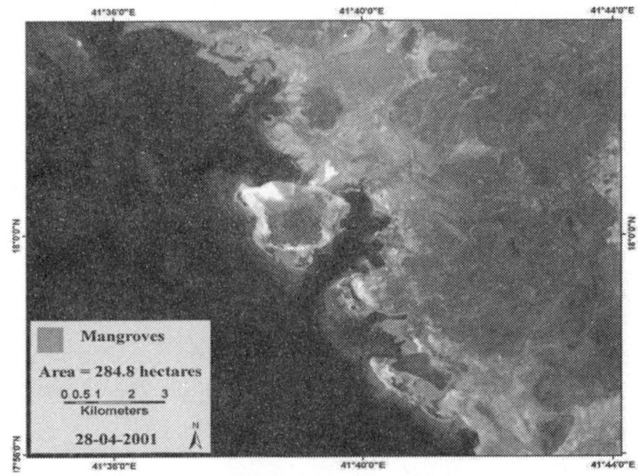

Fig. 11 : Satellite image of mangroves shown at site 28.

Fig. 12 : Satellite image of mangroves shown at site 29.

Fig. 13 : Satellite image of mangroves shown at site 30.

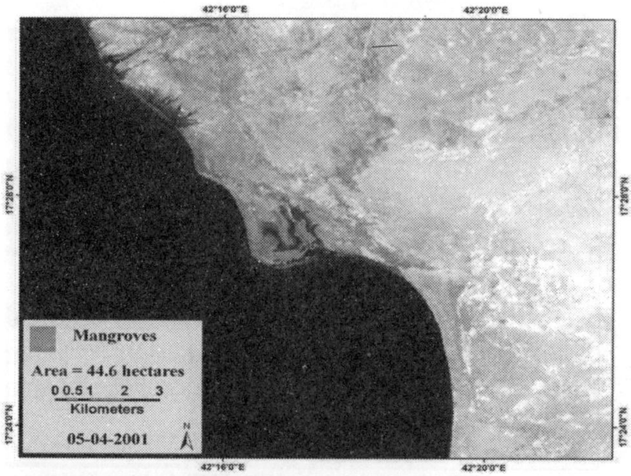

Fig. 14 : Satellite image of mangroves shown at site 31.

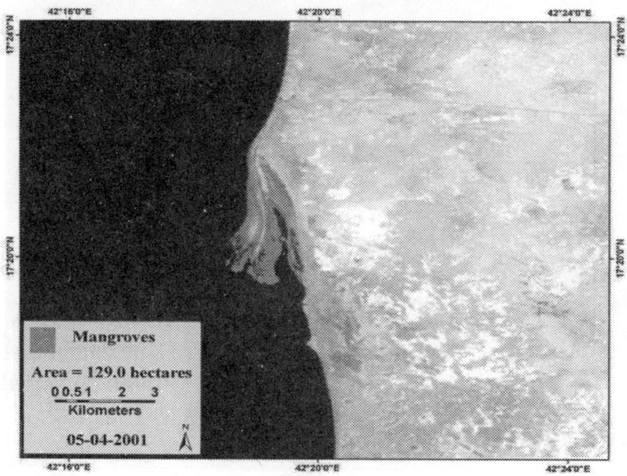

Fig. 15 : Satellite image of mangroves shown at site 32

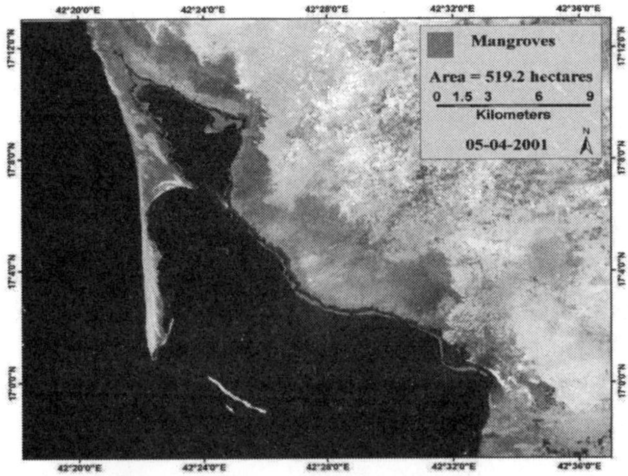

Fig. 16 : Satellite image of mangroves shown at site 33.

Fig. 17 : Satellite image of mangroves shown at site 34.

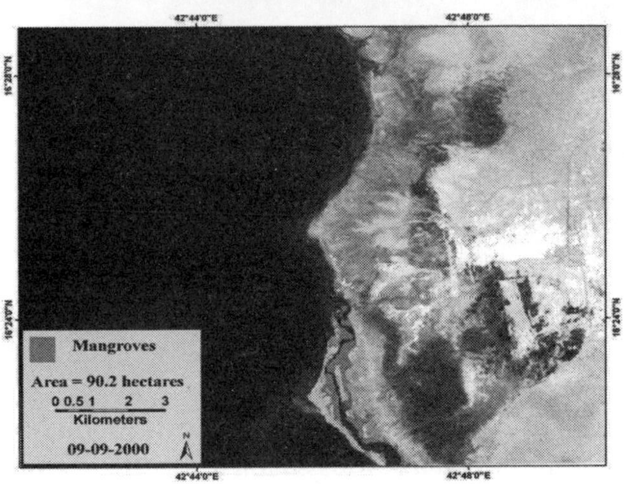

Fig. 18 : Satellite image of mangroves shown at site 35.

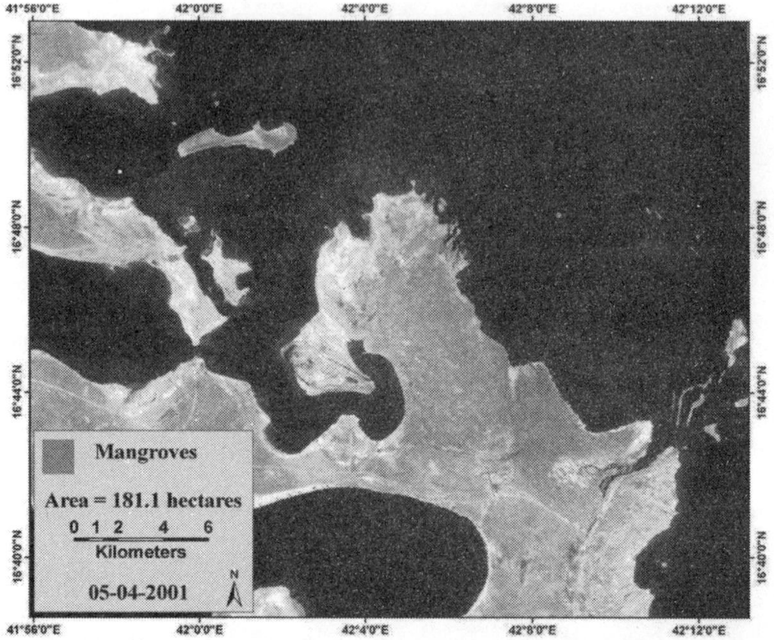

Fig. 19 : Satellite image of mangroves shown at site 36.

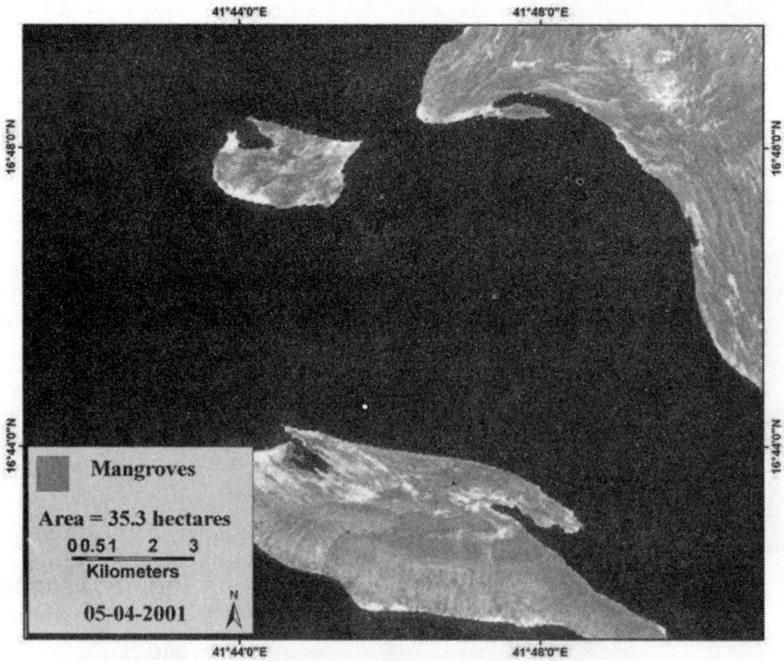

Fig. 20 : Satellite image of mangroves shown at site 37.

Table-4 : Area of mangrove forests (hectares) at the seventeen locations along with their latitude and environmental setting.

Site	Latitude Deg. Min. Sec. (range)	Mangrove area in hectares	Environment
21	20 08 22 to 19 58 40	92.8	Bay, intertidal environment
22	19 49 22 to 19 33 56	211.3	Lagoons, near shore islands
23	19 35 32 to 19 14 25	398.3	Lagoons, near shore islands
24	18 53 43 to 18 32 35	103.8	Bay, intertidal environment
25	18 33 16 to 18 21 08	130.1	Bays and brackish water coastal lakes
26	18 21 19 to 18 09 44	250.8	Lagoons, Bays, near shore islands
27	18 12 30 to 18 03 18	207.4	Lagoons, Islands and intertidal environments
28	18 04 06 to 17 54 55	284.8	Lagoons, Bays, Islands and intertidal environments
29	17 52 47 to 17 43 02	45.1	Bay, intertidal environment
30	17 37 49 to 17 29 43	46.0	Lagoon, Bay, intertidal environment
31	17 31 34 to 17 23 28	44.6	Lagoons and Bays
32	17 24 32 to 17 16 26	129.0	Bays
33	17 12 57 to 16 56 59	519.2	Lagoon, Bay, intertidal environment
34	16 53 08 to 16 37 08	186.1	Lagoon, Bay, intertidal environment
35	16 30 47 to 16 21 37	90.2	Bay, intertidal environment
36	16 52 56 to 16 37 51	181.1	Lagoons, Nearshore islands
37	16 49 43 to 16 41 16	35.3	Nearshore islands

Acknowledgements

We thank King Fahd University of Petroleum and Minerals (KFUPM), Dhahran, Saudi Arabia for permission to publish this paper.

References

Frey, W., Kurschener, H., Sheikh, A. M. E. and Migahid, A. M. (1984) Zonation and photosynthetic pathways of halophytes on the Red Sea coast near Tawwal, Saudi Arabia. Notes from the Royal Botanic Garden, Edinburgh, v. 42(1), pp. 45-56.

Jado, A. R. and Zölt, J. G. (1984) Quaternary Paeriod in Saudi Arabia. Vol. 2. Springer-Verlag, 360p.

Khafaji, A. K., Manfura, A. S. and Saifullah, S. M. (1988) Temporal and spatial variations in the carbohydrate, lipid and protein contents of *Avicennia marina* from the Saudi Arabian Red Sea coast. Proceedings Saudi Biological Society, v. 11, pp. 149-161.

Khafaji, A. K., Manfura, A. S., Saifullah, S. M. and Sambas, A. Z. (1991) Litter production in two mangrove stands in southern Red Sea coast of Saudi Arabia (Jizan). Journal of King Abdulaziz University, Marine Science, v. 2, pp. 93-100.

Kumar, A., Khan, M. A. and Muqtadir, A. (2010) Distribution of mangroves along the Red Sea coast of the Arabian Peninsula: Part-l: the northern coast of western Saudi Arabia. e-Journal Earth Science India, v. 3 (1), pp. 28-42.

Mandura, A. S., Khafaji, A. K. and Saifullah, S. M. (1987) Mangrove ecosystem of southern Red Sea coast of Saudi Arabia. Proceedings Saudi Biological Society, v. 10, pp 165-193.

Mandura, A. S., Saifullah, S. M. and Khafaji, A. K. (1988) Ecology of a mangrove stand of central Red Sea coast area: Ras Hatiba (Saudi Arabia). Proceedings Saudi Biological Society, v. 11, pp.85-112.

Migahid, A. M. (1978) Flora of Saudi Arabia, Two Volumes, 2nd Edition. Riyadh University Publications, Riyadh, 650 p.

Morley, N. F. J. (1975) The coastal waters of the Red Sea. Bulletin Marine Research Centre, Saudi Arabia, v. 5, pp. 8-19.

Saifullah, S. M., Khafaji, A. K. and Mandura, A. S. (1989) Litter production in mangrove stand of the Saudi Arabian Red Sea coast. Aquatic Botany, v. 36, pp. 79-86.

Saifullah, S. M. (1999) Mangrove ecosystem of Saudi Arabian Red Sea coast – an overview. Journal of King Abdulaziz University, Marine Science, v. 7, Special Issue on Red Sea Marine Environment, Jeddah, 1994. pp. 263-270.

Vesey-Fitzgerald, D. F. (1955) Vegetation of the Red Sea coast south of Jedda. Saudi Arabia. Journal of Ecology, v. 43, pp. 477-498.

Vesey-Fitzgerald, D. F. (1957) Vegetation of the Red Sea coast north of Jedda. Saudi Arabia. Journal of Ecology, v. 45, pp. 547-562.

Zahran, M. A., Younes, H. A. and Hajrah, H. H. (1983) On the ecology of mangal vegetation of the Saudi Arabian Red Sea coast. Journal of University of Kuwait, Science, v. 10 (1), pp. 87-99.

□□□

13

Scaling Spectral Analysis : A New Tool for Interpretation of Gravity and Magnetic Data

A.R. BANSAL AND V.P. DIMRI
National Geophysical Research Institute, (CSIR), Hyderabad- 500606

Abstract

The scaling spectral method for gravity and magnetic data is found useful for finding the depth values and statistical properties of the source distribution. The depth values calculated by scaling spectral method are close to the realistic values whereas white noise assumption of sources results in overestimations. The scaling spectral method has been applied to many parts of the world. The power spectrum corresponding to low wave number may be dominated by scaling properties alone rather than the depth values. The scaling distribution is useful for gridding the datasets, delineating the lithological units and enhancing the information of sources. The long non-stationary profiles and larger areas may be divided into piecewise stationary profiles before applying the scaling spectral method. The values of scaling exponents from the field as well source indicate 3- D distribution of susceptibility and density.

Keywords : Scaling spectral method, gridding the datasets, delineating the lithological units, susceptibility, density

Introduction

The interpretation of gravity and magnetic data generally carried out in the space and frequency domains. The interpretation of the gravity and magnetic data is preferred in frequency domain because of simple relation between various source models and field (Naidu and Mathew 1998). The estimation of the depth of anomalous sources is usually carried out by Spector and Grant (Spector and Grant, 1970) method and its variants in frequency domain (Bhattacharyya, 1966; Naidu, 1968; Spector and Grant 1970; Treitel *et al.*, 1971; Hahn *et al.*, 1976; Connard *et al.*, 1983). These methods assume different assemblage of sources like statistical ensemble of prisms (Spector and Grant, 1970), white noise of vertical needles with constant magnetization (Naidu, 1972: Hahn *et al.*, 1976), a sandwich model of uniaxially magnetic sheets (Pedersen, 1991), equivalent density layer (Pawlowski, 1994) etc. These methods are in continuous use since their development because of simplicity (Ofoegbu and Hein, 1991: Cowan and Cowan, 1993: Hildenbrand *et al.*, 1993: Gracia- Abdeslem and Ness, 1994). In case of 2-D data radially averaged power spectrum is proportional to exp (-2dk), where k is wave number, d is depth to anomalous sources.

The assumptions of sources in these methods lead to white noise distribution. The white noise source distribution is assumed because of mathematical simplicity and limited information about the subsurface source distribution. Recent borehole studies from the German Continental Drilling Program (KTB), the Canadian Shield, etc. have shown scaling noise distribution of physical properties rather than white noise distribution (Pilkington and Todoeschuck 1990, 1993, 1995; Maus and Dimri 1995, 1996). The power spectrum of scaling noise can be represented as: $P(k) \sim k^{\hat{a}}$, where $P(k)$, k and â are the power spectrum, wave number and scaling exponent of the source distribution, respectively. The values of scaling exponents indicate the relation between the successive values, a positive value indicate anti-correlated, negative values indicates correlated and zero value corresponds to white noise.

Pilkington and Todoeschuck (1990) have shown that scaling noise models are more suitable for modeling the spatial variation of geophysical parameters. They analyzed six boreholes, three each of 300m depth and 150m depth from Bells Corners, Ottawa, Canada. The data divided into sedimentary and igneous sections for each borehole. The power spectrum of acoustic impedance, resistivity log, density, gamma-ray log, neutron log is calculated using periodogram technique. The values of scaling exponents of density varies from -0.87 to -1.47 for sediments and -0.71 to -1.71 for igneous rocks. In another study of Pilkington and Todoeschuck (1993) the values of scaling exponents are found varying from -1.32 to -1.96 for sedimentary rocks and from -2.08 to -2.72 for igneous rocks. Leonardi and Kumpel (1998) studied the physical properties of the rocks from the KTB borewell data and found the values of scaling exponents between -1.0 to -1.4 for susceptibility and -0.6 to -1.2 for bulk density. Zhou and Thybo (1998) were in the opinion of variation of scaling exponents with depth for the field as well as source distribution. They analyzed the susceptibility data of the KTB super deep main hole and aeromagnetic data over the Southeastern North Sea. The scaling exponent for KTB susceptibility data varies from -0.6 to -2.0 with a decreasing trend with depth. The values of scaling exponents from aeromagnetic data is found -2.3, -1.1 and -0.96 for three different sedimentary basins of different basement depths in the North Sea area (Zhou and Thybo, 1998). The values of scaling exponents may be anisotropic; different in vertical and horizontal directions (Pilkington and Todoeschuck, 1993). Shiomi *et al.* (1997) analyzed well-log data from five deep wells in different tectonic regions in Japan to study the statistical characteristics of P-wave velocity, S-wave velocity and density in the uppermost part of the crust. The three wells are located in sedimentary rocks in the Kanto Plain and two wells are in the Kuju Volcano Group in Kyushu Island. In the Kanto Plain the scaling exponent ranges between -1.1 to -1.3 from the analysis of P-wave velocity and density, whereas the scaling exponent is -1.3 to -1.6 in Kuju Volcano Group for P and S wave velocities and density. This is an indication of less values of scaling exponent in a tectonically stable area as compared to the tectonically active area. Maus and Dimri (1995) found the values of scaling exponent -2.07 to -2.16 over the KTB main hole and -1.8 over the KTB pilot hole from aeromagntic data. Analysis of marine gravity data of Norwegian Coast has shown that the scaling exponent is -4.5 (Maus *et al.*, 1998). An example of susceptibility log measured in the KTB pilot hole (Fig. 1a) and its power spectrum (Fig. 1b) with the values of scaling exponent is shown in Fig. 1 (Maus and Dimri, 1995). The scaling noise has no characteristic correlation length (Feder, 1988).

Pilkington and Todoeschuck (1993) studied the relation between the scaling exponent due to the 3-D susceptibility distribution and 2-D magnetic field. They found that susceptibility distribution with scaling exponent (a) will produce a magnetic field with scaling exponent (a-1). Maus and Dimri (1994) have studied in detail various relationships between the scaling exponents of density and susceptibility distributions and their respective gravity and magnetic fields for different dimensions (Table-1).

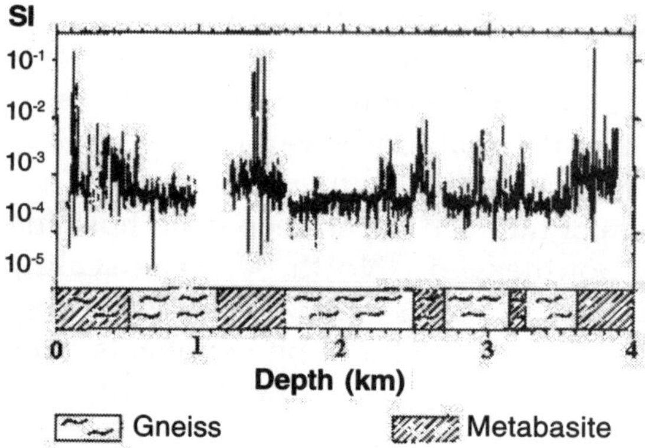

Fig. 1a: Susceptibity log measured in KTB pilot hole with simplified geology (Maus and Dimri, 1995)

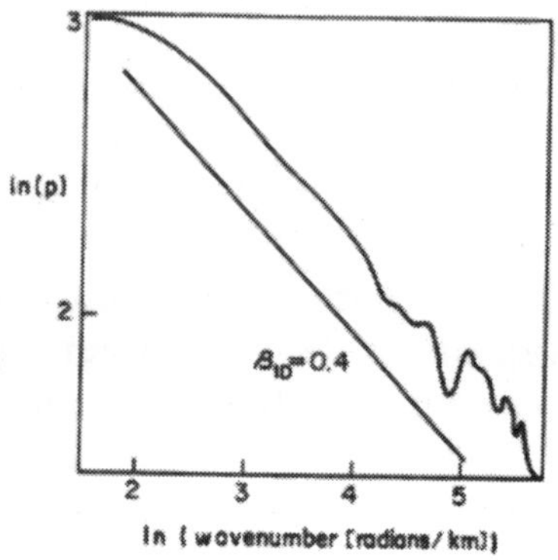

Fig. 1b: An example of calculation of scaling exponent of the susceptibility log from KTB pilot hole.

Table-1 : The relation between the scaling exponent of the density/ susceptibility and gravity/magnetic field in different dimensions (Maus and Dimri, 1994).

	3-D (source)	2-D (source)	1-D (source)	2-D (field)	1-D (field, x, y, z)
Gravity	α	$\alpha-1$	$\alpha-2$	$\alpha+1$	α
Magnetic	α	$\alpha-1$	$\alpha-2$	$\alpha-1$	$\alpha-2$

Calculation of depth values by scaling Spectral Method

The scaling exponent of source distributions can be used as a prior information in term of parameter covariance matrices in linear inversion (Pilkington and Todoeschuck, 1990). Maus and Dimri (1995) suggested the computation of scaling exponents of the field and depth simultaneously from the inversion of power spectrum. In their method power spectrum can be written as:

$$P(k) = Ck^{-a}e^{-2dk} \quad \text{... (1)}$$

where P(k) is the power spectrum, a is the scaling exponent of the field, *k* denotes wavenumber, C is a constant depending on the properties of source and *d* is the depth to the anomalous sources. The parameter a and d are interrelated where lower values of the scaling exponent leads to higher depth and vise versa. The three parameters (C, a, d) can be calculated by the inversion method. The equation (1) may be written in logarithm as:

$$\ln P(k) = C - 2dk - \alpha \ln k \quad \ldots (2)$$

Although Maus and Dimri (1995, 1996) developed the methodology for half space model, the equation (2) can be applied to the multilayer cases with constant values of scaling constant (Fedi *et al.*, 1997).

Applications

Pilkington and Todoeschuck (1990) used the scaling covariance model in the inversion of DC resistivity sounding data. The use of scaling noise covariance model resulted in reduction of the posteriori parameter variances and smoothness of the solution. The value of scaling exponent from aeromagnetic data over the Canadian Shield is found as -3 (Gregotski *et al.*, 1991). The value of scaling exponent -3 is also found from 2-D susceptibility data for the Sierra Nevada Batholith, California and Flin Flon- Snow Lake Belt, Saskatchewan Island Arch. Canada (Pilkington and Todoeschuck, 1995). This value is a representation of 3-D crustal susceptibility with scaling exponent of -4 (Pilkington and Todoeschuck, 1995). The use of scaling exponent values in deconvolution of aeromagnetic data of northeastern Ontario, Canada result in enhancing the magnetic field anomalies and delineating the litological units (Gregotski *et al.*, 1991). Pilkington *et al.*(1994) studied aeromagnetic data from Wollaston Lake and Athabasca Sedimentary Basin in the Canadian Shield. In the Wallston Lake area sources are exposed and depth to magnetic sources is corresponds to the flight altitude of 300m whereas in Athabasca sedimentary basin depth to magnetic basements is established as 1400 m from the seismic refraction surveys and individual magnetic anomalies modeling. The assumption of white noise source distribution result in depth estimation of 1100 m and 2400 m for the Wollastone Lake and Athabasca sedimentary Basin, respectively whereas depth values found from scaling spectral method (scaling exponent -3) are in agreement with seismic refraction studies. On the other hand Maus and Dimri (1996) calculated the scaling exponent -2.2 for the first area (Wollaston Lake) constraining the depth equivalent to flight altitude height and used the values of scaling exponent to calculate the depth values for the second study area (Athabasca Sedimentary Basin) for half space model and found depth values close to the seismic reflection studies (Fig. 2) whereas depth estimation for white noise distribution were overestimated. The fits for different depth values is shown in Fig. 2 for curve (a) for the constant and scaling exponent found from

Wollaston Lake area (curve b). A depth value of 1.7 km is more suitable (Maus and Dimri, 1996). The gridding of magnetic data by using scaling information results in smoothing of the data close to the realistic data as compared to uncorrelated distribution (Pilkington *et al.*, 1994). Maus and Dimri (1996) applied the scaling spectral method to the gravity survey of the Paradox Basin Region in south - eastern Utah. The area was earlier study by the Pawlowski and obtationed depth values of anomalous sources as 5 and 54 km using Spector and Grant Method whereas application of scaling spectrum method indicate dominance of scaling properties of source distribution (Fig. 3). The high values of scaling exponent indicate long range dependency of susceptibility distribution (Maus and Dimri, 1996). Maus and Dimri (1995) applied the method to the aeromagnetic data of the KTB region and found depth value of 61m close to the average flight altitude of 61.7 m. The different values of scaling exponents for different geological units were found from the aeromagnetic data over the KTB (Fig. 4) (Maus and Dimri, 1995).

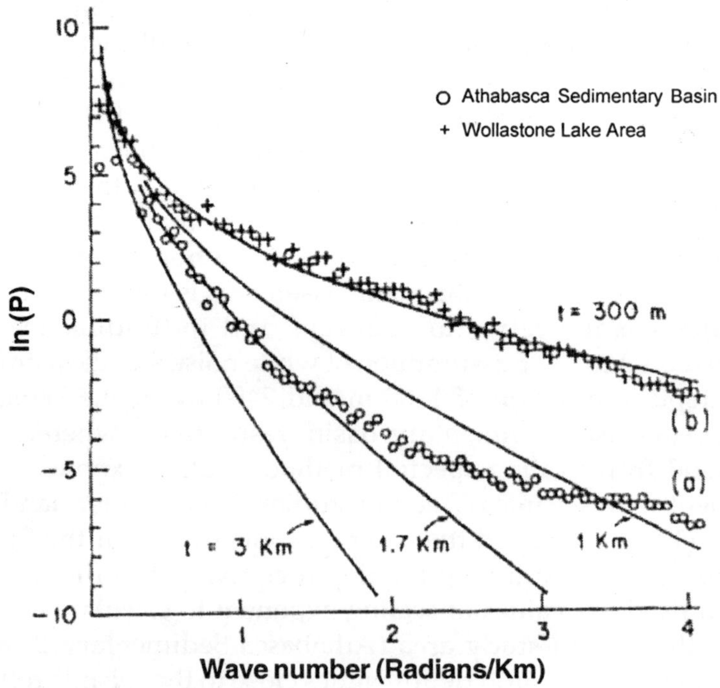

Fig. 2: Radially averaged power spectrum versus wavenumber of aeromagnetic data over the Athabasca sedimentary Basin (a) and Wollastone Lake area (after Maus and Dimri, 2006).

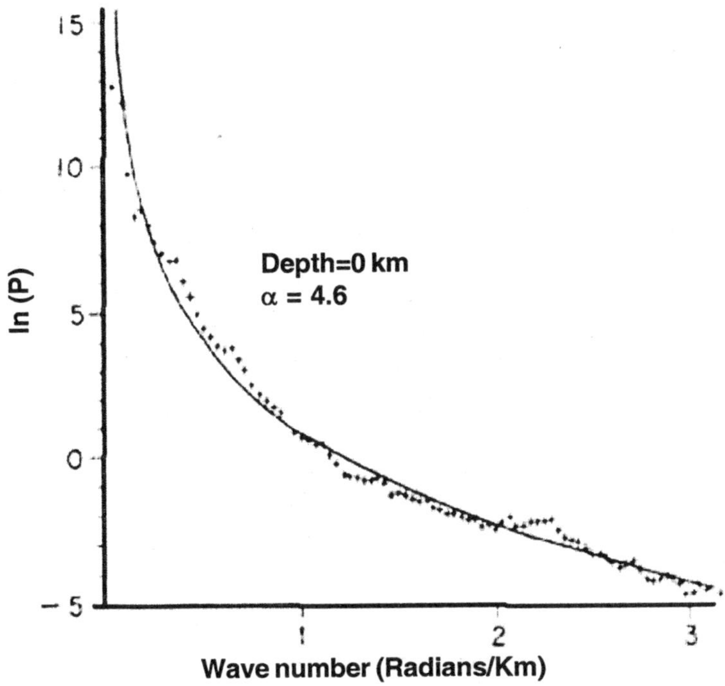

Fig. 3: Depth estimation by scaling spectral method from gravity data of the Paradox Basin(after Maus and Dimri, 2005).

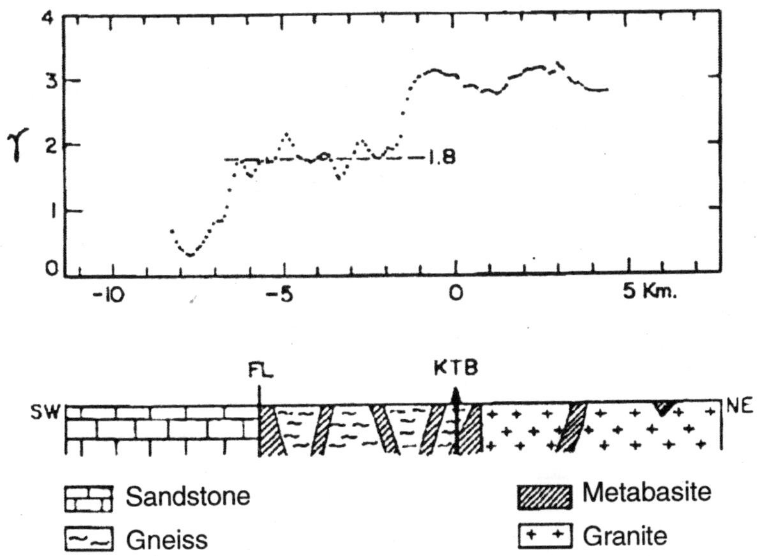

Fig. 4: Variation of scaling exponents with lithology found from aeromagnetic data over the KTB (Maus and Dimri, 1995).

Fedi *et al.* (1997) have shown the inherent power law relation of power spectra of magnetic field. They found the unique scaling exponent -2.9 from aeromagnetic spectra close to -3 (Gregotski *et al.*,1991; Pilkington and Todoeschuck, 1993; Pilkington *et al.*, 1994). In case of synthetic example of two layer assemblage of sources at shallower and deeper depths, shallower depth values found by scaling method is close to the true assumed depths whereas deeper depth values are under estimated (Fedi *et al.*, 1997). The depth values calculated from the non corrected power are over estimated in both the cases. The application of scaling spectral method to the magnetic data of Matonipi Lake, Quebec and Tertiary volcanic lava in Central America have shown two layer and one layer assemblage of sources, respectively. This two layer and one layer case is in confirmation of large and short anomalies in Matinipi Lake area and short wavelength anomalies in Central America map.

Maus and Dimi (1996) applied the half space model with scaling properties to the aeromagnetic data of the Island of Hawaii and found depth values approximately equal to the depth of survey terrain clearance (Fig. 5). The found values of the scaling exponents is high compare with average values of -3 derived by Gregotski *et al.* (1991) from aeromagnetic survey of the North American Continent.

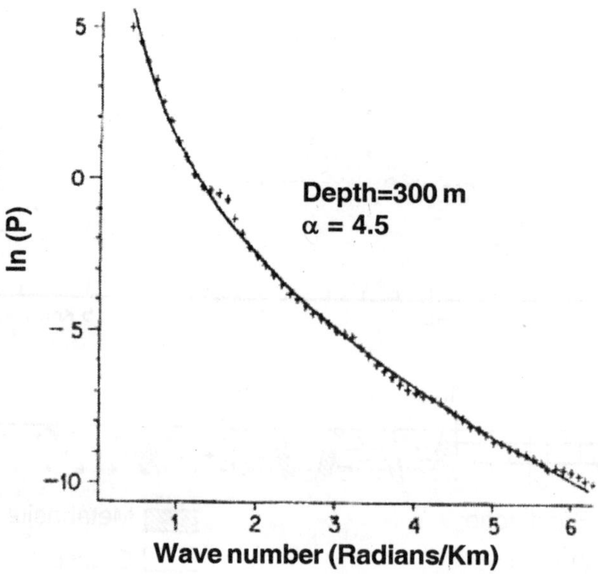

Fig. 5: Calculation of scaling exponent and depth values from aeromagnetic data over the Island of Hawaii (after Maus and Dimri, 1996).

The spectral and scaling spectral methods are applicable to stationary gravity/magnetic data whereas in real field situation gravity and magnetic data is non stationary in nature. Therefore Bansal and Dimri (1999, 2001, 2005) suggested the division of long nonstationary profiles into piecewise stationary profiles by using the Wiener filter theory and technique is called optimum gate length. The technique was earlier developed for seismic traces by Wang (1969). The optimum gate length is selected in such a way that the error between the estimated and true time- varying autocorrelation function is minimum and thus stationarity is maintained in the computed length. The technique of dividing non-stationary gravity and magnetic profiles is consists of following steps:

- First, assume a small reasonable length of the profile.
- Divide the assumed length into three parts and mark the middle of each part.
- Calculate the time-invariant autocorrelation function for each part.
- Write these autocorrelation functions in cosine transform.
- Calculate the coefficients of cosine transform.
- Expand these coefficients in Taylor-series upto second term and calculate the Taylor coefficients.
- Calculate the time- varying auto correlation function, $f_{xx}(t_2,t)$ from following equation :

$$\varphi_{xx}(t_2,\tau)=\frac{1}{L}\sum_{-L/2}^{L/2}x(t+t_2+\frac{\tau}{2})x(t+t_2-\frac{\tau}{2}) \tag{3}$$

where $t_2 = L/2$

- Compute the gate length as:

$$L=144\left[\sum_{\tau=0}^{L-1}\varphi_{xx}(t_2,\tau)^2\sum_{i=1}^{n}\sum_{j=1}^{n}w_iw_j\right]^{1/5} \tag{4}$$

where

- $\phi_{xx}(t_2,t)$ is time-varying autocorrelation function,
- w_i, w_j are Taylor coefficients.
- Calculate the error square between the assumed and calculated length.

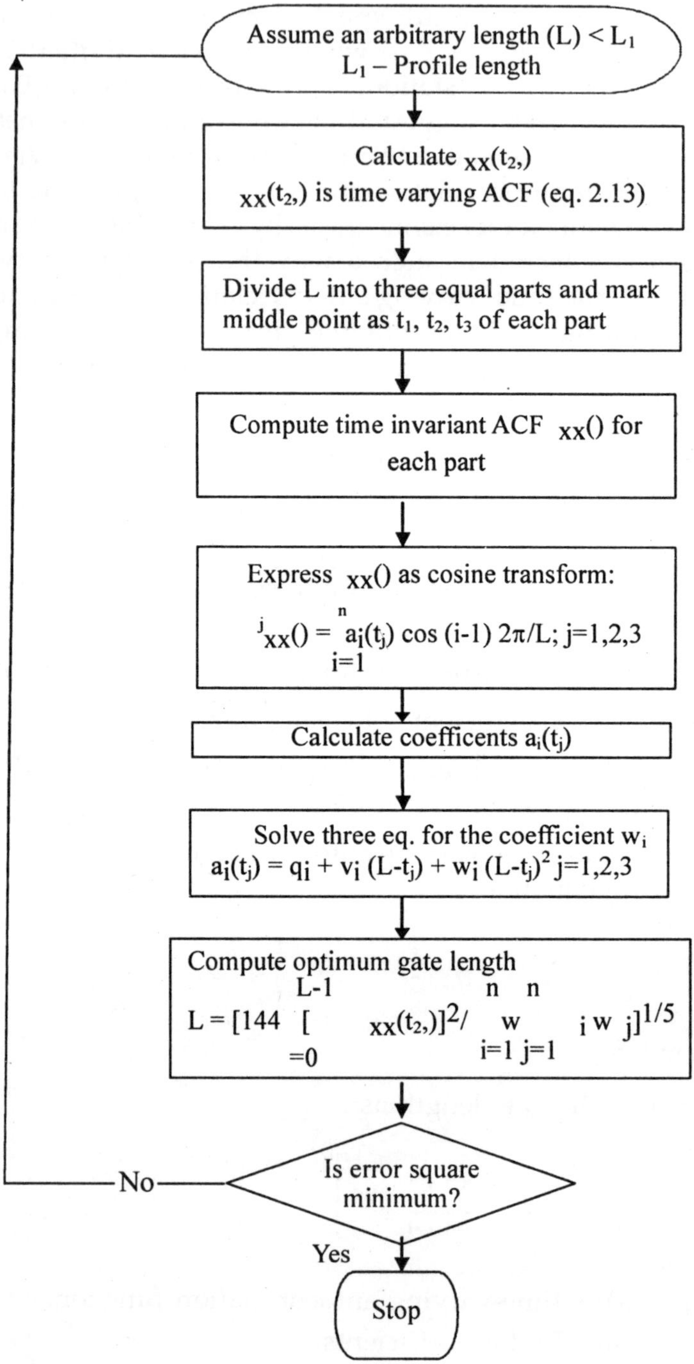

Fig. 6: A flow chart showing the calculation of optimum gate length (after Bansal and Dimri, 2005).

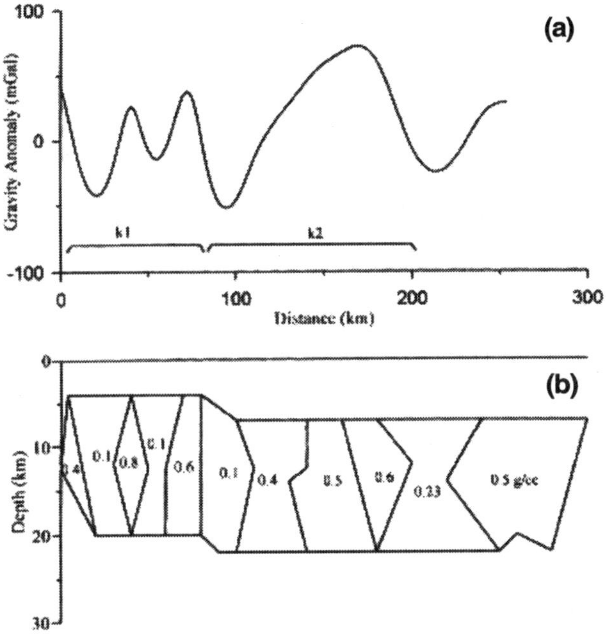

Fig. 7 : (a) Synthetic gravity profile and (b) the assumed subsurface model. K1 and k2 indicate optimum gate length (Bansal and Dimri, 2001).

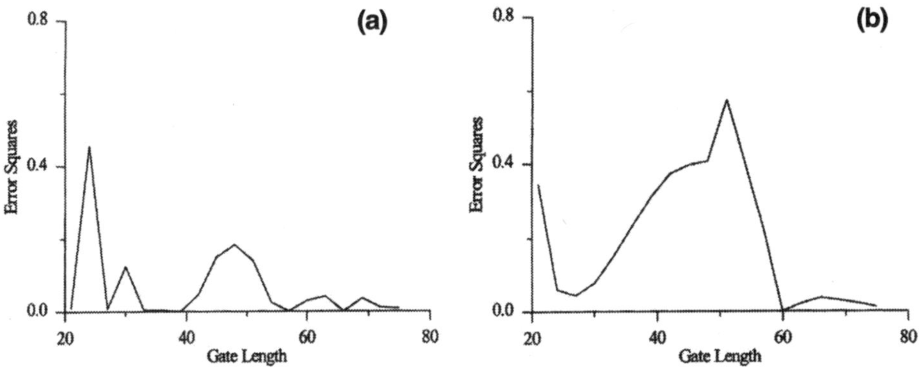

Fig. 8 : Estimation of optimum gate length for (a) k1 and (b) k2 piecewise stationary sub-profiles. The minimum error square for k1 and k2 are found at 39 and 60 points corresponding to the length of 78 and 120 km since sampling interval is 2 km (Bansal and Dimri, 2001).

If the error square between the assumed and computed length is more, then assume a new value of length of the profile and again compute the gate length by above mentioned procedure. Repeat the whole process until a minimum error square is reached.

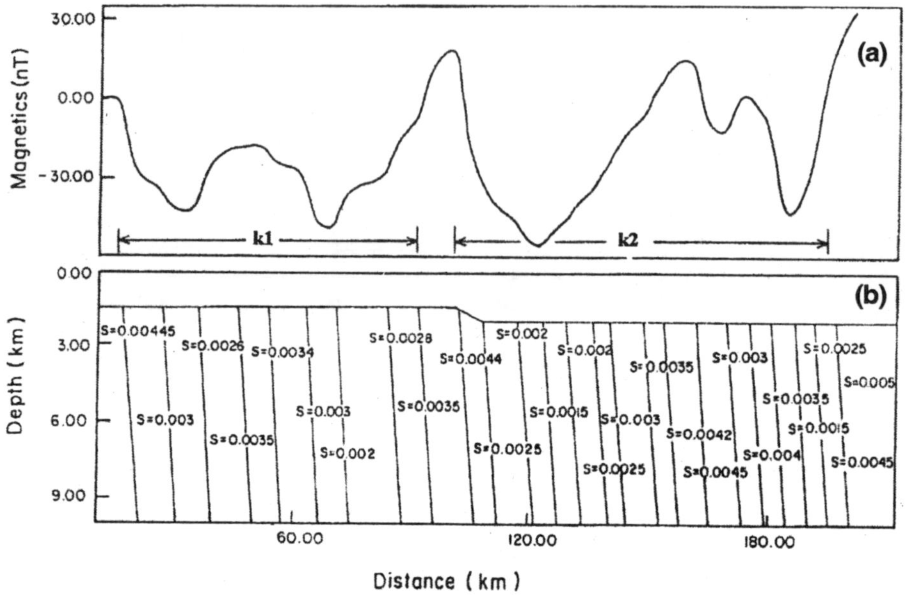

Fig. 9 : (a) Synthetic magnetic profile generated for an assemblage of bodies with varying susceptibility(s) in cgs units (b). The found two optimum gate lengths were shown as k1 and k2 (after Bansal and Dimri, 2005).

After selecting the optimum gate length for the first subsection, repeat the procedure for the remaining profile until whole of the profile is covered. While choosing the optimum gate length one should preserve the wavelength of anomaly. The anomaly of a sub-profile should not be cut in between i.e. the length of section should not be less than the wavelength of anomaly. The procedure of calculating the optimum gate is also presented in Fig. 6. The optimum gate length were further used to calculate depth values from scaling spectral method (Bansal and Dimri, 1999, 2001, 2005). This scheme of two steps is tested on the synthetic gravity (Bansal and Dimri, 2001) and magnetic profiles (Bansal and Dimri, 2005). The synthetic gravity profile was generated for two sets of assemblages of bodies: one shallower with high density contrast and the other deeper with less density contrasts (Fig. 7) (Bansal and Dimri, 2001). The application of optimum gatelength technique results in division of two piece-wise stationary profiles marked as k1 and k2 of length 78 km and 120 km in Fig. 7. The optimum gate length is corresponds to the minimum error square between the assumed and achieved gate length (Fig. 8). The depth values were calculated from these piecewise stationary profiles by applying the scaling spectral approach. The comparison of computed and assumed depth values were presented in Table 2. From the table 2 it can be seen that computed depth values are close to the assumed depth values. A synthetic magnetic profile were generated for an assemblage of sources with top at a depth of 1.5 and 2.0 km extended down up to 30 km for susceptibility distribution shown in Fig. 9 (Bansal and Dimri, 2005). The optimum gate length technique divided it into two stationary sub-profiles of lengths 75 and 93 marked as k1 and k2 in Fig. 9. The depth values calculated from scaling spectral method using maximum entropy method were found as 1.5 km and 1.8 km corresponding to k1 and k2, respectively are close to the assumed depth value of 1.5 and 2.0 km (Bansal and Dimri, 2005). Bansal and Dimri (1999, 2001, 2005) further applied this scheme to real gravity and magnetic data along Nagaur – Jhalawar and Jaipur – Raipur transect. They found calculated depth values are close to the depth values found from the seismic methods.

Table-2 : Comparison of depth values calculated from piece stationary sub profiles (k1 and k2) using scaling spectral method and true assumed depths (Bansal and Dimri, 2001).

k1		k2	
Assumed Depth (km)	Calculated Depth (km)	Assumed Depth (km)	Calculated Depth (km)
4	3.8	7	6.6
20	--	22	23.9

Conclusion

Following conclusions can be made about the use of scaling spectral method for calculating depth values from scaling spectral method:

1. The scaling distribution of the sources is close to the realistic distribution of the sources.
2. The depth estimations of anomalous sources from the scaling spectrum method are close to the real values whereas uncorrected power spectrum results in overestimation of depth values.
3. The power spectrum corresponding to lower wavenumbers may be dominated by the scaling properties rather than depth information.
4. The calculation of scaling exponents and depth from the field data are inter-related. A higher value of scaling exponent results

in lower values of depth and vice versa. Information from the boreholes can be used as preliminary information.

5. The values of scaling exponents may be anisotropic different for horizontal and vertical direction.
6. The values of scaling exponents may vary with depth.
7. The information about the scaling parameter is useful for gridding the data, deriving the lithology and source properties.
8. The longer profiles and larger areas may contain non-stationary gravity and magnetic data. Therefore before applying the scaling spectral method data should be divided into piece-wise stationary data sets.

Acknowledgements

We are thankful to management of NGRI for granting permission for publishing this work and Mr. K. K. Babu for useful discussion.

References

Bansal, A. R. and Dimri, V. P. (1999) Gravity evidence for mid crustal domal structure below Delhi fold belt and Bhilwara super group of Western India. Geophys. Res. Letts., v. 26(18),pp. 2793-2795.

Bansal, A. R. and Dimri, V. P. (2001) Depth estimation from scaling power spectral density of nonstationary gravity profile, Pure appl. Geophys., v. 158, pp. 799-812.

Bansal, A. R. and Dimri, V. P. (2005) Depth determination from non-stationary magnetic profile for scaling geology. Geophys. Prosp., v. 53, pp. 399-410.

Bhattacharyya, B. K. (1966) Continuous spectrum of the total magnetic Field anomaly due to the Rectangular Prismatic Body. Geophysics, v. 31, pp. 97-121.

Connard, G. Couch, R. and Gemperle, M. (1983) Analysis of aromagnetic measurements from the Cascade Range in Central Oregon. Geophysics, v. 48, pp. 376-390.

Cowan, D. R. and Cowan, S. (1993) Seperation filtering applied to aeromagnetic data. Exploration Geophysics, v. 24, pp. 429-436.

Feder, J. (1988) Fractals. Plenum Press, New York. 283p.

Fedi M., Quarta T. and Santis A.D. (1997) Inherent power-law behavior of magnetic field power spectra from a Spector and Grant ensemble. Geophysics, v. 62, pp. 1143–1150.

Garcia-Abdeslem, J. and Ness, G.E. (1994) Inversion of the power spectrum from magnetic anomalies. Geophysics, v. 59(3), pp. 391-401.

Gregotski, M. E., Jensen, O. G. and J. Arkani- Hamed (1991) Fractal stochastic modeling of aeromagnetic data. Geophysics, v. 56, pp. 1706-1715.

Hahn A., Kind E.G. and Mishra D.C. (1976) Depth estimation of magnetic sources by means of Fourier amplitude spectra. Geophysical Prospecting, v. 24, pp. 287–308.

Hildenbrand T. G., J. G. Rosenbaum and J. P. Kauahikaua (1993) Aeromagnetic study of the Island of Hawaii. J. Geophys. Res., v. 98, pp. 4099-4119.

Leonardi, S. and Kümpel, H. J. (1998) Variabiliy of geophysical log data and the signature of crustal heterogeities at KTB. Geophys. J. Int., v. 135, pp. 964-974.

Maus S. and Dimri V.P. (1995) Potential field power spectrum inversion for scaling geology. J. Geophysical Research, v. 100, pp. 12605– 12616.

Maus S. and DimriV.P. (1996) Depth estimation from the scaling power spectrum of potential field? Geophysical J. International, v. 124, pp. 113–120.

Maus, S. and Dimri, V.P. (1994) Scaling properties of potential fields due to scaling sources. Geophys. Res. Lett., v. 21(10), pp. 891-894.

Maus, S., Green, C.M. and Fairhead, J.D. (1998) Improved ocean geoid resolution from retracked ERS-1 satellite altimeter waveform. Geophys. J. Int., v. 134, pp. 243-253.

Naidu, P. S. (1972) Maximum Likelihood (ML) estimation of depth from the spectrum of aeromagnetic field. Pure Appl. Geophys., v. 95, pp. 141-149.

Naidu, P. S. (1968) Spectrum of the Potential Field due to Randomly Distributed Sources. Geophysics, v. 33, pp. 337-345.

Naidu, P.S. and Mathew, M.P. (1998) Analysis of geophysical potential fields, A digital signal processing approach. Elsevier Science publishers, Amseterdam, 298p.

Ofoegbu, C. O. and Hein, K. (1991) Analysis of magnetic data over part of the Yonger Granite Province of Nigeria. Pure Appl. Geoph., v. 136, pp. 173-189.

Pawlowski, R. S. (1994) Green's equivalent- layer concept in gravity band-pass filter design. Geophysics, v. 59, pp. 69-76.

Pedersen, L. B. (1991) Relations between potential fields and some equivalent sources. Geophysics, v. 56, pp. 961-971.

Pilkington, M. and Todoeschuck, J.P. (1990) Stochastic inversion for scaling geology. Geophysical J. International, v. 102, pp. 205–217.

Pilkington, M. and Todoeschuck, J.P. (1993) Fractal magnetization of continenta crust. Geophysical Research Letters, v. 20, pp. 627– 630.

Pilkington, M. and Todoeschuck, J.P. (1995) Scaling nature of crusta susceptibilities. Geophysical Research Letters, v. 22, pp. 779–782.

Pilkington, M., Gregotski, M.E. and Todoeschuck, J.P. (1994) Using fractal crusta magnetization models in magnetic interpretation. Geophysica Prospecting, v. 42, pp. 677–692.

Shiomi, K., Sato, H. and Ohtake, M. (1997) Broad- band power-law spectra o well-log data in Japan. Geophys. J. Int., v. 130, pp. 57-64.

Spector, A. and Grant, F.S. (1970) Statistical model for the interpreting o aeromagnetic data. Geophysics, v. 35, pp. 293–302.

Treitel, S., Clement, W.G. and Kaul, R.K. (1971) The spectral determination o depths of buried magnetic basement rocks. Geophysical Journal of th Royal Astronomical Society, v.24, pp. 415–428.

Wang, R.J. (1969) The determination of optimum gate lengths for time varyin Wiener filtering. Geophysics, v. 34(5), pp. 683-695.

Zhou, S. and Thybo, H. (1998) Power spectra analysis of aeromagnetic data an KTB susceptibility logs, and their implication for fractal behavior of crusta magnetization. Pure and Applied Geophysics, v. 151, pp. 147–159.

14

Study on Bromide in Groundwater in Parts of Nalgonda District, Andhra Pradesh

K. BRINDHA AND L. ELANGO
Department of Geology, Anna University, Chennai, Tamil Nadu, 600 025

Abstract

The quality of groundwater is under constant threat due to urbanization. The purpose of this study is to understand the seasonal and temporal variations with special reference to bromide in groundwater and to determine the reasons behind the presence of this ion in groundwater. This study area is located at a distance of about 135 km towards ESE direction from Hyderabad. The concentration of bromide in groundwater of the study area was analyzed by collecting the groundwater samples from forty six wells once every two months from March 2008 to January 2009. Groundwater level and pH were measured in the field. The concentration of the minor ion namely bromide was determined by analyzing the groundwater samples in the laboratory using ion chromatograph. The bromide concentration ranged from below detection limit to 5.5 mg/l. The permissible limit for bromide in drinking water according to Indian standards specification is not mentioned. However, literature based on toxicity data puts forth 1 mg/l as the maximum admissible level of bromide. In this study, only a few groundwater samples had concentration of bromide above this limit. It is evident from this study that the groundwater quality is good with respect to bromide in this region. By the proper use of fertilizers and pesticides for agriculture, the groundwater quality can further be maintained in its current form or can be improved.

Keywords : Bromide, Groundwater, Hyderabad, Pollution, Drinking water

Introduction

Groundwater is widely distributed than surface water and is used for domestic, industrial and agricultural purposes throughout the world. More than 95% of rural population depends on groundwater for all needs. Because of the rapidly increasing population, several environmental problems are created which includes groundwater quality degradation. Any ion in groundwater present beyond the limit that cannot be admitted by the human body can be considered as pollution of the groundwater. Bromide is a minor constituent in natural waters and toxicity of bromide in human leads to vomiting, nausea, drowsiness, mental dullness and loss of motor coordination. In urban areas, due to treatment of water for drinking by disinfection processes, bromide gets converted into bromate which is a potential carcinogen. There is no standard permissible limit available for bromide in the Indian standards specified for drinking water (BIS, 1992). However, literature based on the toxicity data has set up a groundwater quality criterion of 1 mg/l (Flury and Papritz, 1993). Although there are several studies that have dealt with bromide and bromate concentration in tap water, bottled water and public drinking water supplies (Alessandro *et al.*, 2008) and their health effects, there are no major studies worldwide with respect to bromide content in groundwater. In Nalgonda district, Andhra Pradesh, India, a major groundwater quality investigation is being carried out (Rajesh *et al.*, 2009). The present study forms a part of this major study with the objective of understanding the present status of groundwater quality with respect to bromide and its spatial and seasonal variation.

Study area

The study area forming a part of Nalgonda district, Andhra Pradesh is located at a distance of 135 km ESE of Hyderabad (Fig. 1). The Nagarjuna Sagar reservoir is located in the southeastern side of this area and the southern side of the area is bounded by Pedda Vagu River. The northern boundary is partially along a water divide. The climate varies from arid to semi-arid. The temperature during the summer ranges from 30°C to 46.5°C and in winter it varies between 16°C and 29°C. The rainfall in this region is mostly due to southwest monsoon (June- September) and the average annual rainfall of this area is about 1000 mm. The surface runoff has resulted into the development of dentritic to subdentritic drainage pattern in this area (Fig. 2). The highest elevation in the study area is 348.1 m and the lowest elevation is 169.5 m.

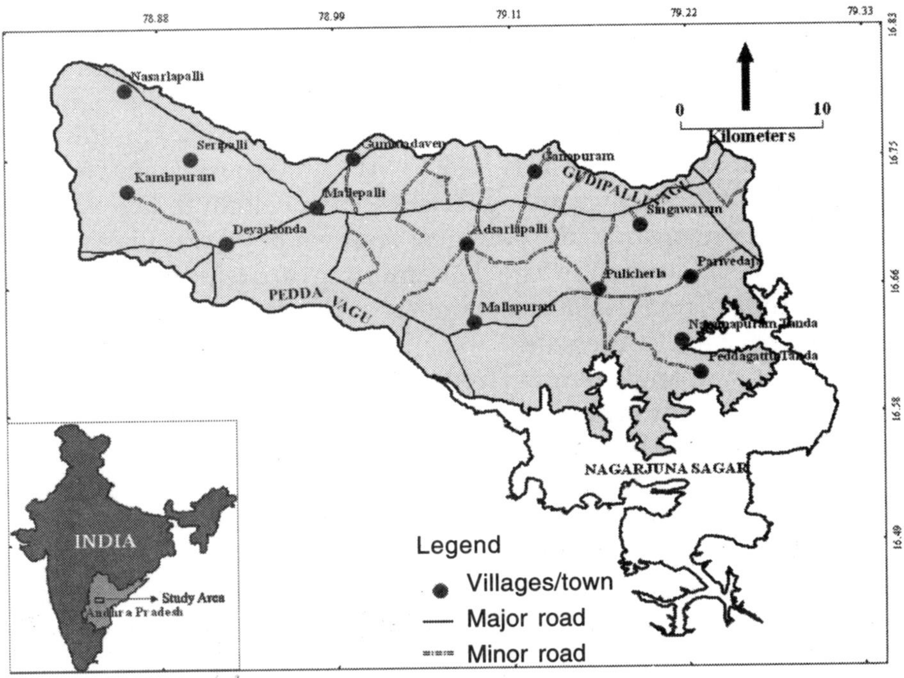

Fig. 1: Location of the study area.

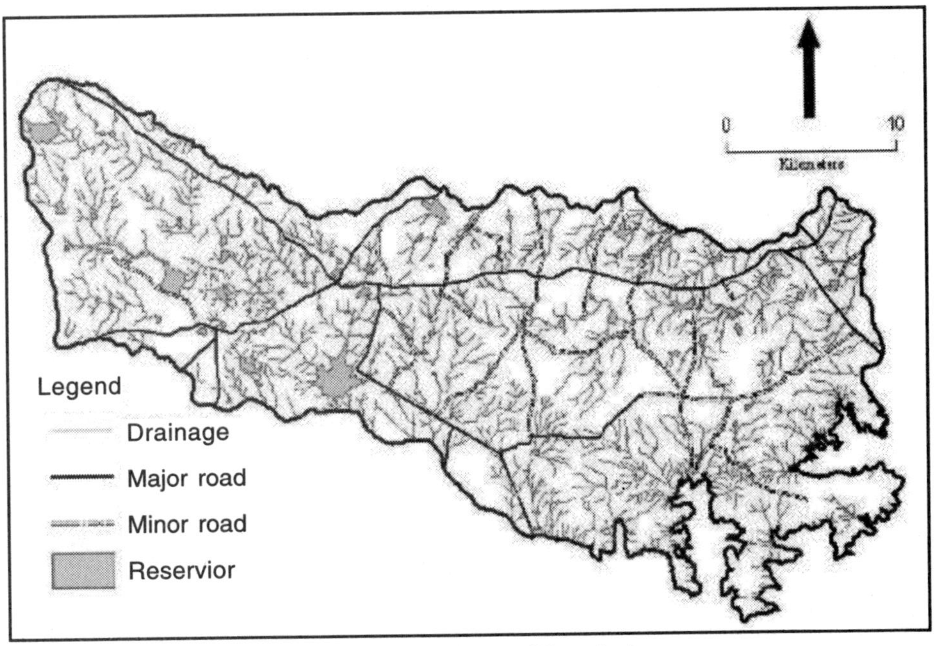

Fig. 2: Drainage pattern of the study area.

Geology

Granite forms the basement which is traversed by numerous dolerite dykes and quartz veins (Figure 3). Granites are generally medium to coarse grained. Rocks such as pink and grey granites occur at most parts of the study area. Most part of the investigated area has exposures of granitic/ granite gneisses which belong to late Archaen. The Srisailam formation, the youngest member of the Cuddapah super group directly overlies the basement granite with a distinct unconformity. The Srisailam metamorphic formation is exposed in the southeastern part of the study area. The litho units of this formation are dipping at an angle ranging from 3° to 5° towards SE. The sediments of Srisailam formation are mainly arenaceous and include pebbly-gritty quartzite shale with dolomitic limestone, intercalated sequence of shale-quartzite and massive quartzite (Fig. 3). The generalized stratigraphic sequence of this area is given in Table-1 (after GSI 1995).

Table-1 : General stratigraphic sequence of the study area (after GSI, 1995)

Cuddapah Supergroup	Massive quartzite
Srisailam Formation	Upper shale
	Quartzite with shale intercalation
	Lower shale with limestone intercalation
	Pebbly and gritty quartzite/ arenite
Eparchean Unconformity	
Late Archean/ Lower Proterozoic	Granite/granitic gneiss intrusion of dolorite dykes and quartz veins

Methodology

An intensive field survey was carried out and nearly 240 wells were considered for sampling. The Electrical Conductivity (µS/cm) was measured in all the wells and a representative well in about every 15km^2 was chosen for the collection of groundwater samples. Based on this survey, a total of 46 wells were selected (Fig. 4) for sample collection from March 2008 to January 2009. Groundwater samples were collected once in two months from these 46 wells. The depth of the wells varied from 1.45 m to about 20 m below ground level. The groundwater samples were collected in clean polyethylene bottles of 500 ml capacity. The bottles were cleaned prior to sampling by soaking them in 1:1 diluted HNO_3 for 24 hours and they were washed thoroughly with tap water. Further, these bottles were washed again before each sampling with the filtrates of the sample. Groundwater level in the wells was recorded using a water level recorder. Groundwater samples collected were transported to the laboratory, and filtered using 0.45 μm Millipore filter paper for determination of bromide concentration of the groundwater samples using Metrohm 861 advanced compact ion chromatograph along with appropriate standards.

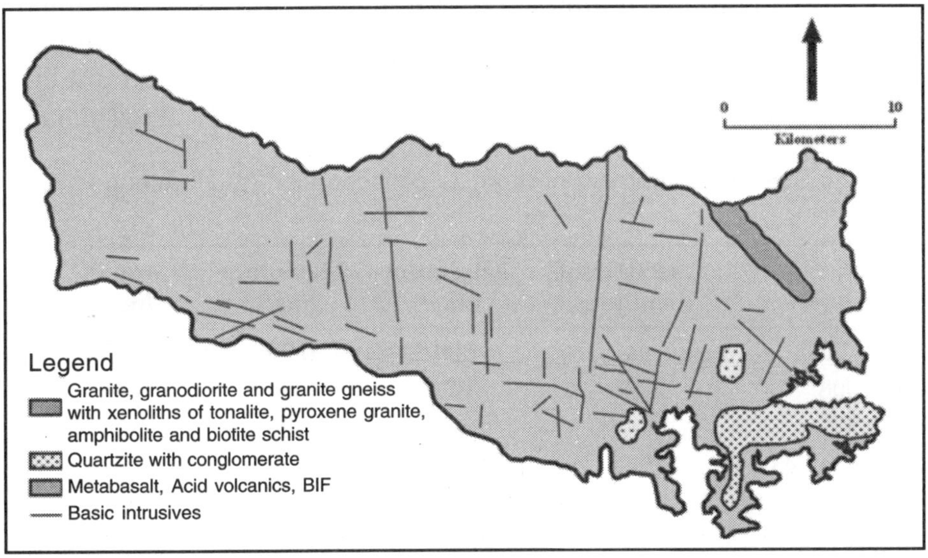

Fig. 3: Geology of the study area (after GSI 1995).

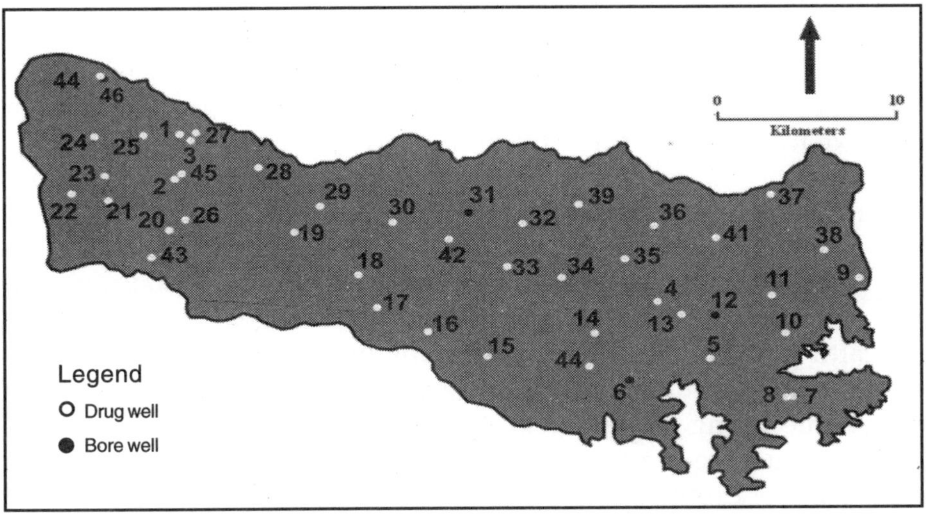

Fig. 4: Location of sampling wells in the study area.

Results and Discussion

The analysis of 251 groundwater samples collected from March 2008 to January 2009 indicate that the concentration of bromide vary with respect to space and time. The detailed description of the causes for this variation is discussed in this section.

The natural source of bromide is usually seawater and dissolution of rocks. Moreover, agricultural fertilizers also contribute to a significant amount of bromide in groundwater. The bromide content in the groundwater of this area was found to be in the range of BDL (Below Detection Limit) to 5.48 mg/l. The minimum, maximum and average bromide concentration in the study area is given in Table-2.

Table-2 : Statistical summary of bromide concentration (mg/l) during various months

Month	Number of samples	Minimum (mg/l)	Maximum (mg/l)	Mean (mg/l)
March 2008	42	BDL*	0.935	0.133
May 2008	38	BDL	5.48	0.505
July 2008	44	BDL	1.252	0.247
September 2008	41	0.001	1.105	0.248
November 2008	43	0.001	0.684	0.19
January 2009	43	0.002	1.655	0.337

*BDL- Below Detection Limit

The Indian standard specification for drinking water (BIS, 1992) does not set a standard for bromide in drinking water. Hence, from the toxicity data put forth by Flury and Papritz (1993), 1 mg/l is considered as the maximum permissible limit of bromide in drinking water. Of the 251 groundwater samples analysed during the entire study, only 14 samples had bromide content above 1 mg/l *i.e.* only 5.51% have bromide above the considered limit of 1 mg/l. Consumption of water with bromide >1 mg/l regularly for a longer period may lead to health problems such as nausea, vomiting, drowsiness and mental illness. However, these healths related problems were not observed in the people living in this area. As disinfection of groundwater for drinking purpose is not carried out in this region, the possibility of conversion of bromide to bromate does not occur. Further, in most of the locations the concentration of bromide is very low in groundwater and the present status poses no major risk to human health. Bromide concentration in groundwater did not show a significant relationship with groundwater level or pH. Spatial variation in groundwater level of the study area for July 2008 and average bromide concentration from March 2008 to January 2009 are shown in Fig. 5.

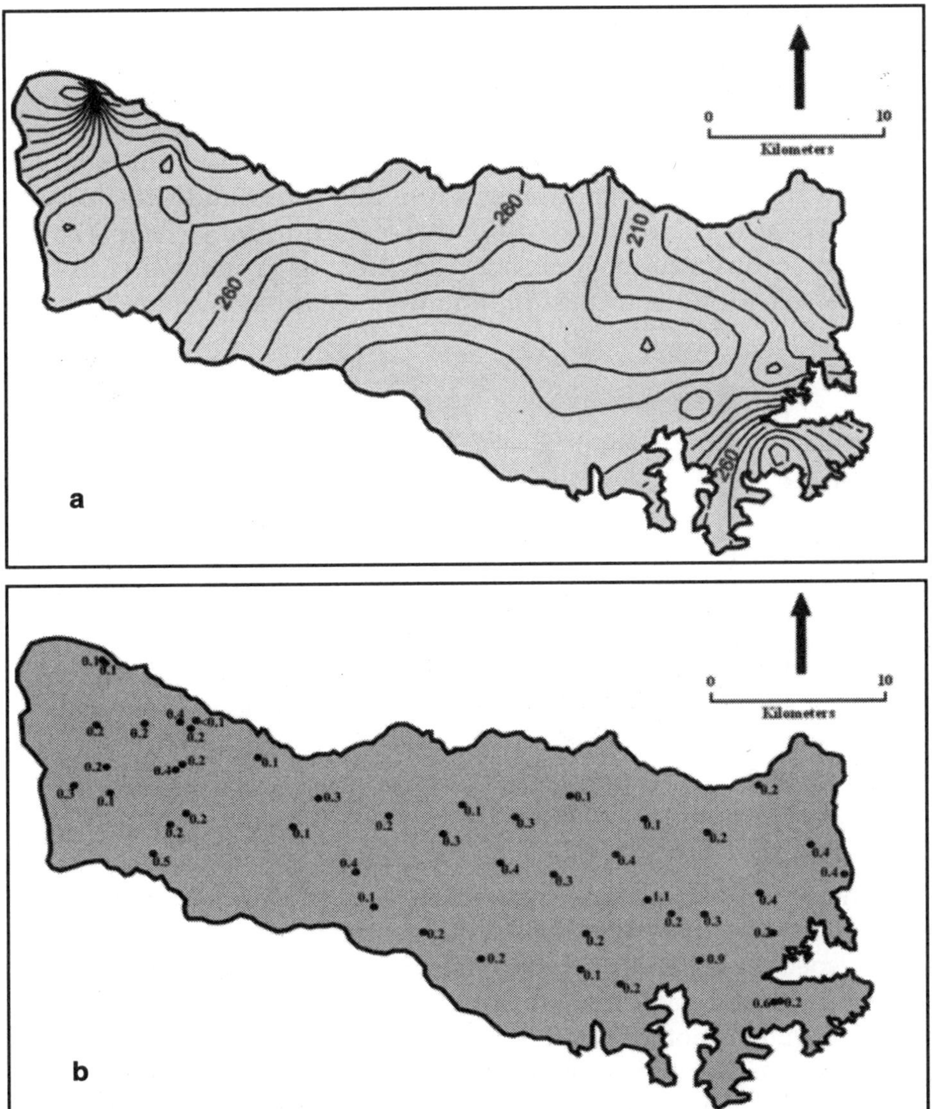

Fig. 5 : a. Spatial variation of groundwater level (msl in m) for July 2008 b. Average concentration of bromide in groundwater (mg/l) samples collected from March 2008 to January 2009.

It is evident from Fig. 5 that the average bromide concentration from March 2008 to January 2009 is high in only one well. Thus, it is very clear that the groundwater of this region is safe for human consumption when considering the bromide concentration in groundwater. Figure 6 shows the average concentration of bromide in wells that are present in irrigation lands and in domestic areas. There seems to be not much variation in the concentration of bromide in both the kind of wells. Only one well in the agricultural area possess average bromide level above 1 mg/l. Further, this well does not contain bromide concentration above 1 mg/l throughout the year. Hence, it is clear that application of fertilizer is not causing increase of bromide in groundwater. The concentration of bromide in groundwater occurs naturally in this area and only in trace amounts. There is no contribution by anthropogenic activities and the groundwater in this region is safe for human consumption with regard to bromide content.

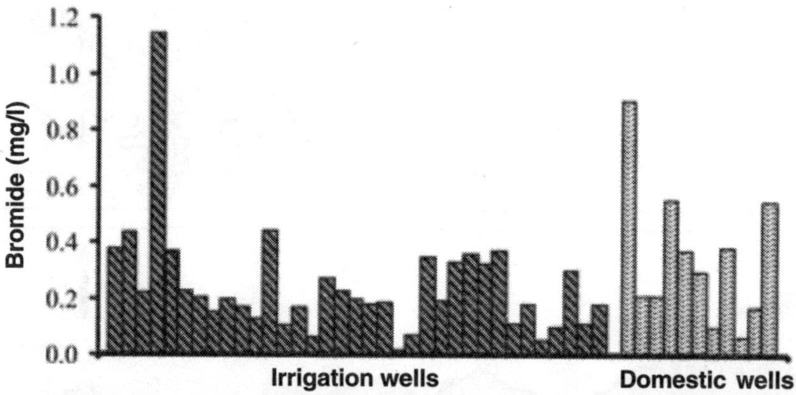

Fig. 6: Average bromide concentration (mg/l) in irrigation and domestic wells.

Conclusion

The present study was carried out with the aim of determining the groundwater quality in parts of Nalgonda district, Andhra Pradesh with respect to bromide. Of the total 251 samples analysed, 5.51% of the groundwater samples had high bromide. Since, disinfection of drinking water is not followed in this area, there is no threat of bromide getting converted into bromate. Hence, groundwater in this area is suitable for human consumption with regard to bromide. This study concludes that there is no major problem with respect to bromide, but proper use of fertilizers and pesticides will help further to improve the quality of groundwater.

Acknowledgement

We would like to thank the Board of Research in Nuclear Sciences, Department of Atomic Energy, Government of India for the financial support. The analytical facilities created using DST-FIST and UGC-SAP grants are also acknowledged.

References

Alessandro, W., Bellomo, S., Parello, F., Brusca, L. and Longo, M. (2008) Survey on fluoride, bromide and chloride contents in public drinking water supplies in Sicily (Italy). Environmental Monitoring and Assessment, v.145, pp.303–313.

BIS, (1992) Indian Standard Specifications for Drinking Water, IS: 10500. http://hppcb.gov.in/EIAsorang/Spec.pdf

Flury, M. and Papritz, A. (1993) Bromide in the natural environment: Occurrence and toxicity. J. Environmental Quality, v.22, pp.747–758.

Geological Survey of India (1995) Geology and minerals map of Nalgonda district, Andhra Pradesh, India.

Rajesh, R., Murugan, R., Brindha, K. and Elango, L. (2009) Hydrogeochemical processes of groundwater in Peddagattu-Lambapur, Nalgonda district, Andhra Pradesh. National Seminar on "Recent Advances in Water Research" (RETWAR- 2009), pp. 32.

❑❑❑

15

Excitation of Hard Rock Aquifers in Southern Peninsular India Associated with the 2004 Sumatra Earthquake

D. MURALIDHARAN, ROLLAND ANDRADE, R. RANGARAJAN AND U. SATHYANARAYANA
National Geophysical Research Institute, Hyderabad- 500 007
(Council of Scientific and Industrial Research)

Abstract

Emergence of springs at surface or in shallow pits reported from discrete locations during the month of January 2005 in the southern Indian peninsular shield area had been an interesting post geo-hydrological phenomenon to investigate. Such freak incidences at places in the interior parts of a continental shield area were reported for the first time. Its connectivity with the 2004 mega thrust great (Mw 9.3) Sumatra earthquake and its multiple aftershocks are studied and analyzed. Resistivity 2D imaging and its hydrogeological modulations revealed perched aquifer conditions at shallow depth (<10 m). There is no geological or geophysical evidence of structurally controlled upward rise of deep groundwater in the area. Electrical conductivity of the emergent spring water and that of deeper (depth >50 m) groundwater support that the spring water is originated due to local effect by the 2004 mega thrust earthquake. The emergence of spring is attributed to ground movement and squeezing of pores in the saturated zone at shallow depth due the mega thrust Sumatra earthquake.

Keywords : Resistivity 2D imaging, perched aquifer, spring, Sumatra earthquake

Introduction

There are many evidences showing significant changes in geological, geophysical and hydrological parameters in response to major earthquakes at a far distance from the epicenter (Amin and Rama, 1982; Muir-Wood and King, 1993; Montgomery and Manga, 2003). In this paper we have reported the anomalous geo-hydrological changes in the interior of Indian continental shield area after the great Sumatra earthquake on 26th December 2004. This earthquake was followed by the tsunami that had left behind a trail of death, destruction and human misery in many Asian countries. The Sumatra earthquake lasted for about 9 minutes, rupturing nearly 250,000 sq. km. of oceanic rock, a large fraction of it slipping 10-15 m across the fault surface and released as much quantum of energy as all other earthquakes combined over the past fifteen years could have generated (Wilson, 2005). Hundreds of aftershocks M >5.0 are registered during the last few years (Preliminary seismological bulletin, GSI). Earthquake induced water level fluctuations in bore wells provide a means to understand the aquifer characteristics through pore pressure changes once the mechanics governing the earthquake occurrence is adequately understood. Similar to the seismograph, the well aquifer system also consists of a mass (the column of water in the well plus some part of the water in the aquifer), a restoring force (the difference between the pressure head in the aquifer and the displaced water level in the well) and a damping force (the friction that accompanies the flow of water through the well and aquifer) (Cooper *et al.*, 1965).

Earthquake and Hydrology

Earthquakes are regarded as ground shaking due to sudden energy release along active faults in tectonic zones at deeper depths of the crust. In the case of Sumatra earthquake, the rupture length was reported to be one of the largest, extending to a length of about 1200 km. and along a gently curving line, following the Indian plate boundary in the North-South direction. Anomalous change in groundwater level is one of the most important signatures associated with earthquakes (Ground Shaking). Water level changes take place not only due to earthquakes, but also from various aseismic effects such as barometric pressure change, tidal variations and precipitation influences (Bredohoeft, 1967; Van der Kamp and Gale, 1983). Several observational attempts are made to confirm the fact that there are certain patterns of water level responses to large earthquakes as distant as several hundreds of kilometers. These responses are purely due to seismic waves but unlike seismic waves water level oscillations are monotonic and persist for days or weeks (Roeloffs, 1998) in certain cases. A well known earthquake related water level change was reported in response to the 1989 Loma Prieta, California earthquake, 157 kms northwest of a monitoring well site (Rojstaczer and Wolf, 1992).

Tectonics and Morphological Setting of the Region

Seismotectonics of the Indian shield region is largely contributed by the Himalayan back-thrust and reactivation of the old dormant faults. The central India constitutes of considerable variance in crustal thickness, tectonic history and thermal structure. The concept of Indian peninsula as a stable Archaean shield region has been changed due to low to moderate earthquakes in the recent years. The energy released in these earthquakes is basically a part of intraplate stress arising due to the collision of the northeastward moving Indian plate with the Tibetan plate (Khattri and Tyagi, 1983; Valdiya, 1976). The observational site on the southern Indian shield (Eastern Dharwar Craton) is characterized by sheared and deformed lithosphere (Pandey and Agrawal, 1999). Basically these K-granite plutons are thought to be emplaced as isolated bodies over the Dharwar Craton. Multi-parametric geological, geophysical and GPS studies over Eastern Dharwar Craton brought out a differential up-liftment of the Hyderabad granite pluton and Bangalore Geotectonic Block between Late Archaean and Early Proterozoic (Catherine and Pandey, 2005). Morphotectonic studies, surface and lake sediments radioactivity, heat flow pattern and tremors indicated persistence of Neotectonic activities in the study region (Pandey *et al.*, 2002). The morphotectonic map of the area with study sites is shown in Fig.1.

The 2004 Sumatra Earthquake

The world's most powerful earthquake of Mw 9.3 had struck on 26th December 2004 near west coast of northern Sumatra Island. The cause was attributed to be thrust faulting at the boundary of the Indian-Burmese plates. The accumulated stress due to ongoing subduction of

Indian Plate beneath the overriding Burma micro-plate and release of strain was the cause of Sumatra earthquake (Purnachandra Rao and Kalpna, 2005). The fault displacement analysis implies that fault rupture propagated to the north from the epicenter. The post-effect of the Sumatra earthquake was generation of giant sea waves (*Tsunami*) and its propagation across the Indian Ocean. Tsunami struck the Indian coast and caused severe destruction along the east and parts of west coast of India, causing heavy loss of property and casualties.

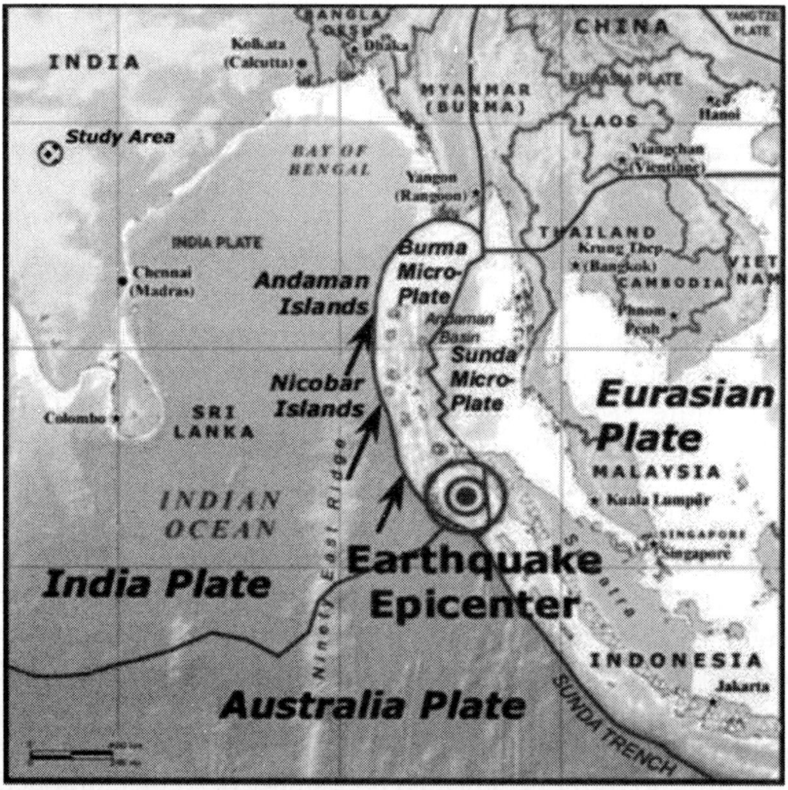

Fig. 1 : Location map of the investigation sites with its morpho-tectonic settings.

Hydrological Response to Earthquakes

Hydrogeological responses to earthquakes are well known phenomena for both near field and far field earthquake sources. Water wells becoming turbid, dry or begin to flow, increase in discharge of springs and forming of new springs have been reported as observations of post earthquake effect. Studies have also indicated that mild earthquakes of M 3 could generate effects on ground water, such as in bore-wells; as far as about 10 miles from the epicenter and effects of magnitude 9 quake could be observed in wells of more than 6000 miles away, as in the case of 1964 Alaska earthquake that registered M 9.2 (Montgomery and Manga, 2003). The observation of 2004 hydrograph data of a well in Hyderabad (NGRI Campus) exhibited the change in hydrostatic pressure from 25th Dec'04 early hours onwards, which is attributed to the stress release before the main shock in the Sumatra-Indonesia subduction zone on 26th Dec'04 (Muralidharan *et al.*, 2005). Similar effects of hydrostatic pressure changes were observed from other places located over Hyderabad granite pluton. These site records were cross-correlated with respect to time and magnitude of water level change. The effects were of the same order in all the well hydrographs.

Emergence of springs at surface or in shallow pits during the month of January 2005 in Ranga Reddy district of Andhra Pradesh was reported in local newspapers (Deccan Chronicle dated 01.02.2005) as a probable post-effect of Sumatra Earthquake of 26th December 2004. However, in a sub-normal rainfall year of 2004, spring emergence at discrete places called for an understanding of the process and causes in interior continental part of southern India, necessitated geophysical and hydrological reconnaissance surveys around these sites, namely Devanur and Mukundapur villages (Fig.1), using resistivity tomography and by water sample collection from all the nearby wells at these sites. Brief report on the outcome of the survey and the inference drawn for the probable cause of spring emergence are presented below.

Resistivity Imaging at Devanur Site

One of the reported sites of spring emergence was near the Devanur village in Ranga Reddy District of Andhra Pradesh and the location of spring emergence is at the foreshore area of an ancient irrigation tank. Discussion with the local villagers yielded information on oozing of water on the surface was first noticed on 16th January 2005 and out of curiosity, the villagers dug out the soil down to 20 to 30 cm and by seeing unusual water seepage in the pits, they linked the incidence to the effects of the 2004 tsunami earthquake which was later reported to the local News Papers. It was an interesting incident to note from such an area where the monsoon over the past few years was below normal. Due to vagary nature and consecutive failure of monsoon, the inflow to the tank reduced over the years, and remained dry. The seepage pits reported were aligned in perpendicular direction (NE-

SW) to the main tank axis and at the mouth of feeder stream channel that brings the runoff water to the tank. In order to understand the cause of emergence of spring, resistivity method was adopted to map the sub-surface nature of the area. Multi-electrode Imaging System with 2 m electrode spacing (a=2m) was deployed in WNW-ESE direction perpendicular to the direction of pits alignment. The center of the image profile was over a pit having water during the time of survey. The sub-surface resistivity image obtained after data processing is shown in Fig.2a.

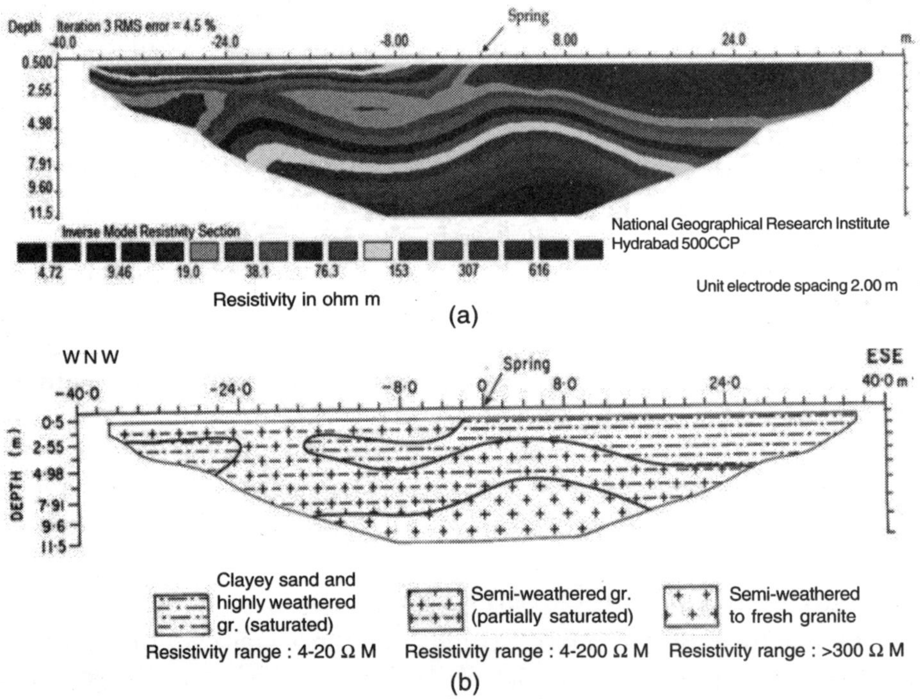

Fig. 2 : (a) 2D resistivity tomography pseudo-section along a profile at Devanur spring emergence site. (b) Subsurface hydrogeological section of the resistivity profile survey at Devanur spring emergence site.

The resistivity section of an 80 m long traverse exhibited a three layered sub-surface formation, of which the top layer with a resistivity range of 4 to 20 Ohm-m extends to a depth of about 5 m clearly depicting a highly conductive nature of the formation and higher degree of saturation. A middle layer with 40 to 200 Ohm-m resistivity followed the top conductive layer from 5 m to 10 m depth marking the semi-weathered part of granite with less saturation. The third geo-electric layer with a resistivity range of 300–650 Ohm-m representing the hard granite found to be occurring below 10 m depth throughout the section. The resistivity imaging results have been modulated to the existing hydrogeological conditions of the site and a section was prepared to understand the prevailing condition. The hydrogeological representation of resistivity image obtained at this site is shown in Fig.2b. The influence of an outcrop of weathered granite seen on the western part of the profile line is getting reflected clearly in the resistivity image by a high resistive zone at shallow depth itself. The eastern side of the profile having the reported spring site reflected a high conductive clayey sand layer with increase in thickness to 5 m on eastern side of the profile. The resistivity range of 4-20 Ohm-m indicates that the geo-electric layer consists of probably clayey sand at the top and followed by highly weathered granite with high degree of saturation. As it is not expected to have such a thick silt deposits in the foreshore area of a tank, it can be visualized that the bottom of the first inferred hydrogeological layer may probably consist of highly weathered granite with saturation as it is reflected with resistivity of 20 Ohm-m.

The hydrogeological layer that follows the first layer from a depth of 5 to 10 m is inferred to be semi-weathered granite with less degree of saturation characterized with resistivity range of 40-200 Ohm-m. A dried nature of a dug well at the time of survey towards NW of the profile line having a depth of 9-10m in the semi-weathered granite located at a distance of 40 m from the center of the image profile affirmed the inferences. The third layer was assumed to be of semi-weathered to fresh granite without saturation, represented by resistivity of more than 300 Ohm-m.

The resistivity imaging and hydrogeological studies collectively indicated that the spring emergence at Devanur site could be due to the formation of pseudo-perched condition of an aquifer system at very shallow depth. The resistivity range obtained for the top layer clearly indicates that the formation is admixture of clay and highly weathered granite materials. The rainwater percolated during the monsoon of 2004 probably saturated the soil zone. Due to poor rainfall during later monsoon months, the percolated water front has not further moved into deeper zones under the normal process of natural recharge. The dried nature of dug well near the site at the time of spring emergence at very shallow level should of localized and not of regional phenomenon. As the resistivity imaging studies did not revealed any signatures of structural features at the site spring emergence, upward movement of ground water from deeper horizons to shallow levels is being questionable. Also, the very low order of electrical conductivity of 200 to 275 micro-mhos of the seepage pit water indicate that the water is not from deeper groundwater source as the electrical conductivity of groundwater source collected from boreholes of 60 m depth was measured to be varying from 800 to 880 micro-mhos. Reoccupation of the same site with the resistivity imaging profile in the month of February 2007 for validating the sub-surface feature seen with same electrode spacing of 2 m confirmed the persistence of shallow highly saturated soil zone without any spring emergence in spite of being normal rainfall year of 767 mm.

Resistivity Imaging at Mukundapur Site

The emergence of spring reported from another site near Mukundapur village in the same district was noticed in the form of more wetness in rain fed agricultural field before harvesting itself. After harvesting, in the month of January 2005, they dug a pit at the place of more wetness and found that the groundwater was oozing and flowing out of the pit. As the water was continuously coming out of the pit, the villagers attributed the observation to be the effect of the 2004 tsunami earthquake and reported to the News Papers. The location of the site is on a sloping ground (*agricultural terraces*) of a broad valley. The area is underlined by granites similar to that of the Devanur site. A stream channel was noticed on the downstream of site. The excavated pit was at the edge of an agricultural field at higher elevation. The geomorphologic condition suggested that the site was having a good catchment area on the upstream with a topographical slope of roughly about 5-10%.

Resistivity tomography survey was carried out at this site keeping the orientation of the profile (N-S) parallel to the field terrace border. The orientation was chosen so that the study is made perpendicular to the general groundwater flow direction. The center of the profile line was shifted by 17 m towards north of the spring emergence site due to site constraints. The profile length of 80 m was laid with the electrodes placed at 2 m interval. The processed image obtained for the study profile is shown in Fig.3a.

The hydrogeological interpretation of the resistivity image obtained to a depth of 11.5 m revealed that the sub-surface nature of the investigated area can be grouped into three layered structure similar to that of the Devanur site. The top layer of conductive nature extending to a depth of 5 m with a resistivity range 6-30 Ohm-m was found to be uniformly present through out the study traverse line. However, a bowel shaped anomalous zone was reflected just at/near the location of spring site. This layer followed by a semi-weathered formation with a resistivity range of 50-230 Ohm-m forming the middle layer extending to a depth of 8-9 m. The third layer with resistivity more than 300 Ohm-m marked the basement.

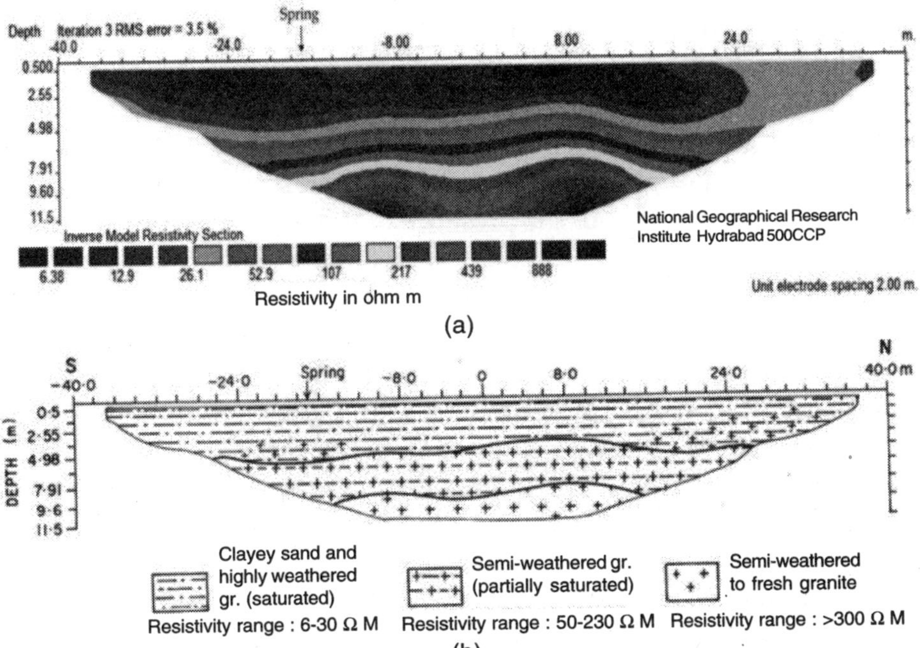

Fig.3: (a) 2D resistivity tomography pseudo-section along a profile at Mukundapur spring emergence site. (b) Subsurface hydrogeological section of the resistivity profile survey at Mukundapur spring emergence site.

Similar to the Devanur site, a section was constructed transforming the resistivity imaging results into a hydrogeological representation and shown in Fig.3b. The top layer extending to a depth of 5 m through out the section is attributed to clayey sand layer at shallow and highly weathered granite, at the bottom part with high degree saturation. However, on the northern side of the section, this layer is with very less topsoil as evidenced in the field with exposure of highly weathered granite. This was well supported by higher resistivity over this area when compared to the areas on the southern side for the same hydrogeological layer. A semi-weathered granite layer with partial saturation follows this highly conductive top layer and its thickness extends to a depth of 8-9 m. The third layer is deciphered to be semi-weathered to fresh granite layer representing the basement. The high order resistivity of more than 300 Ohm-m for this layer indicates that the rocks are compact and devoid of saturation.

The resistivity imaging and other hydrogeological observations at the Mukundapur site brings out the plausible reasons for the emergence of spring may be attributable to the presence of conductive formation at shallow depth with saturation may be the source for the spring. The bowl shaped anomalous zone at the spring emergence location is probably due to excavation and filling with sediments. Recharge from the upland area might have helped in increase in water accumulation of water at this site due to the presence of terrace cue nearby. Presence of conductive layer throughout the section at shallow depth with saturation supports the movement of water from the upland area as a sheet flow following the topographical slope and the terrace cut surface boundary near the site favored the emergence of spring. The electrical conductivity of spring water of 163 micro-mhos exhibits the recent origin. The electrical conductivity of groundwater of the region of 700 of a well near the stream on the downstream and 2270 micro-mhos of village hand pump on the upstream clearly points out that the spring water is not from deep groundwater source. The practice of keeping the agricultural field dry before harvesting supports that the water emerged as a spring is probably due to the movement of upland monsoon recharge as a sheet flow under the influence of topographic slope.

Conclusion

The hydrogeological conditions and resistivity imaging at the two sites in Ranga Reddy district of Andhra Pradesh points out that the emergence of spring was of localized nature. However, the mobilization of water and emergence of spring might have been caused due to ground movement associated with the 2004 Sumatra earthquake and hundreds of aftershocks. Continuous monitoring at these two sites by having few observation wells of shallow depth for next hydrological cycle is required to better understand the subsurface mechanism.

References

Amin, B.S. and Rama (1982) A search for correlation between seismicity and radon anomaly in hot springs. Earth Planetary Sciences, v. 91(1), pp.15-19.

Bredohoeft, J.D. (1967) Response of well-aquifer systems to earth's tides. J. Geoph. Res, v.72, pp. 3075-3087.

Catherine, J.K. and Pandey, O.P. (2005) Differential uplift Between Hyderabad and Bangalore Geotectonic Blocks of Eastern Dharwar Craton, South India. Geology, v.65, pp. 493-496.

Cooper, Hilton H. Jr., Bredehoeft, John D., Papadopulos, Istavros S. and Bennett, Robert R. (1965) The response of well aquifer system to seismic waves. J. Geoph. Res, v.70, pp.3915 - 3926.

Khattri, K.N. and Tyagi, A.K. (1983) The Transverse tectonic features in the Himalaya. Tectonophysics, v. 96, pp.19-29.

Montgomery, David R. and Manga, Michael (2003) Stream flow and Water Well Responses to Earthquakes. Science, v. 300, pp.2047-2049.

Muir-Wood, R. and King, G.C.P. (1993) Hydrological signatures of earthquake strain. J. Geophysical Research, v. 98, pp. 22,035–22,068.

Muralidharan, D., Andrade, Rolland, Laxminarayana, G., Swathi, T. and Swetha, T. (2005) Imprint of 26th December 2004 Sumatra Earthquake in Aquifers of Hyderabad Granite Pluton. Current Science, v. 89(7), pp.1083 - 1086.

Pandey, O.P. and Agrawal, P.K. (1999) Lithospheric mantel deformation beneath Indian craton. Geology, v.107, pp.683-692.

Pandey, O.P., Agrawal, P. K. and Chetty T. R. K. (2002) Unusual lithospheric structure beneath the Hyderabad granitic region, Eastern Dharwar Craton, South India. Phys. Earth Planet. Int., v.130, pp.51-69.

Preliminary seismological bulletin for seismic events recorded and registered at broad band seismic observatory, Nagpur, Maharashtra, India, for Sumatra earthquake of 26th December 2004 and the aftershocks. GSI; Broad Band Seismic Observatory, Nagpur.

Purnachandra Rao, N. and Kalpna, G. (2005) Deformation of the subducted Indian Lithospheric slab in the Burmese Arc. Geoph. Res. Let, v.32, pp.1-5 LO5301.

Rojstaczer, Stuart and Wolf, Stephen (1992) Permeability changes associated with large earthquakes: An example from Loma Prieta, California. Geology, v.20, pp. 211-214.

Roeloffs, Evelyn A. (1998) Persistent water level changes in a well near Parkfield, California, due to local and distant earthquakes. J. Geoph. Res, v.103, pp. 869-889.

Valdiya, K.S. (1976) Himalayan transverse faults and folds and their parallelism with subsurface structures of north Indian plains. Tectonophysics, v.32, pp. 353-386.

Van der Kamp, G. and Gale, J.E. (1983) Theory of Earth tides and barometric effects in porous formation with compressible grains. Water Resources Research, v.19, pp. 538-544.

Wilson, Mark (2005) Modeling the Sumatra-Andaman Earthquake reveals a complex, non-uniform rupture. J. Geol. Soc. India, v. 66(4), pp. 499 – 503.

□□□

16

Use of Meridional Wind Index at East Coast of India for Monitoring Drought and Flood Monsoon Conditions over Indian Region

S. S. DUGAM
Indian Institute of tropical Meteorology, Pune-411008

Abstract

In this paper, the relationship between daily meridional winds (v) at 200 hPa taken from NCEP/NCAR, over the domain (15°N to 25°N and 80°E to 95°E) at east coast of India and daily rainfall during the Indian Summer Monsoon (June-September) has been studied. Analysis is carried out for fifty three (1951-2003) years of data. The twenty days composite rainfall and meridional wind for the 10 drought and 8 flood monsoon years have been analyzed. The twenty days variations of the southerlies (northerlies) meridional wind in the upper troposphere during the monsoon at east coast the of India and twenty days rainfall activity goes hand in hand in drought (flood) monsoon years. From the analysis, it is observed that during the drought (flood) years southerlies (northerlies) persist over this domain in monsoon season.

Keywords : Iron oxide/hydroxide coating, Bulk sample, Bulk leach gold, Stream sediment survey.

Introduction

A major proportion of the annual rainfall over most parts of India is received during a short span of 4 months from June to September, known as the Southwest or the summer monsoon season. The summer monsoon rainfall averaged over the whole of India is found to be stable over the past 125 years, with no long-term trend, but is dominated by high inter-annual variability. This variability in monsoon rainfall occasionally leads to large-scale droughts and floods over different parts of the country, resulting in serious reduction in agricultural output and affecting the national economy. In view of the critical influence of such variability, adequate forecasting of monsoon rainfall would be of profound importance for policy-making and planning of mitigatory efforts. This variability is generally attributed to the slowly varying surface boundary forgings of sea surface temperature, soil moisture, snow cover and the circulation features of the upper-troposphere (200 hPa May meridional wind over India) and the mid-troposphere (April 500 hPa ridge along 75° E).

Insufficient, untimely and unevenly distributed monsoon rainfall leads to meteorological drought over the country during the main kharif cropping season. It is known that the occurrence of drought is due to insufficient rainfall in time and space. However, it is still difficult to quantify the conditions that lead to the onset of drought. The early prediction of drought and its spatial extent has a very important role in Indian economy. No known method exits to predict reliably occurrence, continuation, cessation or recurrence of drought. The present scientific knowledge is insufficient for accurate and early prediction of drought, so it is necessary to understand the causes of drought and hence lead to better predictions. Many scientists studied the extreme monsoons and their association with large-scale regional and global perturbation in atmospheric and oceanic circulation. Sikka (1980, 1999); Rasmusson and Carpenter (1983); Rajeevan (1993); Goswami and Xavier (1966); Goswami *et al.*(2003) Fasullo and Webster (2002); Gadgil *et al.* (2002); Gadgil *et al.* (2004); Kalsi *et al.* (2004); Krishnan *et al.* (2006); Joseph *et al.* (1981;1999) and Bhat (2006).

Data and Methodology

Daily grid point meridional wind (v) data at 200 h Pa level taken from NCEP/NCAR for a period from 1 June to 30 September (122 days) for 53 years (1951-2003) are considered. The data over the domain15°N to 25°N and 80°E to 95°E is used for the analysis. Daily All India monsoon Rainfall (DAIMR) are taken from daily gridded rainfall data published by Indian Metrological Department (IMD) based on 1803 stations (Rajeevan *et al.*, 2006) for the same period and years. The simple moving average and correlation analysis methodology is used for the analysis. Daily averages meridional winds in the domain are calculated for 122 days (1 June-30 September) for the 53 years. Daily meridional wind anomalies for each day and for each year are calculated by taking mean from the base period 1951-2003.

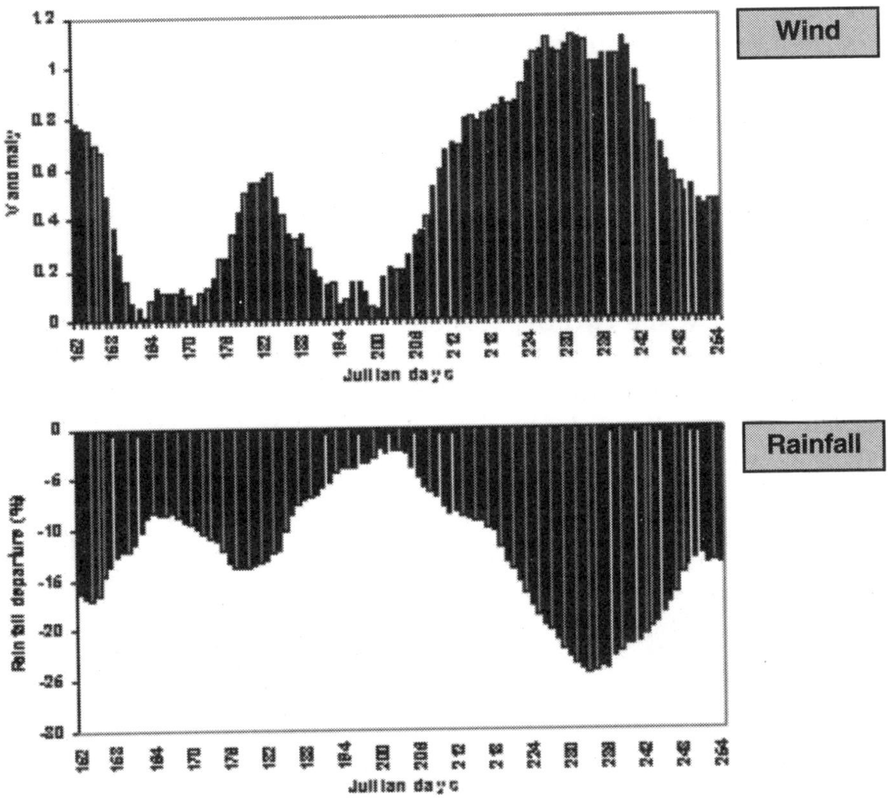

Fig. 1 (a): 20 days running meridional wind (V) anomalies at 200 h Pa level for 10 drought monsoon years.

Discussions

To see the daily variation of meridional wind anomaly in flood and drought years the 20 days moving averages of ECMWI anomalies for the period 1 June to 30 September calculated for 10 drought (1951; 65, 66; 68; 72; 74; 79; 82; 85; 86; 87 and 2002), and 8 flood years (1956; 59; 61; 70; 75; 83; 88 and 94). Similarly, daily all ISMR anomalies for same drought and flood years are also calculated. Comparisons of this analysis are shown in Fig. 1a and 1b. From the figure, it is seen that the trends in southerlies and the subdued activity of rainfall goes hand in hand in almost all the drought years. It may be seen that the southerlies occurring in the upper troposphere in July and August in the large- scale monsoon failure show considerable persistence. Therefore, the persistence of the southerlies meridional at 200 h Pa components of winds in the upper troposphere may be useful for monitoring the large-scale monsoon failures. The composite meridional wind anomalies at 200 h Pa for these drought and flood years also calculated over the domain mention above. It is observed that during the drought (flood) years southerlies (northerlies) persist over this domain in monsoon season. Composite divergence field anomaly at 200 h Pa for these extreme years also calculated over the above domain. It is seen that the large scale convergence (divergence) is observed in the southern Bay in drought (flood) years which suppressed the synoptic scale activity and that may leads to major drought condition vice versa. No depression formed in the year 2002 for the first time in past 180 years while two depressions were occurred in 2003 may be contributing normal rainfall. In major drought 2002 total number of synoptic scale disturbances (including low pressure systems, upper air cyclonic circulations; depressions etc.) are 21 and in normal monsoon 2003 they are 24 included 2 depressions.

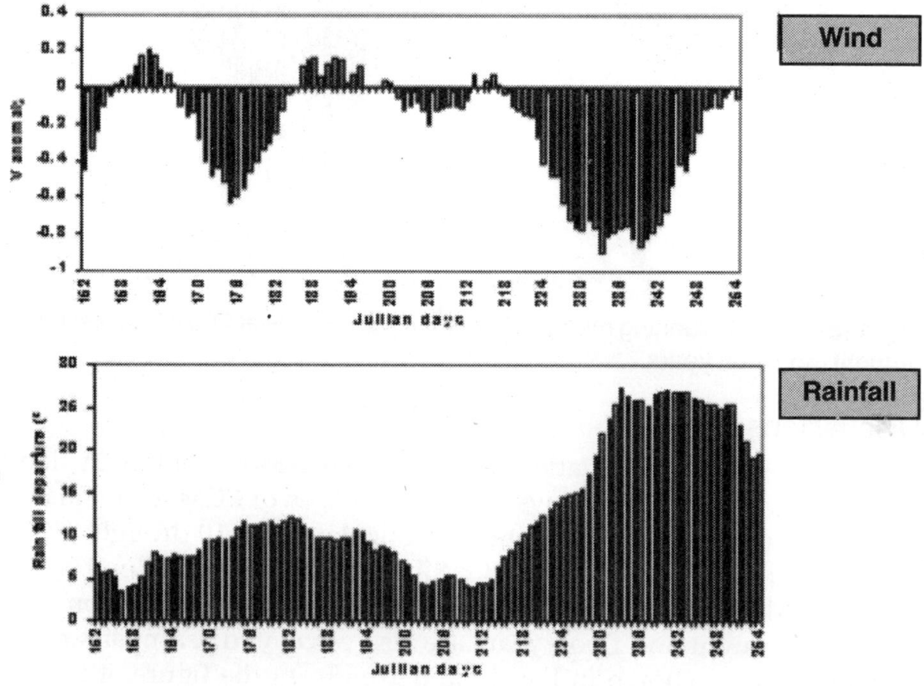

Fig.1 (b): 20 days running meridional wind (V) anomalies at 200 h Pa level for 8 flood monsoon years.

Conclusion

1. The twenty days variations of the southerlies (northerlies) meridional wind at 200 hPa during the monsoon period (June-September), at east coast in the upper troposphere and twenty days rainfall activity goes hand in hand in drought (flood) years. Similar result also holds good for individual drought and flood years.

2. The twenty day tendency of the variation of meridional wind at 200 hPa in domain mention above will gives some clue for next coming twenty days rainfall activity.

Acknowledgments

The authors are grateful to Prof. B. N. Goswami Director, I.I.T.M. and Dr P. N. Mahajan, Head, Forecasting Research Division for his encouragement and valuable suggestion. The author is thankful to DST Government of India for funding.

References

Bhat, G. S. (2006) Indian drought of 2002– a sub-seasonal phenomenon? Quart. J. Roy. Meteor. Soc., v. 132, pp. 2583–2602.

Fasullo, J. and Webster, P. J. (2002) Hydrological signatures relating the Asian summer monsoon and ENSO. J. Clim., v. 15, pp. 3082–3095.

Gadgil, S., Srinivasan, J., Nanjundiah, R. S., Krishna, Kumar K., Munot, A. A., Rupa and Kumar K. (2002) On forecasting the Indian summer monsoon: the intriguing season of 2002. Curr. Sci., v. 83, pp. 394–403.

Gadgil, S., Vinayachandran, P. N., Francis, P. A. and Gadgil, S. (2004) Extremes of the Indian summer monsoon rainfall, ENSO and equatorial Indian Ocean oscillation. Geophys. Res. Lett, v. 31 L12213 doi 10.1029/2004GL019733

Goswami, B. N. and Xavier, P.K. (2003) Potential predictability and Extended Rang prediction of Indian summer monsoon breaks. Geophys. Res. Lett., v. 30,doi:101029/2003GL017,810-2003

Goswami, B. N., Ajay Mohan, R.S., Xavier, P.K. and Sengupta, D. (2002) clustering of low pressure systems during the Indian summer monsoon by intraseasonal Oscillations. Geophys. Res. Lett., v. 30, doi: 10,1029/ 2002GLO16, 734-2003.

Joseph, P. V., Mukhopadhyaya, R.K., Dixit, W.V. and vaidya, D. V. (1981) Meridional wind index for long-range forecasting of Indian summer monsoon rainfall. Mausam, v. 32, pp. 31-34.

Joseph, P.V. and Srinivasan, J. (1999) Rossby waves in May and the Indian summer monsoon rainfall. Tellus, v. 51A, pp. 854-864.

Kalsi, S.R., Hatwar, H. R., Jayanthi, N., Subramanian, S. K., Shyamala, B., Rajeevan, M. and Rajendra, Kumar R. K. (2004) Various aspects of unusual behavior of monsoon 2002. India Meteorol. Dep. Monogr No.2, 97 p.

Krishnan, R., Ramesh, K. V., Samala, B. K., Meyers, G., Slingo, J. and Fennessy, M. J. (2006) Indian Ocean-Monsoon coupled interactions and impending monsoon droughts. Geophy. Res. Lett. V. 33 L08711 doi 10.1029/2006GL025811

Rasmusson, E.M. and Carpenter, T.H. (1983) The relationship between eastern equatorial Pacific sea surface temperature and rainfall over India and Sri Lanka. Mon. Wea. Rev., v. 111, pp. 515–528.

Rajeevan, M., Bhate, J., Kale, J. D. and Lal, B. (2005) Development of a high resolution daily gridded rainfall data for the Indian region. Met. Monograph Climatology 22/2005, IMD Pune, 27 P.

Krishnan, R., Ramesh, K. V., Samala, B.K., Meyers, G., Slingo, J. and Fennessy, M. J. (2006) Indian Ocean-Monsoon coupled interactions and impending monsoon droughts. Geophy. Res. Lett., v. 33 L08711 doi 10.1029/2006GL025811

Sikka, D. R. (1980) Some aspects of the large scale fluctuations of summer monsoon rainfall over India in relation to fluctuations in the planetary and regional scale circulation parameters. Proceedings of Indian Academy of Sciences (Earth & Planetary Sciences), v. 89, pp. 179-195.

Sikka, D. R. (1999) Monsoon drought in India-joint COLA/CARE. Technical report No.2, Center for Ocean- Land Atmosphere Studies, Calverton, MD 20705, USA.

□□□

17

Thunderstorms/Lightning Generated Sprite and Associated Phenomena

DEVENDRAA SIINGH[1], SANJAY KUMAR[2] AND A. K. SINGH[2]
[1]Indian Institute of Tropical Meteorology, Pune-411008
[2]Department of Physics, Banaras Hindu University, Varanasi-211 005

Abstract

The study of the thunderstorms has advanced rapidly during the past century and lots of efforts have been made towards the understanding of lightning, thunderstorms and their associated optical phenomena such as sprites, elves, blue get etc. Even though, thunderstorms and lightning are well understood, but our knowledge on sprites and associated phenomena, is limited. In this paper, we review the distribution of thunderstorm/lightning and their association with optical emissions. Detailed discussions on lightning generated sprites and various natural phenomena associated with them are also included. Recent results in this emerging field are summarized.

Keywords : thunderstorms, lightning, sprites, elves, optical emissions

Introduction

A thunderstorm is characterized by strong winds in the form of squall, heavy precipitation and low-level wind shear. The formation, intensification and propagation of thunderstorms are mostly governed by the synoptic and thermodynamic conditions of the atmosphere; their microphysical and electrical characteristics are known to significantly affect the formation and intensity of precipitation. These are the deepest convective clouds caused by buoyancy forces set up initially by the solar heating of the Earth's surface. A number of theories have been proposed to explain how the thunderstorms get electrified (Uman, 1987). Several field and laboratory experiments have been conducted to determine the electrical nature of storms. Possible electrification processes have been simulated in the laboratory and also by theoretical modeling (Saunders, 2008; Yair, 2008).

MacGorman and Rust (1988) reviewd the complete understanding, up to late 20th century of the electrification of thunderstorm, also Rakov and Uman (2003) have provided the similar review for lightning. Recent work by Yair (2008), Saunder (2008) and Davydenko *et al.* (2004) have discussed how the complex and variable structure and charge distribution in cloud leads to a variable contribution to the current flow in global circuit. The anomalous electrification is observed in super cell and other hail- and severe-weather producing storms and is surprisingly common, particularly in the large storms of the western and central Great Plains in the United States.

Thunderstorms exhibit cloud (including intra-cloud, cloud-to-cloud, and cloud-to-air), cloud-to-ground and cloud-to-ionosphere lightning discharges. Cloud-to-ground (CG) discharges are the most studied as a good part of them is observed from the Earth's surface. The discharges occur mostly between the main negative or positive charge center and the ground. Each flash consists of several strokes, with each stroke consisting of a leader and a return stroke; thus, negative or positive charges are brought to the ground. On the other hand, most of the intra-cloud (IC) discharges occur between the positive and negative charge centers of the main dipole. The upward discharges from cloud to the ionosphere may occur as a result of electrical breakdown between the upper storm charge and the screening charge attracted at the cloud top. They could also occur due to electrical breakdown between the main mid-level charge and a screening depleted upper-level charge that continues to propagate out of the top of the storm (Krehbiel *et al.*, 2008). The first process has been used to explain blue jets, while the second one could explain gigantic jets (Krehbiel *et al.*, 2008). Thus, recently observed optical emissions such as sprites, elves, jets, blue starters, etc.. are associated with thunderstorms (Rodger, 1999; Barrington-Leigh *et al.*, 2002; Su *et al.*, 2003). Recent studies established a link between individual positive ground flashes that stimulated sprites and the excitation of global Schumann resonances within the Earth-ionosphere cavity (Boccippio *et al.*, 1995).

Research in the past two decades has identified a surprising variety of "Transient Luminous Events" (TLEs). Amongst them the most common is the so called sprite, which is a manifestation of electrical breakdown of the mesosphere at 40 - 90 km altitude (Sentman *et al.*, 1995). Sprites are associated with positive cloud to ground lightning discharges which lower positive charges from a cloud to the ground. Blue jets are discharges propagating upwards into the stratosphere from cloud tops in a similar way to classical lightning, consisting of leaders and a return stroke. They may or may not be associated with cloud to ground lightning activity (Wescott *et al.*, 1995). Elves are concentric rings of optical emissions propagating horizontally outwards at the bottom edge of ionosphere at ~ 90 km altitude (Fukunishi *et al.*, 1996), which are caused by the electromagnetic pulse radiated by the cloud to ground discharge current of cloud to ground lightning of either polarity (Cho and Rycroft, 1998).

Gigantic jets seem to be a discharge where a blue jet triggers a sprite, creating electrical breakdown of the atmosphere from the thunderstorm clouds directly up to the bottom of the ionosphere (Su *et al.*, 2003; Pasko *et al.*, 2002). Another related event is the Terrestrial Gamma-ray Flash (TGF) with energies up to 20 MeV, which is observed in association with lightning onboard a satellite (Fishman *et al.*, 1994; Smith *et al.*, 2005; Ostgaaard *et al.*, 2008). TGFs may be bremsstrahlung radiation from upward propagating relativistic electron beams generated in a runaway discharge process powered by the transient electric field in the stratosphere and mesosphere following a lightning event. The runaway discharge process has also been suggested for the initiation of lightning and sprites (Roussel-Dupre and Gurevich, 1996), but so far no evidence of a direct connection between sprites and TGFs could be obtained.

Lightning discharges in thunderstorms radiate powerful radio noise bursts over a wide frequency range from a few Hz to several Megahertz. Radio noise in the ELF/VLF frequency range can propagate over long distances through the Earth-ionosphere waveguide. The waves with the frequency range less than 50 Hz can propagate globally with extremely low attenuation rates, allowing these radio waves to propagate a few times around the globe before dissipating into the atmosphere, whose interference results in the Earth-ionosphere cavity resonance known as Schumann resonance (SR) (Cummar *et al.*, 1996) which are not prone to interference sources such as an electric railways, mechanical vibrations of the antennas, surrounding vegetation, drifting electrically charged clouds, power line transients etc. Presently, Schumann resonances are being used to monitor global lightning activity, global variability of lightning activity and sprite activity (Siingh *et al.*, 2007). The relations between lightning and ELF noise levels on the global basis have been analyzed to study the space-time dynamics of world-wide lightning activity (Magunia, 1996). Schlegel and Fullekrug (1999) showed that solar proton events cause increase in frequency, Q-factor and amplitude of the SR modes.

The SR and GEC have been suggested to be closely linked to weather and climate (Williams, 1992; Price, 1993; Price and Rind, 1994; Harrison and Carslaw, 2003). Williams (1992) reported a positive correlation between the monthly means of the tropical surface-air-temperature anomaly and the magnetic field amplitude for the fundamental SR mode. He also demonstrated that the SR phenomena is a sensitive measure of temperature fluctuations in the tropical atmosphere (and thus generally of global climate change). Buoyancy of thunderclouds is closely linked with the precipitation and electrification processes in thunderclouds. Temperature perturbations are nonlinearly related with the cloud electrification and thus to the lightning activity in clouds.

Thunderstorms directly couple the atmosphere to the magnetosphere through electromagnetic radiation from lightning (whistlers) precipitated fluxes of the energetic electron from the radiation belts into the atmosphere. Sprites may assist the reverse process, injecting MeV electrons into the magnetosphere during the relativistic breakdown mechanism. Recently, Siingh *et al.* (2008) have summarized the phenomena taking place from the troposphere to the magnetosphere and discussed the current status of understanding of transient luminous events in the mesosphere, whistler mode signals in the ionosphere/magnetosphere, its parent-lightning discharges and thunder storms.

A thunderstorm is a complex phenomenon which has been studied during the last century and still there are various aspects which have not been explored properly. In this review paper we have tried to present the update of different phenomena associated with it and wherever possible some suggestions are also given for further studies. The importance of the subject is further increased because it controls the weather of the biosphere and is controlled by the space weather variables. Thunderstorm couples the atmosphere, ionosphere and magnetosphere through the electromagnetic processes and associated energy. The electrification of cloud and hence thunderstorm is a complex phenomena and microprocesses involved also control cloud development and hence climate (Harrison and Carslaw, 2003). Recently, observed optical phenomena above thundercloud and below the ionosphere is another aspect, which causes interest in this subject.

Thunderstorm/Lightning

The major source of the dc energy is the thunderstorm/lightning discharge. In recent years some progress has been made towards understanding thunderstorm and lightning. To explain lightning phenomena, Kasemir (1983) introduced the concept of the bidirectional uncharged leader, emphasizing that the essential factor in maintaining the lightning discharge is the continuing breakdown at the tip of the positive or negative ends of the lightning leader that extends the channel into new regions with stored electrostatic energy. The widely accepted model of thunderstorm electrification, shown in Fig. 1, is one in which conduction, displacement and precipitation current densities below the negative layer of charge in the thundercloud vary with altitude. The same figure also shows the various Transient Luminious Emissions (TLEs) of the stratosphere and mesosphere. In the region between the bottom and the top of thunderstorm cloud charging, conduction, displacement and precipitation current densities vary in space and time. In the region above the thunderstorm only conduction and displacement currents are considered. All lightning currents are considered as discontinuous charge transfers. In the fair-weather regions far away from thunderstorms, only conduction currents flow.

Spatial, Temporal and Global Distribution of Thunderstorm

Lightning activity is concentrated in three distinct zones - East Asia, Central Africa and America and it is more prevalent in the northern hemisphere than the southern hemisphere and mostly occurs over the land surface. The observation of lightning activity from space shows that two of every three lightning flashes occur in tropical region (Schlegel and Fullekrug, 1999). In addition to the tropical lightning, extra-tropical lightning activity plays significant role in the summer season in the northern hemisphere, resulting in the global lightning activity having a maximum from June to August. Fig. 2 shows the potential gradient as a function of universal time. The upper panel shows an annual diurnal variation of the potential gradient (V/m) on the oceans as measured by the research vessel Carnegie and Maud expeditions (Parkinson Torrensen, 1931). The lower panel shows the annual diurnal variation of global thunderstorm activity (Whipple and Scrase, 1936). The similarity of the diurnal variations of the electric field over the oceans and of the worldwide thunderstorm activity supports the hypothesis that thunderstorms are the main electrical generators in the GEC. The largest of the three maxima occurs at the time of maximum thunderstorm activity over the America, although this is weaker than that over Africa. This paradoxical effect has been explained by the fact that South American thunderstorms are close to the magnetic dip equator, whereas most African thunderstorm occur over the Congo at higher (southern) dip latitude (Clayton and Polk, 1977). About 200 thunderstorms are active at any time, which are mainly concentrated over the tropical land masses during the local afternoons and cover about 10 % of the earth's surface (Sato *et al.*, 2008). On the remaining 90% of the earth's surface return current (fair weather current) ~1000A (~1 pA/m^2) from the ionosphere to the earth's surface flows. There is also good relation between ac component of the global circuit (Schuman resonance) and global lightning activity (Zipser, 1994). All these studies show that the global circuit has a maximum at ~1800 hrs and a minimum at 0300 UT (Price, 1993). The correlation of Latitudinal and longitudinal distributions of lightning activity and electromagnetic VLF wave activity in the VLF range recorded at Earth surface. Recently Sato *et al.*, 2008 studied the temporal and regional variation of lightning occurrence and their relation to sprite activity and climate variability based on 1–100 Hz ELF magnetic field data obtained at the Syowa (Antarctica), Onagawa (Japan) and Esrange (Sweden) for the period from September 2003 to August 2004. They found that in the northern summer season (June to August) the lightning occurrence rates are higher in the northern hemisphere than in the southern hemisphere with large enhancements in North America, South-East Asia and the northern part of Africa. On the other hand, in the northern winter season (December to February) these rates are higher in the southern hemisphere, with large enhancements in South America, Australia and the southern part of Africa.

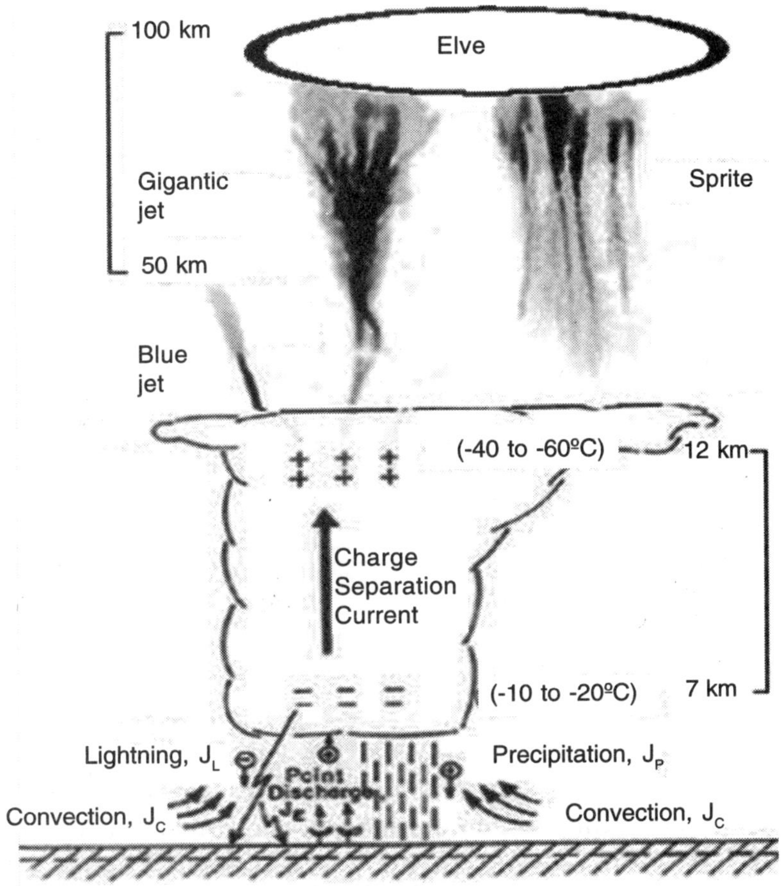

Fig. 1: Various currents that flow in the vicinity of an active thundercloud; there are five contributions of the total current. In the same figure are shown the various Transient Luminious Emissions (TLEs) of the stratosphere and mesosphere (Siingh *et al.*, 2008).

Thunderstorm / Lightning and Precipitation

Lightning and convective processes are two related phenomena of the thunderstorm. The lightning is more severe in highly convective clouds. As the electrification increases with the height of the convective clouds, the tall cumulonimbus clouds can produce severe lighting. Several workers have explored the technique to estimate rainfall directly from cloud-to-ground (CG) lightning observations. Zipse (1994) has used the ratio of monthly rainfall to number of thunderstorm day to study the rainfall and thunderstorm relation for West African region. Petersen and Rutledge (1998) used the total rain mass and CG flash

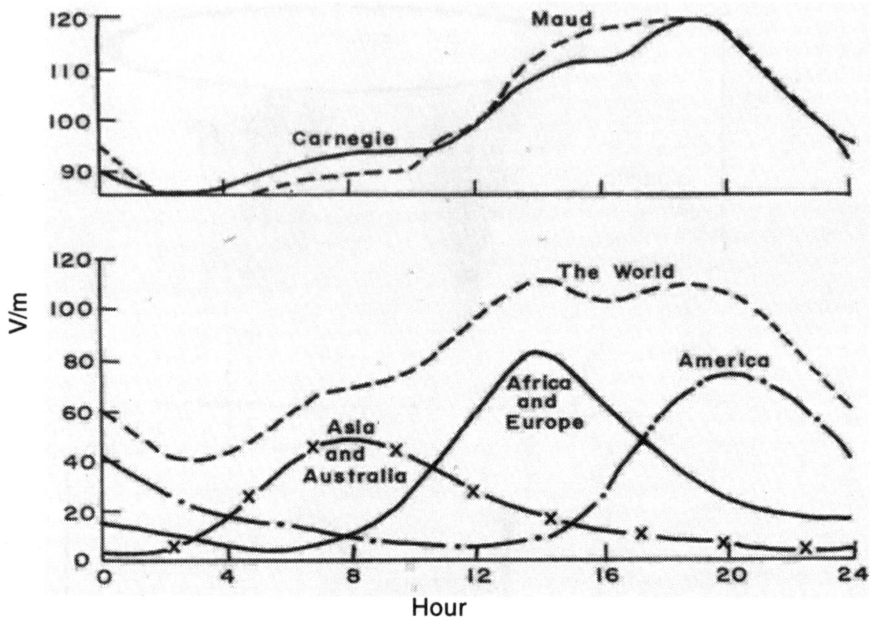

Fig. 2: Annual diurnal variation of the potential gradient measured on the surface of the oceans (Kartalev *et al.*, 2006).

density to examine the relationship over large spatial and temporal scales for several different parts of the globe. The relationship between lightning and rainfall may vary significantly, depending on air-mass characteristics and cloud microphysics. Earlier studies showed some positive relationship between lightning and area averaged convective rainfall (Chze and Sauvageot, 1997; Tapia *et al.*, 1998), and total lightning flash rate and convective rainfall (Buechlers and Goodman, 1990; Williams *et al.*, 1989). The relation between rainfall and lightning is generally expressed in terms of the Rainfall-Lightning ratio (RLR). This ratio estimates the convective rainfall volume per cloud-to-ground lightning (CG) flash. The RLR depends on thermal and microphysical characteristics of thunderstorm, its location, local climatology and convective regime (Seity *et al.*, 2001; Lang and Rutledge, 2002; Williams *et al.*, 1992; Williams *et al.*, 1993) and varies over a wide range. The nature of the relationship is less certain over the oceans and in particular, over the tropical oceans (Piepgrass *et al.*, 1982; Buechlers and Goodman, 1990). Rakov and Uman, 2003 have shown that the RLR ranges from 2×10^4 m^3 per CG flash to 2×10^7 m^3 per CG flash on various scales of space and time in variety of geographical locations. It has been also observed that severe storms produce lower RLR values than ordinary thunderstorms.

The lightning-rainfall relationship has been used by some researchers to improve the skill of numerical forecast. For example, Alexander *et al.* (1999) found relatively good correlation between convective rainfall and lightning rates on 1993 Super storm and showed improved numerical forecast by assimilating latent heating rates derived from lightning data. On the basis of the comparison made between the lightning rate measured by Pacnet and convective rainfall data obtained from TRMM's (Tropical Rainfall Measuring Mission) microwave sensors for a variety of storm systems over the central north Pacific. Pessi *et al.* (2004) suggested that the lightning data over the Pacific can be assimilated into numerical models as a proxy for latent heat release in deep convective clouds.

Thunderstorm and Atmospheric Aerosoles

The aerosol particles in the atmosphere partly act as nucleation centers for cloud formation and partly as ice forming nuclei. The number of aerosol particles which can serve as cloud condensation nuclei (CCN) increases with increasing super saturation because even smaller particles participate in nucleation at higher super saturation value. Observations show that continental air masses are generally richer in CCN than the maritime air masses. Some fraction of the aerosol particles also serves as ice forming nuclei.

Aerosol particles are generated by gas-to-particle (GPC) or drop-to-particle conversion processes. Trace gases also play a significant role through the aquous-phase chemistry of the atmosphere and contribute to the generation and destruction processes of aerosols. The Aitken nuclei produced by gas-to-particle reactions under the super saturation condition are likely to become cloud nuclei (Heintzenberg *et al.,* 1989). *Hegg et al.,* 1980 observed a shift of the particle size spectrum towards larger size on passing from the upstream to downstream side of a cloud. A numerical simulation of the growth and subsequent evaporation of a convective cloud also produced similar result. Heintzenberg *et al.* (1989) observed a pronounced shift of the size distribution to the larger size due to the processing of the clear air particles by the cloud. Hoppel *et al.* (1994) proposed that drop-to-particle conversion is the cause for double maximum found in maritime aerosol particle spectra.

Variety of aerosol particles and trace gases vastly differing in their physical and chemical characteristics are removed from the atmosphere across of nucleation scavenging (incorporated in cloud drops, raindrops or ice crystals), impact scavenging (collected by collision with the cloud drops, raindrops or ice crystals) and gas scavenging (absorption of trace gases by cloud particles and raindrops). As a result of various scavenging processes and because of varying solubility of different particles and trace gases components at different temperatures, both gas-phase and aquous-phase chemistry of the atmosphere becomes complex. The whole variety of anthropogenic aerosol particles and trace gases injected into the atmosphere from man-made sources makes the situation still more complex. Chemically active constituents of the system undergo various chemical reactions and significantly change the condensation, freezing, coalescence and breakup, properties of the cloud drops, raindrops and ice particles in the cloud. Thus, the chemical characteristics of the air and the cloud particles can significantly alter the microphysical properties of the cloud and their study can help in understanding important phenomenon like acid rain, which adversely affects the life, structure and agriculture on the surface of the earth.

It is difficult to distinguish the effects of aerosols from that of thermodynamic/dynamic on the thunderstorm activity effects. Williams *et al.* (2002) conducted experiments to verify the enhanced lightning activity predicted by the aerosol hypothesis and arrived at different results during the lightning-active pre-monsoon and during the less active wet season in Brazil. Williams and Stanfill (2002) supports the thermal hypothesis to explain the lightning activity variation with island area acting in oceans as the heat islands in the continent. Lyons *et al.* (1998) and Murray *et al.* (2000) attribute the enhancement in the positive cloud-to-ground flashes in North America to incursion of smoke from fires in Mexico and subsequent ingestion by these thunderstorms. However, the effect of smoke from biomass burning on the thunderstorm activity observed in Brazil does not support the above result (Williams *et al.*, 2002). Mac Gorman and Burgess (1994) reported clustering of positive ground flashes below storms which developed in strong instability.

Thunderstorm and Global Electric Circuit

The main source of electric fields and currents in the global electric circuit (GEC) are thunderstorms in the troposphere and dynamo situated in the ionosphere and magnetosphere. For detailed study about the GEC and thunderstorm, the readers may refer the recent reviews by Siingh *et al.*, 2005, 2007 and Tinsley, 2008.

Thunderstorm/Lightning and Climate

Recently, the importance of lightning in climate is recognized. For detailed study in this area readers may refer the recent reviews by Williams (2005) and Siingh *et al.* (2007).

Recent researches show a warmer climate in the future (Price, 2006). The key questions are related to the impact of global warming on thunderstorms, and severe weather. Will thunderstorm activity increase in a warmer world? Since the majority of global thunderstorm activity occurs in the tropics, changes in future thunderstorm activity will depend on changes in the tropical climate. We would expect a drier climate to produce fewer thunderstorms, and less lightning. However, experimental and modeling studies have shown that as tropical regions dry in the present climate, they experience greater lightning activity. This paradox may be explained by noting that while drier climate conditions result in fewer thunderstorms and less rainfall, the thunderstorms that do occur are more explosive, resulting in more lightning activity. The predicted drying of the Mediterranean region in the future may also imply increase in the intensity of thunderstorm and lightning activity in the future.

Sprites

The sprites phenomena are one of four currently known forms of TLEs occurring at high altitude in the Earth atmosphere, which are related to the lightning activity in underlying thunderstorms. Sprites may occur in clusters of two, three or more "carrot" shaped emissions of ~1 km thickness over a horizontal distance of 50-100 km, with the separation between sprite elements of ~10 km (Neubert *et al.*, 2008). Sprites may also occur as single luminous columns termed C-sprites (Westcott *et al.*, 1998). For example, a group of carrot sprites image observed at Hungary during the ILAN Campaign (2005-2006) on January 14, 2006 at 20:12:08.0602 UT are shown in Fig. 3. Telescopic imaging revealed sprites with intertwined discharge channels and beads to scales down to 80 m diameter and smaller (Gerken *et al.*, 2000; Mende *et al.*, 2002). The optical intensity of a sprite cluster as observed by TV-frame rate cameras is comparable to that of a moderately bright auroral arc (Sentman *et al.*, 1995). The brighter region is in the altitude range 65-85 km with most of the intensity in the red, and with wispy faint blue tendrils extending down to 40 km or at times as low as the cloud top (Wescott, 2001). High speed photometry shows that the duration of sprites is from a few ms to ~200 ms (Armstrong *et al.*, 1996; Winckler *et al.*1998). High speed imaging reveals that the discharges are initiated at ~65 km altitude; they propagate downwards and shortly after upwards (Stanley *et al.*, 1999; Moudry *et al.*, 2003). Recent imaging at 10 kHz frame rates shows the formation and propagation of streamers and resolves the streamer heads (McHarg *et al.*, 2007). Stenback-Nielsen *et al.* (2007) showed that sprite emission rate on the shortest time scale can reach 1 - 500 GR.

Fig. 3 : A group of carrot sprites image as was captured on January 14, 2006, 20:12:08.0602 UT at Hungary during the ILAN campaign 2005-2006 (Greenberg *et al.*, 2007).

The generation of sprites depends on the critical level of charge moment change of thunderstorms (Huang *et al.*, 1999; Williams, 2001; Lyons, 2008; Sato *et al.*, 2002). The electric field from upward discharges (Krehbiel, 1986) decreases as the inverse of the square of distance where as the dielectric strength follows the density of the air which decreases exponentially. Under this circumstances there may exist an altitude range, well above thunderstorms, where the imposed electric field exceeds the dielectric strength and initiates sprites (Williams, 2001; Wilson, 1925). The conductive atmosphere above 90 km prevents luminous breakdown and hence sprites may not be observed above 90 km.

Based on ELF/VLF observations, Ohkubo *et al.* (2005) suggested that IC discharges play a significant role in sprite generation. The IC flash and sferic activity is caused by breakdown processes inside the clouds feeding the continuing currents of +CG discharges. Using the Euro-sprite data, this point has been examined by Neubert *et al.* (2008), who showed that column sprites are generally characterized by a short time delay relative to the causative +CG discharge with high peak current, little IC activity and short, intense bursts of broadband VLF radio wave activity, whereas carrot sprites are associated with longer time delays, large IC activity and weaker, longer lasting bursts of lightning discharges. This shows that IC discharges play an important role in the generation of carrot sprites but may not play significant role for the impulsive column sprites. Neubert *et al.* (2008) have presented a sprite which is laterally displaced from the +CG and have proposed a process in which a +CG discharge may transfer charge from a remote region of the storm to the ground. A laterally displaced sprite was also reported by Mazur *et al.* (1998).

Fig. 4 shows the illustration of different phenomena (top panel) and theoretical mechanisms (bottom panel) of lightning-ionosphere interactions operating at different altitudes and producing optical emissions (Chern *et al.*, 2003).

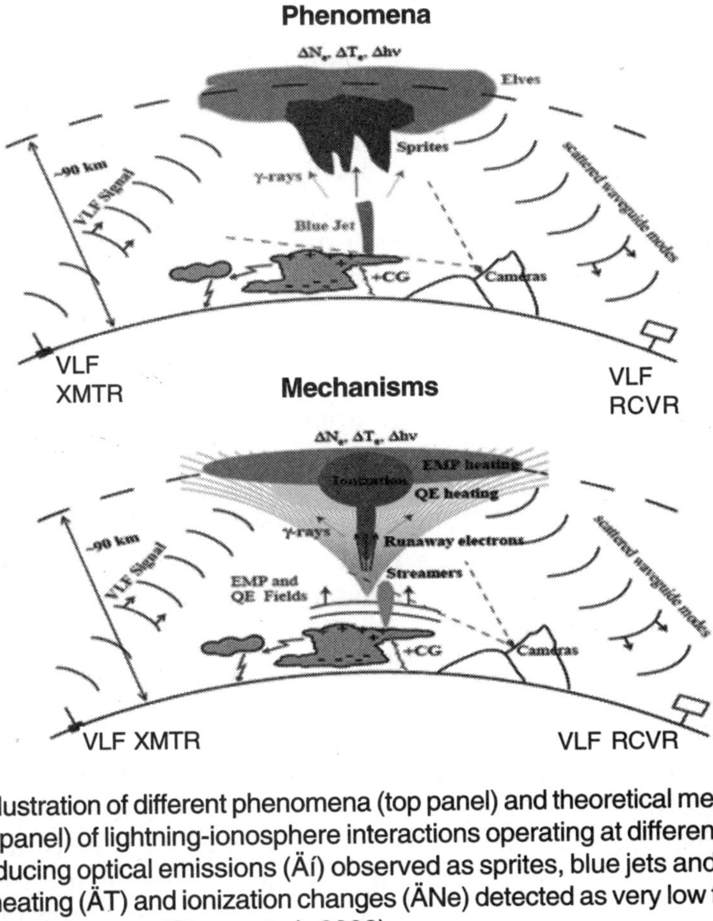

Fig. 4: Illustration of different phenomena (top panel) and theoretical mechanisms (bottom panel) of lightning-ionosphere interactions operating at different altitudes and producing optical emissions (Äí) observed as sprites, blue jets and elves, as well as heating (ÄT) and ionization changes (ÄNe) detected as very low frequency (VLF) signal changes (Chern *et al.*, 2003).

The experimental evidence indicating strong upward electrodynamics coupling of thunderstorms to the mesosphere and lower ionosphere includes early/fast perturbations of sub-ionospherically propagating VLF signals associated with lightning discharges in underlying thunderstorms (Inan *et al.*, 1988), and optical emissions in clear air above thunderstorms associated with sprites (Sentman *et al.*, 1995; Lyons, 1994, 1995,1996; Boeck *et al.*, 1994; Rairden and Mende, 1995; Winckler *et al.*, 1996), blue jets (Wescott *et al.*, 1995), gigantic jets (gigantic jets are special phenomena of the oceanic thunderstorms) (Su *et al.*, 2003) and airglow enhancements (Winckler *et al.*, 1996).

Rowland (1998) has studied lightning driven electric fields at high altitudes and has shown that the electric field is sufficient to cause thermal breakdown and runway breakdown at the height corresponding to the observation of sprites and elves. The latter are believed to be closely associated with recently discovered rapid (<1 ms) optical emissions at 80-95 km altitudes with lateral extents up to 300 km preceding sprites and referred to as "elves" (Rairder and Mende, 1995; Fukunishi, *et al.*, 1996; Inan *et al.*, 1995). Sprites are clearly associated both temporally and spatially with intense lightning discharges in underlying thunderstorms (Boccippio *et al.*, 1995) as well as with early/fast VLF events (Inan *et al.*, 1995) providing strong evidence of significant changes in the mesospheric electrical conductivity. Sprites are almost uniquely associated with positive cloud-to-ground lightning discharges (Boccippio *et al.*, 1995; Lyons, 1996; Winckler *et al.*, 1996). The upper portion of the sprite is red, with wispy, faint blue tendrils extending to 40 km or lower. Boccippio *et al.* (1995) showed that about 80% of sprites are associated with ELF transient events and +ve CG lightning return strokes having large peak current (>35 kA)[96] and large total charge moment change of the thunderstorm ($ÄM_Q$) values. Some sprites associated with –ve CG lightning have also been observed (Barrington-Leigh *et al.*, 1999) and explained by Williams *et al.* (2007). Recently, Soula *et al.* (2009) have reported observation of sprites during the summers of 2003 to 2006 over thunderstorms in France. They further reported that all sprites were associated with +CGs except one which was observed after –CG discharge. The first observations of gamma ray bursts of terrestrial origin (Fishman *et al.*, 1994; Fishman and Inan, 1988) associated with positive cloud-to-ground lightning discharges (Inan *et al.*, 1996), and intense wideband VHF bursts (Holden *et al.*, 1995) constitute additional new examples of upward coupling of energy originating in thunderstorms.

Pasko *et al.* (2002) have reported a video recording of a blue jet propagating upwards from a small thundercloud cell to an altitude of about 70 km. As relatively small thundercloud cells are very common in the tropics, it is probable that optical phenomena from the top of the clouds may constitute an important component of the GEC. This becomes more pertinent in view of the fact that power supplied by thunderstorms is insufficient to maintain a field of the magnitude observed in fair-weather regions. However, the inclusion of the effect of sprites and other optical phenomena also could not be explained because, optical phenomena occur in the upward branch of the global electric circuit above the thunderstorms and they are likely to influence only the upper atmosphere conductivity. Moreover, since they occur much less frequently (only one sprites out of 200 lightning) because of their association with intense lightning discharges (Siingh *et al.*, 2007; Rycroft *et al.*, 2000), they may not play a major role in GEC (Rycroft *et al.*, 2000). Since optical emissions could change electrical properties of the atmosphere and influence processes related with weather and climate, intense research activity in this area is required. Further, thunderstorms are also the source of Schumann resonance, which control the electrodynamics of the lower atmosphere.

The sprites provide a link between tropospheric processes in the thunderstorms and mesospheric processes in the upper atmosphere. Hiraki *et al.*, 2002 suggested that sprites may change the concentration of NO_x and HO_x in the mesosphere and lower atmosphere. These chemical changes have impact on the global cooling or heating in the middle atmosphere (Galloway *et al.*, 2004; Schumann Huntriese, 2007). Nitrogen oxides are critical components of the troposphere which directly affect the abundance of ozone and hydroxyl radicals (Crutzen, 1974). Ozone absorbs solar ultraviolet radiation and controls the dynamic balance of the atmosphere. NO_x creates ozone in the troposphere and destroys it in the stratosphere and mesosphere. The concentration of NO_x in the mesosphere is enhanced by transient events such as auroras and solar proton events (Crutzen and Solomon, 1980). The vertical transport of NO_x by the neutral wind is an important process to control NO_x concentration in the thermosphere (Saetre *et al.*, 2007). TLEs also affect the concentration of NO_x in the stratosphere and mesosphere and the processes involved could be similar to those in the aurora. The production of NO_x is through energetic neutral atom conversion where the heating of the atmosphere by the continuing current in the ion channel of the lightning stroke allows suprathermal oxygen atoms to react directly with N_2 (Balakrishnan and Dalgarno, 2003).

The observations listed above have led to the discovery of new interaction mechanisms, which result in troposphere-mesosphere/ lower ionosphere coupling. These mechanisms are based on heating of the ambient electrons by electromagnetic pulses generated by lightning discharges (Inan *et al.*, 1991; Milikh *et al.*, 1995; Rowland *et al.*, 1995) or by large quasi-electrostatic thundercloud fields (Boccippio *et al.*, 1995; Rairden and Mende, 1995; Pasko *et al.*, 1995), and on runaway electron processes (Rairden and Mende, 1995; Bell *et al.*, 1995; Roussel-Dupre and Gurevich, 1996; Tarenenko and Roussel-Dupre, 1996). It has been proposed by Inan *et al.*, 1996 that quasi-electrostatic thundercloud fields are capable to maintain the ionospheric electrons at a persistently heated level well above their ambient thermal energy. Changes in the thundercloud charge (*e.g.*, in lightning discharges) lead to heating/cooling above/below this quiescent level, and are registered as early/fast VLF events (Inan *et al.*, 1996). The simultaneous observations of early/fast VLF events and sprites (Inan *et al.*, 1995) indicate that these two effects may be a manifestation of the same physical process consistent with model predictions of both optical emissions, heating and ionization changes associated with quasi-electrostatic thundercloud fields (Inan *et al.*, 1995, 1996). Some of these physical processes are illustrated in Fig. 4 (Pasko, 1996). It should be noted that in existing scientific literature the possibility of breakdown ionization of the upper atmosphere by thundercloud fields was first mentioned in 1925 by *C.T.R. Wilson*. He recognized that the relation between the thundercloud electric field which decreases with altitude z as $\sim z^{-3}$ and the critical breakdown field which falls more rapidly (being proportional to the exponentially decreasing atmospheric density) leads to the result that "there will be a height above which the electric force due to the cloud exceeds the sparking limit" (Wilson, 1925). Three decades later *Wilson* speculated that a discharge between the top of a cloud and the ionosphere might often be the normal accompaniment of a lightning discharge to ground (Wilson, 1956). Pasko (1996) investigated in detailed the heating, ionization, attachment and optical emission effects associated with intense quasi-electrostatic thundercloud fields existing above large thunderstorms before (producing blue jets) and after (generating sprites) lightning discharges.

From the above discussion on the sprites, mainly focused in mid latitude thunderstorms, while the large majority of thunderstorms occur in the tropics. Also few studies have been performing the global sprites distribution from the space and satellite (Chen *et al.*, 2003; Yair *et al.*, 2004). The field experiments needs to be expanded to additional regions of the globe to examine how does tropical sprites activity differ from mid latitude sprites activity and how do the parent lightning flashes that trigger sprites differ between regions. Winter thunderstorms are generally smaller than the summer thunderstorms, but winter thunderstorms are also producing the sprites. What is special about the winter storms that produced the sprites? Generally, it was found that positive polarity lightning produced the sprites on land as well as on ocean. Is this always true? Studies showed that many negative discharges over the oceans (Fullerkrug *et al.*, 2002; Greenberg and Price, 2004), few sprites observations have been targeted above oceanic thunderstorms. Is any physical characteristics of sprites differs due to the type of thunderstorm, region, season, and types of lightning? Finally the main important issue for scientific community is how important are sprites and their associated lightning to the GEC? Are sprites producing storms, the main contributor to the GEC? These are some unresolved problems of the sprites and related phenomena.

The readers may refer the recent article by Lyons (2006) for many more outstanding research questions related to TLEs.

Global Occurrence of Sprites

Global occurrence rate of sprite in the summer, fall and winter seasons shows that the high occurrence regions of sprites are concentrated in the major lightning source regions (Sato and Fukunishi, 2003). In the summer season, sprite occurrence rates are higher in the northern hemisphere than in the southern hemisphere with large enhancements in North America and South-East Asia. On the other hand, these rates are higher in the southern hemisphere in the winter season with large enhancements in South America and Africa. It is also estimated that sprites occur with rates of ~113 and ~190 events/day in North America and South-East Asia in summer season, respectively, while they occur with a rate of ~240 events/day in Africa in the winter season (Williams, 2007 a,b).

Sprites and Schumann Resonance

Natural electromagnetic field variations in the 3-120 Hz transition zone, characterized by interference of propagating wave in the Earth-ionosphere cavity, called Schumann Resonance (SR) are mainly excited by globally occurring cloud to ground and cloud to ionosphere discharges. The vertical electric and horizontal magnetic fields measurements at a single station are being used to locate the source. Since it is well recognizes that Schumann Resonances are accompanied with sprite lightning (Fig. 5, Rycroft *et al.*, 2000), the measurement of the former at a long distance can yield accurate information about the latter. During the explanation of the results of day-night asymmetry, hypothesis of cavity has been introduced to be responsible for systematic errors in the source-receiver distance measurements based on uniform cavity model. The uniform cavity model itself is under question because of local/transient charge fluctuations at the boundary and day-night asymmetry. Even the transient natures of Schumann Resonance wave-fields are not properly used. Recent measurements show polarity asymmetry of sprite producing lightning (Williams *et al.*, 2007a, b). This also needs a well focused study of probing potentiality of Schumann resonance phenomena.

Considerable evidence has accrued that a consideration of the day-night asymmetry is needed in integrating local field intensities for the global invariants. Sentman and Fraser (1991) showed substantial improvements in the summation of magnetic intensities at two distant SR stations when corrections were made for the variations of local ionospheric height. Sátori *et al.* (2008) have shown that a proper ordering of the lightning source strengths in the three dominant tropical chimneys also requires consideration of the day-night asymmetry.

The accurate treatment of this asymmetry with the 2D transmission line model (Kirillov *et al.*, 1997) requires values for the two frequency-dependent characteristic heights (Greifinger *et al.*, 1878). The lower characteristic height for daytime and nighttime conditions has been worked out (on the basis of measured ionospheric profiles) by Greifinger *et al.* (2007).

Forward problem in SR is the algorithm used to compute field intensities worldwide, given a complete knowledge of the lightning sources. The forward problem for a uniform cavity is analytic in nature. However, the real cavity is demonstrably non-uniform and problem becomes complex. But using the transmission line approach, accurate

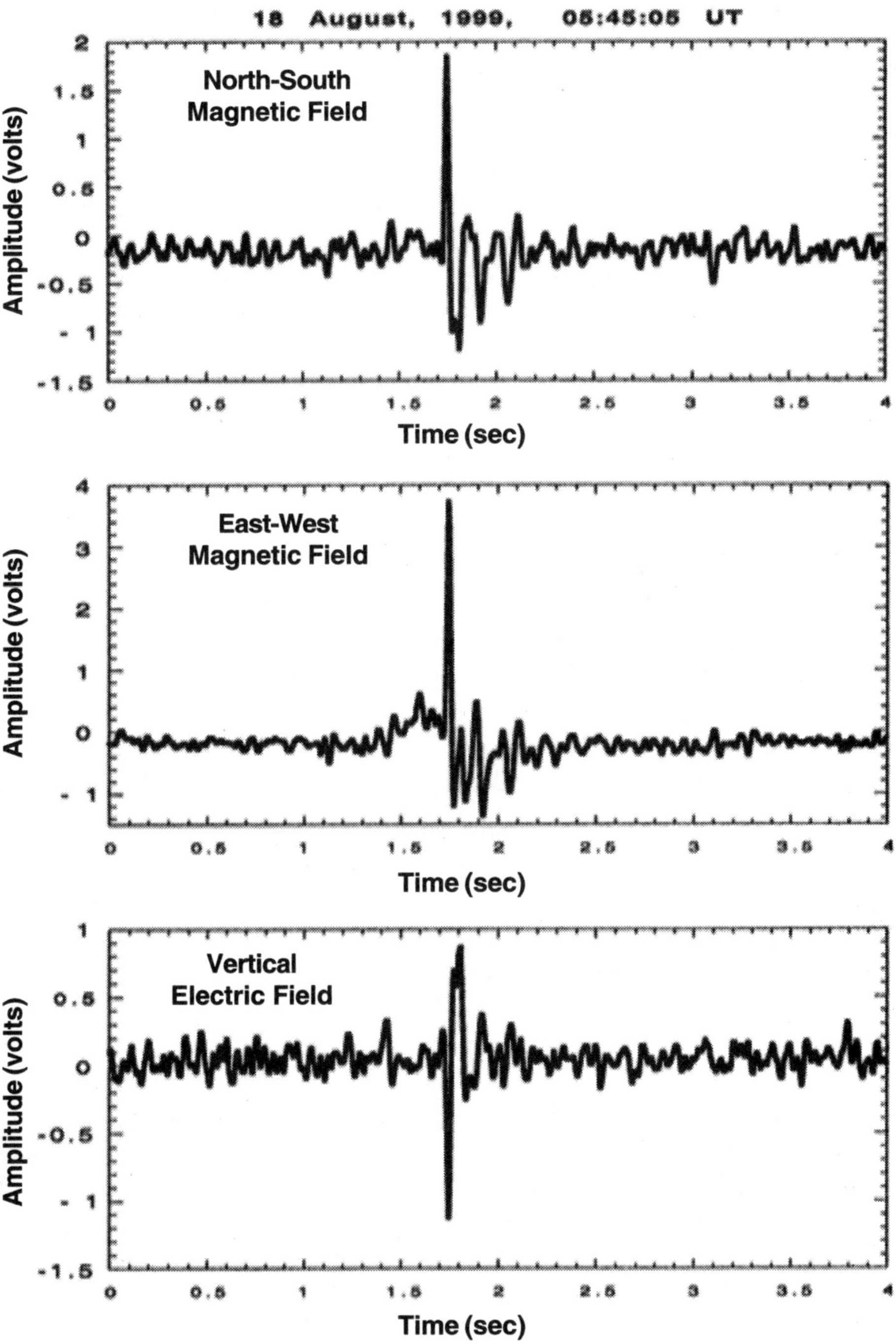

Fig. 5: Example of the ELF radio signal (SR mode) observed in Israel (associated with the sprites generated in United State (Galloway *et al.*, 2004).

models are available as equivalent networks (Madden and Thompson, 1965), and as such analytical results are available (Kirillov *et al.*, 1997). Recent results of forward simulations support the accuracy of the latter non-uniform models (Sátori *et al.*, 2008).

Electromagnetic transients have been recorded at the Massachusetts Institute of Technology field station in West Greenwich, Rhode Island (USA), coinciding with sprite producing lightning flashes in northern Australia, at a distance of 16.6 Mm. Single-station Schumann resonance methods have been used to locate the parent lightning flashes and to evaluate their vertical charge moments (Williams, 2007a). The charge moment thresholds for sprite production are consistent with similar measurements with identical methods made at considerably closer range (~2 Mm). The use of a uniform model for the Earth-ionosphere waveguide can produce systematic errors in the source-receiver distance, of the order of 1 Mm. Schumann resonance ELF methods have been used to measure the charge moments of millions of flashes worldwide with sufficient accuracy.

The measurements of Schumann Resonance wave fields can also be used to study its effects on the ionospheric height variations, on the eigen frequencies from transient perturbation to the ionosphere, contribution from excitation sources other than cloud to ground lightning, apart from locating the lightning/sprites (Williams, 2007a,b).

Transient optical emissions in the mesosphere and ionosphere denoted sprites and elves respectively are also associated with strong positive cloud-to-ground discharge (Boccippio *et al.*, 1995). Sprites are electrical discharges in the mesosphere induced by energetic lightning flashes in the troposphere have been observed in the ELF range (Cummer and Inan, 1997), in the lower ELF range (Boccippio *et al.*, 1995) and in the ELF/ULF transition range (Fullekrug *et al.*, 1996). Now the question is "Is there any relation between the global excitations of the Earth-ionosphere cavity resonance and sprites? Fullerkrug and Reising (1998) and Bell *et al.* (1998) reported sprites above the mesoscale convective system in the Midwestern United State on 1 August, 1996 by low light level TV camera observations, Stanford university Fly's Eye experiment at Yucaa Ridge, Colorado. During the time interval 0630 -0813 UT, 30 lightning flashes met this criterion and excited Earth-ionosphere cavity resonance, verified by the National Lightning Detection Network (Fullerkrug and Reising, 1998). These lightning flashes are mainly associated with positive cloud discharges with peak currents 20-70 kA and 80% of these lightning flashes were associated with sprites. The probability of sprites detection by global excitation of Earth-ionosphere cavity resonances can be explained by high current associated with those lightning discharge with together excite Earth-ionosphere cavity resonance and sprites (Fullerkrug and Reising, 1998). From this result we can say that its relation of global excitations of Earth-ionosphere cavity resonance and sprites.

Sprites and Global Electric Circuit

The impact of sprites on the GEC has not yet been quantified, even though the quasi-static atmospheric electric field plays an important role in the global climate system (Carslaw *et al.*, 2002). The most global DC atmospheric electric field ~150 V/m is maintained by thunderstorm electric field. The sprites contribution to the global DC atmospheric electric field may be similar to the contribution from particular intense lightning discharges ~5-120mV/m (Fullekrug, 2004). For measuring the contribution of sprites towards the global DC atmospheric electric field it is required very high sensitivity electric field mills. Fullekrug and Rycroft (2006) proposed a new methodology to infer the global DC atmospheric field of individual sprites from conventional global dynamic (AC) electric field measurements in the ULF/ELF range made with radio wave antennas at frequencies = 4Hz. They found that global atmospheric electric field from individual sprites is = 44 mV/m or ~ 3×10^{-4} E_z, where E_z = 150V/m is the total global DC atmospheric electric field.

Is there any effect of sprites in the GEC or just beautiful natural phenomena like rainbow? The question is a challenge to our scientific community to find out the possible influence of sprites on GEC or our environment. This question also have large issue to address the possible connection between atmospheric electrical properties and climate and also how modulated the electrical conductivity and cloud nucleation rates by cosmic radiation (Carslaw *et al.*, 2002).

Conclusion

The modern developments experimental and computational facilities have enhanced our understanding of the complex and exciting field of thunderstorm, lightning and sprites activity and its impact on atmospheric phenomena. In this short review, we have attempted to discuss our present understanding of sprites and its related phenomena in the lower atmosphere. It is emphasized that major experiment should be carried out in low latitude where the occurrence of thunderstorm is maximum, to study sprites and other Transient Luminous Events.

Acknowledgments

The work is partly supported by ISRO, Bangalore and partly by DST, New Delhi. We are thankful to reviewer for their valuable comments and suggestions.

References

Alexander, G. D. Weinman J. A.,Karmpudi, V. M., Olson, W.M., Lee A.C.L. (1999) The effect of assimilating rain rate derived from satellites and lightning on forecasts of the 1993 superstrom. Mon Weather Rev., v. 127, pp. 1433-1457.

Armstrong, R. A., Shorter, J. A., Taylor, M. J., Suszcynsky, D. M., Lyons, W. A. and Jeong, L. S. (1998) Photometric measurements in the SPRITES '95 and '96 campaigns: nitrogen second positive (399.8 nm) and the first negative (427.8 nm) emission. J. Atmos. Terr. Phys. v. 60, pp.787-799.

Baker, M. B., Blyth, A. M., Christian, H. J., Latham, J. Miller, K. L. and Gadian, A. M. (1999) Relationships between lightning activity and various thunderlcoud parameters: Satellite and modeling studies. Atmos. Res., v. 51, pp. 221-236.

Balakrishnan, N. and Dalgarno, A. (2003) Nitric oxide production in collisions of hot O (3P) atoms with N_2. J Geophys. Res., v. 108 (A2) doi:10.1029/2002JA009566.

Barr, R. Llanwyn, J. D. and Rodger, C. J. (2000) ELF and VLF radio waves. J. Atmos. Solar-Terr Phy., v. 62, pp. 1689-1718.

Barrington-Leigh C. P., Inan, U. S., Stanley, M. and Cummer, S. A. (1999) Sprites triggered by negative lightning discharge. Geophys. Res. Lett., v. 26, pp. 3605-3608.

Barrington-Leigh, C. P., Pasko, V. P. and Inan, U. S. (2002) Exponential relaxation of optical emissions in sprites. J. Geophy. Res., v.107(A5), pp.1065 doi:10.1029/2001JA900117.

Bell, T. F., Pasko, V. P. and Inan, U. S. (1995) Runaway electrons as a source of Red Sprites in the mesosphere. Geophys. Res. Lett., v. 22, pp. 2127.

Bell, T. F., Reising, S.C. and Inan, U. S. (1998) Intense continuing currents following positive cloud-to-ground lightning associated with red sprites. Geophys. Res. Lett., v. 25 , pp.1285–1288.

Boccippio, D. J., William, E. R., Heckman, S. J., Lyons, W. A., Baker, I. T. and Boldi, R. (1995) Sprites ELF transients, and positive ground strokes. Science, v.269, pp.1088-1091.

Boeck, W. L., Vaughan, O. H., Blakeslee, R. J., Vonnegut, B., Brook, M. and McKune, J. (1994) Observations oflightning in the stratosphere. J. Geophys. Res., v. 100 (D1, pp.1465-1475).

Buechlers, D. E. and Goodman, S. J. (1990) Echo size and asymmetry; impact on NEXRAD storm identification. J. Appl. Meteoro., v. 29, pp. 962-969.

Carslaw, K. S., Harrison, R. G. and Kirkby, J. (2002), Cosmic rays, clouds, and climate. Science (USA), v. 298 , pp. 1732-1737.

Chern, J. L., Hsu, R. R., Su, H. T., Mende, S. B., Fukunishi, H., Takahashi, Y., Lee L, C. (2003) Global survey of upper atmospheric Transient Luminous Events on the ROCSAT-2 satellite. J. Atmos. Solar. Terr. Phys., v. 65, doi.1029/ 2002JD002347.

Chze, J. L. and Sauvageot, H. (1997) Area average rainfall and lightning. J. Geophys. Res., v. 102, pp. 1707-1716.

Clayton, M. D. and Polk, C. (1977) Diurnal variation and absolute intensity of worldwide lightning activity, September 1970 to May 1971'. *In: H. Dolezalek and R. Reiter (eds.)* Electrical Processes in Atmospheres (Verlag, Darmstadt, Germany), pp.423- 438.

Crutzen, P. J. and Solomon, S. (1980) Response of mesospheric ozone to particle precipitation. Planet Space Sci., v. 28, pp.1147-115.

Crutzen, P. J. (1974) Photochemical reactions initiated by and influencing ozone in unpolluted tropospheric air. Tellus, v. 26, pp. 47-57.

Cummer, S.A. and Inan, U. S. (1997) Measurement of charge transfer in sprite producing lightning using ELF radio atmospheric. Geophys. Res. Lett., v. 24, pp. 1731-1734.

Cummer, S. A., Inan, U. S., Bell, T. F. and Barrington-Leigh, C. P. (1998) ELF radiation produced by electric currents in sprites. Geophys. Res. Lett., v. 25, pp.1281-1285.

Cho, M. and Rycroft, M. J. (1998) Computer simulation of electric field structure and optical emission from cloud top to ionosphere. J. Atmos. Solar-Terr Phys., v.60, pp.871-888.

Davydenko, S. S., Mareev, E. A., Marshall, T. C. and Stolzenburg, M. (2004) On the calculation of electric fields and currents of mesoscale convective systems. J. Geophys. Res., v.109, 10.1029/2003JD003832.

Fullekrug, M., Price, C. and Williams, E. R. (2002) Intense oceanic lightning. Ann. Geophys., v. 20, pp. 133.

Fukunishi, H. Y., Takahashi, Y., Kubota, M., Sakanoi, K., Inan, U. S. and Lyons, W. A. (1996) Elves: lightning-induced transient luminous events in the lower ionosphere. Geophys. Res. Lett., v. 23, pp. 2157-2160.

Fukunishi, H. Y., Takahashi, Y., Kubota, M., Sakanoi, K., Inan, U. S. and Lyons, W. A. (1996) Elves: lightning-induced transient luminous events in the lower ionosphere, Geophys. Res. Lett., v.23, pp.2157-2160.

Fishman, GJ., Bhat, P.N., Mallozzi, R., Horack, J.M., Koshut, T., Kouveliotou, C., Pendleton, G.N., Meegan, C.A., Wilson, R.B., Paciesas, W.S., Goodman, S.J., Christian, H.J. (1994) Discovery of intense gamma-ray flashes of atmospheric origin. Science, v. 164, pp.1313-1317.

Fullekrug, M., Reising, S. C. and Lyons, W. A. (1996) On the accuracy of arrival azimuth determination of sprite associated lightning flashes by Earth-ionosphere cavity resonances. Geophys. Res. Lett., v. 23(25) , pp.3691.

Fullekrug, M. and Reising, S.C. (1998) Excitation of Earth–ionosphere cavity resonances by sprite-associated lightning flashes. Geophys. Res. Lett., v. 25 (22), pp. 4145.

Fullekrug, M. (2004) The contribution of intense lightning discharges to the global atmospheric electric circuit during April 1998. J. Atmos. Solar-Terr Phys., v. 66, pp. 1115 -1119.

Fullekrug, M. and Rycroft, M. J. (2006) The contribution of sprites to the global atmospheric electric circuit. Earth Planets Space, v. 58, pp. 1193-1196.

Galloway J. M. Dentener, F .J. Capone, D. G. Boyer, E. W. Howarth, R. W. Seitzinger, S. P. Asner, A. G. Cleveland, C.Green, P. Holland, E. Karl, D. M. Michaels, A. F. Porter, J. H. Townsend, A. and Vorosmarty, C. (2004) Nitrogen cycles: Past, Present and Future. Biogeochemistry, v. 70, pp. 153-226.

Gerken, E. Inan, U. and Barrington-Leigh, C. (2000) Telescopic imaging of sprites. Geophys. Res. Lett., v. 27, pp. 2637-2640, doi:10.1029/2000GL000035.

Greenberg, E. and Price, C. (2004) A global lightning locations algorithm based on the electromagnetic signature in the Schumann resonance band. J. Geophys. Res., v.109 doi:10.1029/2004JD004845.

Greenberg, E., Price, C., Yair, Y., Canot, M., Bore, J. and Satori, C. (2007) ELF transients associated with sprites and elves in Eastern Mediterranean winter thunderstorm. J. Atmos. Solar Terr. Phy., v. 69, pp 1569-1589.

Greifinger, C. and Greifinger, P. (1978) Approximate method for determining ELF eigenvalues in the Earth-ionosphere waveguide. Radio Science, v.1 , pp.831-837.

Greifinger, P. S., Mushtak, V. C. and Williams, E. R. (2007) On modeling the lower characteristic ELF altitude from aeronomical Data. Radio Science, v. 2 RS2S12, doi: 10.1029/2006RS003500.

Harrison, R. G. and Carslaw, K. S. (2003) Ion-aerosol-cloud processes in the lower atmosphere. Rev. Geophys., v. 41, doi:10.1029/2002RG000114.

Hegg, D. A., Hobbs, P. V. and Radke, L. F. (1980) Observations of the modification of cloud condensation nuclei in wave cloud. J. Rech. Atmos., v. 14, pp. 217-222.

Heintzenberg, J., Ogren, J.A.,Noone, K.J.,Gardneus, L.(1989) The size distribution of submicrometer particles within and about stratocumulus cloud droplets on Mt Areskutan, (Sweden).v. 24, pp.89-101.

Hiraki, Y., Lizhu, T., Fukunishi, H., Nambu, K. and Fujiwa, H. (2002) Development of a new numerical model for investigation the energetic of sprites. Eos. Tran. AGU 83(47) O Fall Meet. Suppl. Abstract A11 60105.

Holden, D. N., Munston, C. P. and Davenport, J. C. (1995) Satellite observations of transionospheric pulse pairs. Geophys. Res. Lett., v. 22, pp. 889-892.

Hoppel, W. A., Frick, G.M., Fitzgerald, J.W., Wattle, B.J. (1994) A cloud chamber study of the effect that non-precipitating water clouds have on the aerosol size distribution. Aerosol. Sci. Techno., v.20, pp. 1.

Huang E., Williams, E., Boldi, R., Heckman, S., Lyons, W., Taylor, M., Nelson, T. and Wong, C. (1999) Criteria for sprites and elves based on Schumann resonance observations. J. Geophys. Res., v. 104, pp. 16943-16964.

Inan, U. S., Bell, T. F., Pasko, V. P., Sentman, D. D., Wescott, E. M. and Lyons, W. A. (1995) VLF signatures of ionospheric disturbances associated with sprites. Geophys. Res. Lett., v. 22, pp. 3461-3464.

Inan, U. S., Bell, T. F. and Rodriguez, J. V. (1991) Heating and ionization of the lower ionosphere by lightning. Geophys. Res. Lett., v. 18, pp. 705.

Inan, U. S., Sampson, W. A. and Taranenko, Y. N. (1996) Space-time structure of optical flashes and ionization changes produced by lightning EMP. Geophys. Res. Lett., v. 23, pp. 133-136.

Inan, U. S., Shafer, D. C., Yip, W. Y. and Orville, R. E. (1988) Subionospheric VLF signatures of nighttime D-region perturbations in the vicinity of lightning discharges. J. Geophys. Res., v. 93, pp. 11455.

Kartalev, M. D., Rycroft, M. J., Fullekrug, M., Papitashvili, V. O. and Keremidarska, V. I. (2006) A possible explanation for the dominant effect of South American thunderstorms on the Carnegie curve. J. Atmos. Solar-Terr. Phys., v. 68, pp. 457-468.

Kasemir, H. W. (1983) Static discharge and triggered lightning, Colorado Scientific Research Corp Berthoud, Report.

Kirillov, V. V., Kopeykin, V. N. and Mushtak, V. C. (1997) ELF electromagnetic waves within the Earth-ionosphere waveguide (in Russian). Geomag. Aeron., v. 37, pp.114-120.

Krehbiel, P. L., Riousset, J. A., Pasko, V. P., Thomas, R. J., Rison, W., Stanley, M. A. and Edens, H. E. (2008) Upward electrical discharges from thunderstorms. Nature geoscience, v.1, pp.233-237, doi:10.1038/ngeo162.

Krehbiel, P. R. (1986) The electrical structure of thunderstorms, in The Earth's Electrical Environment. National Academy Press, Washington, D.C (USA), pp. 90–113.

Lang, T. J. and Rutledge, S. A. (2002) Relationship between convective storm kinematics, microphysics and lightning. Mon. Weather Rev., v.130, pp. 2492-2506.

Lyons, W. A. (1994) Characteristics of luminous structures in the stratosphere above thunderstorms as imaged by low-light video. Geophys. Res. Lett., v.21, pp. 875-878.

Lyons, W. A. (1995) Characteristics of luminous structures in the stratosphere above thunderstorms as imaged by low-light video. Geophys. Res. Lett., v. 21, pp. 875.

Lyons, W. A. (1996) Sprites observations above the US high plains in relation to their parents thunderstorm systems. J. Geophys. Res., v. 10129, 641-29, pp.652.

Lyons, W. A., Cummer, S. A., Stanley, M. K., Huffiness, G. R., Wiens, K. C. and Nelson, T. E. (2008) Supercells and sprites. American Meteorological Society, BAMS (August issue), pp.1165-1174.

Lyons, W. A. (2006) The meteorology of transient luminous events-an introductory and overview, in Sprites Elves and intense lightning discharges. *In: Martin Füllekrug, Eugene A. Mareev and Michael J. Rycroft (eds.)* Sprites, Elves and Intense Lightning Discharges, pp.19-25,

Lyons, W. A., Uliasz, M. and Nelson, T. E. (1998) Large peak current cloud-to-ground lightning flashes during the summer months in the Contiguous United States. Mon. Weather Rev., v.126, pp. 2217-2234.

Markson, R. (1978) Solar modulation of atmospheric electrification and possible implications for the Sun-weather relationship. Nature, v. 273, pp. 103-109.

Murray, N. D., Orville, R. E. and Huffines, G. R. (2000) Effect of pollution from Central American fires on cloud-to-ground lightning in May 1998. Geophys. Res. Lett., v. 27, pp. 2249.

Madden, T. R. and Thompson, W. B. (1965) Low-frequency electromagnetic oscillations of the Earth-ionosphere cavity. Rev. Geophys., v. 3, pp. 211.

Mazur, V. Shao, X. M. and Krehbiel, P. R. (1998) Spider, lightning in intracloud and positive cloud-to-ground flashes. J. Geophys. Res., v. 103(D16), 19,811-19,822, doi:10.1029/98JD02003.

McHarg, M. G., Stenbaek-Nielsen, H. C. and Kammae, T. (2007) Observation of streamer formation in sprites. Geophys. Res. Lett., v. 34 L06804, doi:10.1029/2006GL027854.

Milikh, G., Papadopoulos, M. and Chang, K. C. L. (1995) On the physics of high altitude lightning. Geophys. Res. Lett., v. 22, pp. 85.

MacGorman, D.R. and Rust, W. D. (1998) The Electrical Nature of Storms, Oxford University Press. New York, USA, pp. 422.

MacGorman, D. R. and Burgess, D. W. (1994) Positive cloud-to-ground lightning in tornado storms and hailstorms. Monthly Weather Rev., v. 122, pp. 1671-1697.

Magunia, A. (1996) The thunderstorm- driven diurnal variation of the ELF electromagnetic activity level. J. Atmos. Solar-Terr. Phys., v. 56, pp.1683 - 1698.

Mende, S. B., Frey, H. U. R., Rairden, L., Han-Tzong, Su, Hsu,R-R., Allin, T. H., Neubert, T., Gerken, E. A., Inan, U. S. (2002) Fine structure of sprites and proposed global observations, Cospar Colloquia Series. *In: L.H. Liu (ed.),* Space Weather Study using Multipoint Techniques, 12. Pergamon Elsevier Science, pp. 275-282.

Moudry, D., Stenbaek-Nielsen, H., Sentman, D. and Wescott, E. (2003) Imaging of elves, halos and sprite initiation at 1 ms time resolution. J. Atmos. Sol.-Terr. Phys., v.65, pp.509-518.

Neubert, T., Rycroft, M., Farges, T., Blanc, E., Chanrion, O., Arnone, E., Odzimek, A., Arnold, N., Enell, C.-F., Turunen, E., Bösingèr, T., Mika, Á., Haldoupis, C., Steiner,R. J., van der Velde, O., Soula, S., Berg, P., Boberg, F., Thejll, P., Christiansen, B., Ignaccolo, M., Füllekrug, M., Verronen,P. T., Montanya, J., and Crosby, N. (2008) Recent results from studies of electric discharges in the mesosphere. Surv. Geophys., v.29, pp.71-137.

Ohkubo, A., Fukunishi, H., Takahashi, Y. and Adachi, T. (2005) VLF/ELF sferic evidence for in-cloud discharge activity producing sprites. Geophys. Res. Lett., v. 32 L04812, doi:10.1029/2004GL021943.

Ostgaaard, N., Gjesteland, T., Stadsnes, J., Connell, P. H. and Carlson, B. (2008) Production altitude and time delays of terrestrial gamma flashes: Revisiting the Burst and Transient Source Experiment spectra. J. Geophys. Res., v. 113, A02307, doi:10.1029/2007JA012618.

Pasko, V.P. (1996) Dynamic coupling of quasi-electrostatic thunderstromcloud fields to the mesosphere and lower ionosphere, sprites and jets. Ph.D thesis, Stanford University (USA).

Pasko, V. P., Stanley, M. A., Mathews, J. D., Inan, U. S. and Wood, T. G. (2002) Electrical discharge from a thundercloud top to the ionosphere. Nature, v.416, pp.152-154.

Pasko, V. P., Stanley, M. A., Mathews, J. D., Inan, U.S. and Wood, T. G. (2002) Electrical discharge from a thundercloud top to the ionosphere. Nature, v. 416, pp. 152-154.

Pasko, V. P., Inan, U. S., Taranenko, Y. N. and Bell, T. F. (1995) Heating, ionization and upward discharges in the mesosphere due to intense quasi-electrostatic thundercloud fields. Geophys. Res. Lett., v. 22, pp. 365.

Parkinson, W. C. and Torrensen, O. W. (1931) The diurnal variation the electrical potential of the atmosphere over ocean. Compt. Ren de l'Assemblee de Stockholm, 1930, IUGG Sect Terres. Magn Electr Bull., v. 8, pp 340-345.

Pessi, A. S., Businger, K. L. and Cummins, T. (2004) Tuner On the relationship between lightning and convective rainfall over the central Pacific Ocean. 18th International lightning Detection Conferences, 7-9 June 2004, Helsinki, Finland.

Petersen, W. A. and Rutledge, S. A. (1998) On the relationship between cloud-to-ground lightning and convective rainfall. J. Geophys. Res. v. 103, pp.14025-14040.

Piepgrass, M. V., Krider, E. P. and Moore, C. B. (1982) Lightning and surface rainfall during Florida thunderstorms. J. Geophys. Res., v. 87, pp. 11193-11201.

Price, C. (2006) Global thunderstorm activity, Sprites, Elves and Intense Lightning Discharge. *In: M. Fullekrug et al. (eds.)* Sprites, Elves and Intense Lightning Discharges, pp. 85-99.

Price, C. (1993) Global surface temperature and the atmospheric electric circuit. Geophys. Res. Lett., v. 20, pp. 1363-1366.

Price, C. and Rind, D. (1994) Possible implication of global climate change and global lightning distributions and frequencies. J. Geophys. Res., v. 99, pp. 10823-10831.

Rakov V. A. and Uman, M. A. (2003) Lightning: Physics and Effects. Cambridge University Press, Cambridge, UK, pp. 687.

Roussel-Dupre, R. A. and Gurevich, A. V. (1996) On runaway breakdown and upward propagating discharges. J. Geophys. Res., pp. 1012297.

Rodger, C. J. (1999) Red sprites, upward lightning and VLF Perturbations. Review Geophysics, v.37, pp.317-336.

Roussel-Dupre, R. A. and Gurevich, A. V. (1996) On runaway break-down and upward propagating discharges. J. Geophys. Res., v. 101, 2297, doi:10.1029/95JA03278.

Rowland, H. L., Fernsler, R. F., Huba, J. D. and Bernhardt, P. A. (1995) Lightning driven EMP in the upper atmosphere. Geophys. Res. Lett., v. 22, pp. 361.

Rowland, H. L. (1998) Theories and simulations of elves, sprites and blue jets. J. Atmos. Solar-Terr. Phys., v.60, pp. 831-844.

Rutledge, S.A., Williams, E. R. and Keenan, T. D. (1992) The down under doppler and electricity experiment (DUNDEE) overview and preliminary results. Bull. Ameri. Meteorological Soc., v. 73, pp. 3.

Rycroft, M. J. Israelsson, S. and Price, C. (2000) The global atmospheric electric circuit, solar activity and climate change. J. Atmos. Solar-Terr Phys., v. 62, pp.1563-1576.

Rairden, R. and Mende, S. (1995) Time resolved sprite imagery. Geophys. Res. Lett., v. 22, pp. 3465-3468.

Saetre, C., Barth, C. A., Stadsnes, J., Ostgaard, N., Ailey, S. M., Baker, D. N., Germany, G. A., Gjerloev, J. W. (2007) Thermospheric nitric oxide at higher latitudes: Model calculations with auroral energy input. J. Geophys. Res., v. 112 A08306, doi:10.1029/2006JA012203.

Sato, M. and Fukunishi, H. (2003) Global sprites occurrence locations and rates derived from triangular of transient Schumann Resonance events. Geophys. Res. Lett., v. 30 (16), pp. 1859.

Sato, M. Fukunishi, H. Takahashi, Y. Adachi, T. Okubo, A. Lyons, W. A. (2002) Charge moment differences in winter of Japan and summer of the United States. EOS, v. 83 A11C-0103.

Sato, M., Takahashi, Y., Yoshida, A. and Adachi, T. (2008) Global distribution of intense lightning discharges and their seasonal variations. J. Phys. D. Appl. Phys., v. 41, 234011 (10pp), doi:10.1088/0022-3727/41/23/234011.

Sátori, G., Mushtak, V. and Williams, E. R. (2008) Schumann resonance signatures of global lightning activity, Lightning: Principles, Instruments and Applications. *In: H.-D Betz, U. Schumann and P. Laroche, (eds.)* Title Lightning: Principles, Instruments and Applications, Springer.

Saunders, C. (2008) Charge separation mechanisms in clouds. Space Sci. Rev., v. 137, pp. 335-353.

Schumann, U. and Huntrieser, H. (2007) The global lightning-induced nitrogen oxides source. Atmos. Chem. Phys., v. 7, pp. 3823-3907.

Schlegel, K. and Fullekrug, M. (1999) Schuman resonance parameter changes during high energy particle precipitation. J. Geophys. Res., v. 104, pp. 10111-10118.

Seity, Y., Soula, S. and Sauvageot, H. (2001) Lightning and precipitation relationship in costal thunderstorms. J. Geophys. Res., v. 106, pp. 22801-22816.

Sentman, D. D. and Fraser, B. J. (1991) Simultaneous observations of Schumann resonances in California and Australia: Evidence for intensity modulation of the local height of the D-region. J. Geophys. Res., v. 96, pp. 15973.

Sentman, D. D., Wescott E. M., Osborne, D. L., Hampton, D. L. and Heavner, M. J. (1995) Preliminary results from the sprites 94 aircraft campaign 1 red sprites. Geophys. Res. Let., v.22, pp.1205.

Siingh, Devendraa, Singh, R. P., Kamra, A. K., Gupta, P. N., Singh, R., Gopalakrishnan, V. and Singh, A. K. (2005) Review of electromagnetic coupling between the Earth's atmosphere and the space environment. J. Atmos. Solar-Terr. Phys. v. 67, pp. 637-658.

Siingh, Devendraa, Gopalakrishnan, V., Singh, R. P., Kamra, A. K., Singh, S., Pant, V., Singh, R. and Singh, A. K. (2007) The atmospheric global electric circuit: An overview. Atmos. Res., v. 84, pp. 91-110.

Siingh, Devendraa, Singh, A.K., Patel, R.P., Singh, R., Singh, R. P., Veenadhari, B. and Mukherjee, M. (2008) Thunderstorms, lightning, sprites and magnetospheric whistler-mode radio waves. Surv. Gephys., v. 29, pp. 499-551.

Smith, D. M., Lopez, L. I., Lin, R. P. and Barrington-Leigh, C. P. (2005) Terrestrial gamma-ray flashes observed up to 20 MeV. Science, v. 307, pp.1085-1088, doi:10.1126/oiscience.1107466.

Soula, Serge, Van Der Velde Oscar, Montanya Joan, Neubert Torsten, Chanrion Olivier, Ganot Michal (2009) Analysis of thunderstorm and lightning activity associated with sprites observed during the Eurosprites campaigns: Two case studies. Atmos. Res., v.91, pp.514-528.

Stanley, M., Krehbiel, P., Max Brook, Charles, M. Rison, W. (1999) High speed video of initial sprite development. Geophys. Res. Lett., v. 26, pp. 3201-3204.

Stenbaek-Nielsen, M. G., McHarg Kammae, T. and Sentmann, D. D. (2007) Observed emission rates in sprite streamer heads. Geophys. Res. Lett., v. 34(11) L11105, doi:10.1029/2007GL029881

Su, H. T., Hsu, R. R., Chen, A. B., Wang, Y. C., Hsiao, W. S., Lai, W. C., Sato, M. and Fukunishi, H. (2003) Gigantic jet between a thundercloud and the ionosphere. Nature, v.423, pp.974-976.

Taranenko, Y. N. and Roussel-Dupre, R. A. (1996) High altitude discharges and gamma-ray flashes:a manisfestation of runaway air breakdown. Geophys. Res. Lett., v. 23, pp. 571.

Tinsley, B. A. (2008) The global atmospheric electric circuit and its effects on cloud microphysics. Rep. Prog. Phys., v.71, 066801 doi:10.1088/0034-4885/71/6/066801.

Tapia, A., Smith, J. A. and Dixon, M. (1998) Estimation of convective rainfall from lightning observations. J. Appl. Meterol., v. 37, pp.1497-1509.

Uman, M. (1987) The lightning discharge. Academic Press, Orlando, Florida, USA. pp. 377.

Vohra, K. G., Vasudevan, K. N. and Nair, P. V. (1970) Mechanisms of nucleus forming reactions in the atmosphere. J. Geophys. Res., v. 75, pp. 2951.

Wescott, E. M., Sentman, D., Osborne, D., Hampton, D. and Heavner, M. (1995) Preliminary results from the sprites 94 aircraft campaign: 2, blue jets. Geophys. Res. Lett., v.22, pp.1209-1212.

Wescott, E. N., Stenback-Mielsen, H. C., Sentman, D. D., Heavmer, M. J., Moudry, D. and Sabbas, F. T. S. (2001)Triangulation of sprites, associated halos and their possible relation to causative lightning and micrometeors. J. Geophys. Res., v. 106, pp. 10467-10478.

Wescott, E. M., Sentman, D., Heavner, M., Hampton, D., Lyons, W. A. and Nelson, T. (1998) Observations of 'Columniform' sprites. J. Atmos. Terr. Phys., v. 60, pp. 733-740.

Whipple, F. J. W. and Scrase, F. J. (1936) Point discharge in the electric field of the earth. Geophys. Mem., v. VIII (68), pp. 20.

Wilson, C. T. R. (1956) A theory of thundercloud electricity. Proc. R. Soc. London, Ser A (UK), pp.236 297.

Winckler, J., Lyons, W., Nelson, T. and Nemzek, R. (1996) New high-resolution ground-based studies of sprites. J. Geophys. Res., v. 101, pp. 6997-7004.

Winckler, J. R., Lyons, W. A., Nelson, T. E. and Nemzek, R. J. (1996) New high-resolution ground-based studies of sprites. J. Geophys. Res., v. 101, pp. 6997-7004.

Williams, E. R. Mushtak, V.C., Boldi, R., Dowden, R.L. and Kawasaki, Z.I. (2007b) Sprite lightning heard round the world by Schumann resonance methods. Radio Sci., v. 42, RS2S20, doi:10.1029/2006RS003498.

Williams, E. R., Downes, E., Boldi, R., Lyons, W. A. and Heckman, S. (2007) Polarity asymmetry of sprite-producing lightning: A paradox? Radio Sci., v. 42, RS2S17, doi:10.1029/2006RS003488.

Williams, E. R. (2001) Sprites, elves and glow discharge tubes. Phys. Today (September issue), pp. 1-7.

Williams, E. R. and Stanfill, S. (2002) The physical origin of the land-ocean contrast in lightning activity. Competes. Rendus., Physique, v. 3, pp.1277.

Wilson, C. T. R. (1925) The electric field of a thunderstorm and some of its effects. Proc. R. Soc. Lond, v. 3732D, pp. 534.

Williams, E. R. (2005) Lightning and climate: A review. Atmos. Res., v. 76, pp. 272-287.

Williams, E. R. and Stanfill, S. (2002) Contrasting convective regions over the Amozon: Implications for cloud Electrification. J. Geophys. Res. v. 107, D20.8082doi101029/2001 JD 000380.

Williams, E. R., Rutledge, S. A., Geotis, S. G., Renno, N., Rasmussen, E. and Rickenbach, T. A. (1992) radar and electrical study of tropical 'hot towers'. J. Atmos. Sci., v. 49, pp. 1386-1345.

Williams, E. R. and Renno, N. O. (1993) An analysis of the conditional instability for the tropical atmosphere. Mon. Weather Rev., v. 121, pp. 21-36.

Williams, E. R., Weber, M. E. and Oville, R. E. (1989) The relationship between lightning type and convective state of thunderclouds. J. Geophys. Res., v. 94, pp. 13213-13220.

Williams, E. R. (1992) The Schumann Resonance: a global thermometer. Science, v. 256, pp. 1184-1187.

Yair, Y. (2008) Charge generation and separation processes. Space Sci. Rev., v.137, pp.119-131.

Yair Y. Israelevich, P. Devir, A.D., Moalem, M. Price, C., Oseph Joachim, H. Levin, Z. Baruch, Z. Sternlieb, A. Telle, A. (2004) New observations of sprites from the Space Shuttle. J. Geophys. Res., v. 109(D), doi: 10.1029/ 2003JD004497.

Zipser, E. J. (1994) Deep cumulonimbus cloud system in the tropics with and without lightning. Mon. Weather Rev., v. 122, pp.1837-1851.

□□□

18

Long Term Trend, Diurnal and Seasonal Variations of Atmospheric Ozone at Indian Antarctic Station Maitri

NANDITA D. GANGULY AND VINCENT JOEL
Department of Physics, St. Xavier's College, Ahmedabad-380009,Gujarat

Abstract

The long term trend, diurnal and seasonal variations of atmospheric ozone at the Indian Antarctic station Maitri has been examined using ground based and satellite data. Total ozone column (TOC) at Maitri exhibits a clear seasonal variation, which is found to be maximum in January and minimum in September/October while the surface ozone does not exhibit any significant diurnal variation. Both, the monthly mean TOC and minimum TOC during the ozone hole period exhibits a sharp decreasing trend for the period 1979-1988, which is observed to slow down during the period from 1989-2001. The observed weakening in ozone trends are discussed in the light of the possible influence of Montreal protocol and other natural and anthropogenic factors, which also influence the recovery in ozone levels.

Keywords : Maitri, seasonal variation, diurnal variation, trend, polar vortex, ozone hole.

Introduction

Antarctica is the southernmost continent of the world located between 65°S and 90°S latitudes. Every year about 60% of the total amount of overhead ozone gets depleted over some parts of Antarctica during spring time. This phenomenon known as the Antarctic ozone hole was first discovered in the mid eighties (Farman *et al.*, 1985) and it led to the triggering of intense scientific research and signing of international agreements like Montreal Protocol.

The main factors affecting the Antarctic ozone loss are high chlorine and bromine levels, extremely low temperature during winter, and relative isolation of the polar region from the mid latitudes by the polar vortex. However, in the recent years, it has been established that the development of the Antarctic ozone hole is strongly influenced by photochemical balance (Crutzen *et al.*, 1986), transport of ozone by winds and meteorological conditions in the stratosphere (Jain *et al.*, 2008). In the light of this background, the diurnal and seasonal variations and long term trend of atmospheric ozone at Maitri has been studied using ground based and satellite data.

Measurement site and Data

The Indian Antarctic research station Matri (70°46′S, 11°45′E) is located in the coastal area of Queen Maud Land on a rocky area and is devoid of snow throughout the year (Bhattacharya, 1987). The monthly mean total ozone column (TOC) data for Maitri has been obtained from Nimbus-7 Total Ozone Mapping Spectrometer (Nimbus-7 TOMS) and Earth Probe Total Ozone Mapping Spectrometer (EP-TOMS). This data has been used to study the trend and seasonal variability of TOC at Maitri for the period from 1979– 2001. Since TOMS data are either incomplete or unavailable for the years 1993-1996, they have been neglected. Further, due to continuous changes in the optical properties of the front scan mirror on EP-TOMS, a latitude dependent error was observed, which could not be corrected by a simple calibration correction. Because of this error, data since the year 2002 has not been considered for trend analysis as per the recommendation from NASA (http://toms.gsfc.nasa.gov/news/news.html#18-Nov 2004). TOMS provides daily global measurement of the TOC of the atmosphere by measuring solar ultraviolet radiation back scattered from the atmosphere. The surface ozone data and vertical ozone profiles at Maitri have been obtained from India Meteorological department (IMD). The wind direction has been obtained from NCEP/NCAR Reanalysis (Kalnay *et al.*, 1996).

Results and Discussions

1. Diurnal variation of surface ozone

A comparison of surface ozone measured in units of parts per billion by volume (ppbv) on four different days at Maitri is shown in Fig. 1. Surface ozone at Maitri exhibits an insignificant diurnal variation of around 5 ppbv. This is because, since Maitri lies in the polar region, the sun shines for all the 24 hours every day during October to February. Thus, the solar radiation does not vary much from morning to night hours, resulting in almost constant ozone concentration throughout the day. Apart from this, surface ozone is produced by the photo oxidation of pollutants (carbon mono oxide & hydrocarbons) in the presence of adequate amount of nitrogen oxides during daytime. As vehicular traffic and industrial pollution levels are almost negligible at Maitri, insitu photochemical production of ozone may not be very significant. The general wind direction during all these four days was from the south east direction (NCEP/NCAR Reanalysis; Kalnay *et al.*, 1996), which rules out the possibility of transport of surface ozone and its precursors from polluted continental areas to this region. Moreover, as ozone is destroyed efficiently during transport over the sea/ocean due to relatively high insolation and humidity (Roelofs and Lelieveld, 2000), transport of air over sea/ocean do not cause the ozone levels to increase. Similar results were observed by Tripathi and Jain (2000), who have studied the diurnal variation of total ozone at Maitri during ozone hole period. They have observed that TOC decreased slightly from morning to evening hours.

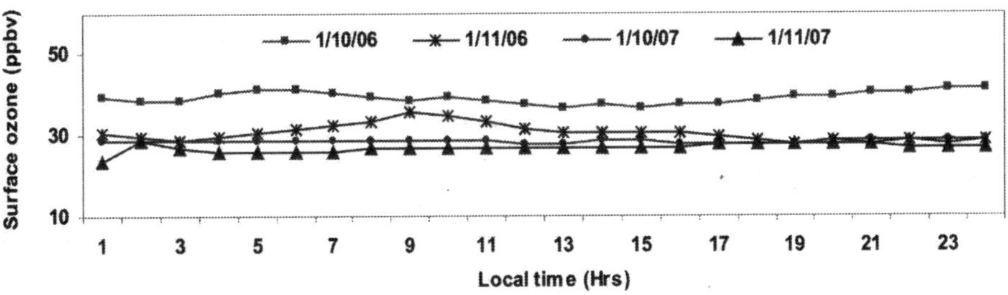

Fig. 1: Diurnal variation of surface ozone (ppbv) at Maitri on four different occasions (Source of data: India Meteorological Department).

2. Seasonal variation of total ozone column (TOC)

The monthly mean TOC in Dobson unit (DU) obtained for Maitri (70°46′S, 11° 45′E) located at the fringe of the polar vortex, and Station

X (65°S, 11°E) located outside the polar vortex from TOMS data were averaged for the period from 1997 - 2001 to study seasonal variation of ozone (Fig. 2). Since TOMS determines the TOC by measuring the solar ultraviolet radiation back scattered from the atmosphere, TOC data are not available at Maitri during winter darkness (Polar night period). TOC at Maitri is found to be maximum in January (283 DU). It decreases thereafter and reaches a minimum in September (170 DU). It again increases gradually from October to December and reaches a peak in January. A clear seasonal variation of around 113 DU is observed.

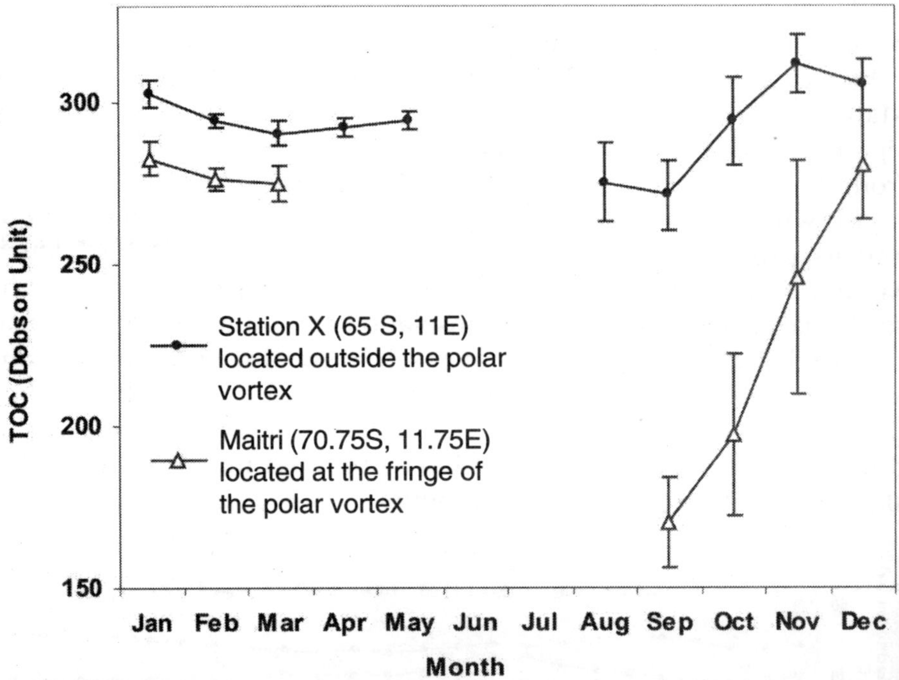

Fig. 2 : Seasonal variation of total ozone (Dobson unit) at Maitri and station X for the period 1997-2001. Bars indicate 2σ deviation (Source: EP TOMS).

Unlike Maitri, TOC at Station X is found to be maximum in November (312.2 DU). It decreases thereafter and reaches a minimum in September (271.4 DU). A clear seasonal variation of around 40.8 DU is observed. The observed seasonal variation at Maitri is because during Antarctica's dark winter, air temperatures above the South Pole drops to very low values. This forms a ring of high-altitude winds around Antarctica, which traps the ozone and ozone-depleting substances inside the Antarctic region and blocks outside air from entering into the Antarctic region. With the return of sunlight in September, the chemical reactions that destroy ozone are energized, resulting in low ozone values (<220 DU) in September/October. In November, as the air warms, the vortex breaks down and ozone-rich air from elsewhere mixes with the ozone-poor polar air, thus resulting in an increase in ozone concentration and disappearance of the hole. Unlike station X, which exhibits a standard deviation of around 5.36 DU from August to December, the ozone concentration at Maitri exhibits a high standard deviation of around 36 DU from September to December. The large amplitude of seasonal variation and large standard deviation observed at Maitri confirms the observation of Carswell *et al.* (1996) that ozone at Matri is strongly affected by the dynamical behavior of the polar vortex and planetary wave conditions. As Maitri is located on the fringe of the vortex (Tripathi and Jain, 2000), the amount of exchange between vortex and mid latitude air influences the amount of chemical processing of ozone and hence TOC (Schoeberl *et al.*, 1991). Singh and Sharma (2008) have studied the seasonal variation of TOC at Maitri in 2003 using Brewer Spectrophotometer. They have observed maximum TOC in the month of January and minimum in October.

3. Seasonal variation in the vertical ozone profile

A comparison between ozone levels at Maitri during the year 1997 for the months March (which can be considered as a normal ozone profile), August (before the ozone hole started forming) and October (when the ozone hole was at its maximum) is shown in Fig. 3.

It is observed that the ozone peak is higher during March (summer), but its altitude decreases in August (winter). This may be because during winter as the air begins to cool, ozone rich air grows denser and sinks to lower altitudes, resulting in a lowering of the ozone peak. The October profile indicates that ozone has almost disappeared from about 16-21 km altitude range which would be where the polar stratospheric clouds that enhance ozone destruction are mostly located.

4. Long term trend of total ozone column (TOC)

The variability in the TOC at Maitri has been studied by several Indian scientists by using satellite and ground based observations (Tripathi and Jain; 2000; Singh and Sharma, 2008). However, it may be noted that these studies are either restricted to coordinated

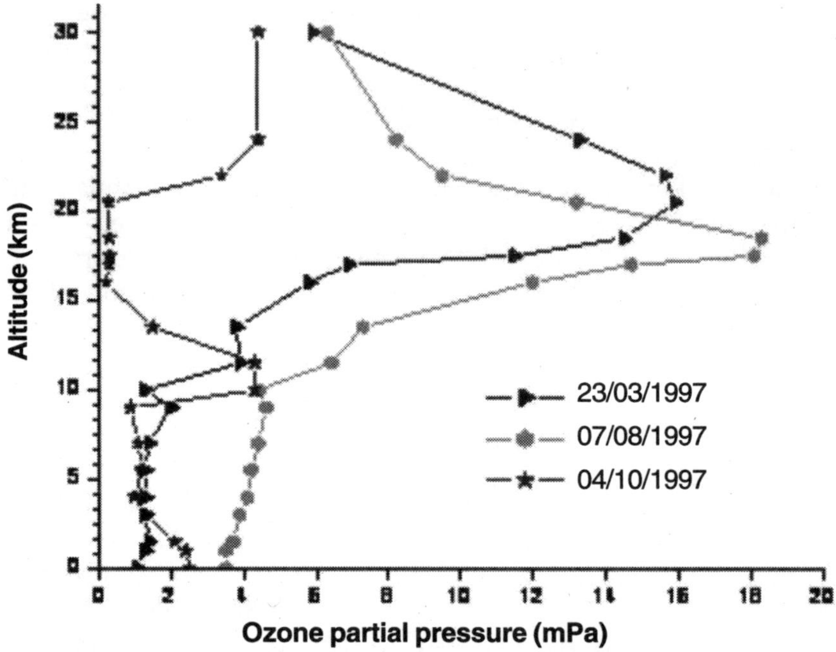

Fig. 3: Comparison between ozone levels at Maitri during the year 1997 for the months March, August and October (Source of data: India Meteorological Department).

campaigns or encompasses data measured over a continuous period of time, which is much before and after the implementation of the Montreal Protocol, and therefore cannot delineate the real effects of the Protocol. Since the implementation of Montreal Protocol entered into force from January 1989, the entire period of available data set is separated into two segments: 1979-1988 (before the implementation of Montreal Protocol) and 1989 - 2001 (after the implementation of Montreal Protocol) with a view to study the impact of Montreal protocol and other factors on ozone levels. The linear trend of monthly mean TOC from 1979-1988 and 1989-2001 at Maitri is plotted in Fig. 4.

The monthly mean TOC at Maitri exhibits a small decreasing trend of 0.331 DU yr^{-1} for the period 1979-1988. Thereafter, during the period from 1989-2001, the depletion in ozone levels is observed to slow down and a decreasing trend of 0.2015 DU yr^{-1} is observed. The observed weakening in ozone trends is +0.1295 DU yr^{-1}.

Similarly, the minimum TOC during ozone hole period at Maitri (Fig. 5) exhibits a sharp decreasing trend of 5.99 DU yr^{-1} for the period

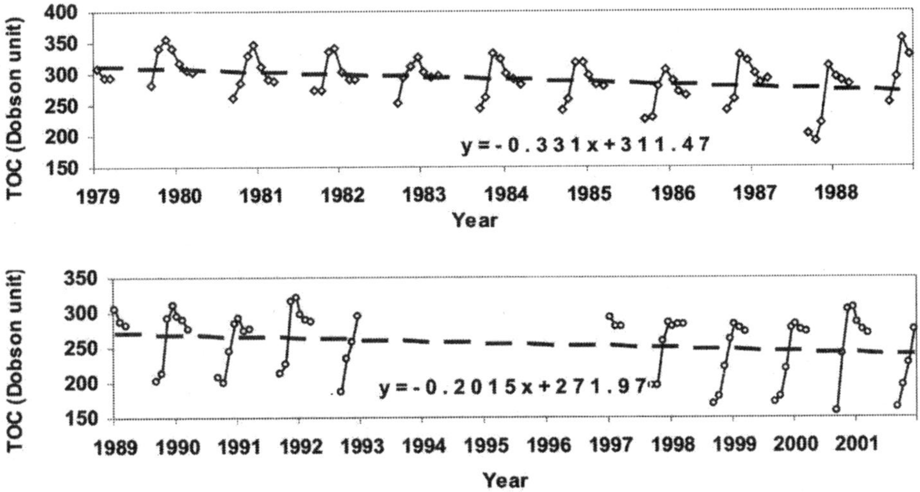

Fig. 4: Trend of TOC (DU) at Maitri for the period 1979-1988 and 1989-2001 (Source of data: Nimbus- 7 TOMS and Earth Probe TOMS)

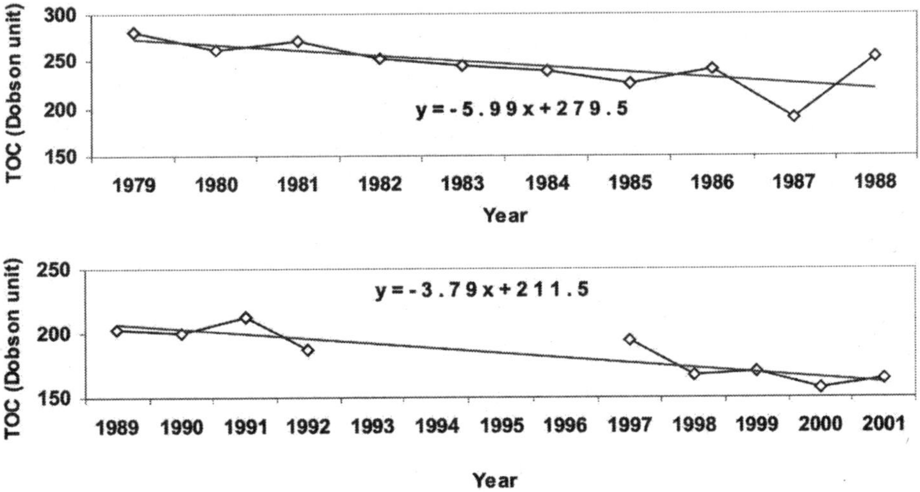

Fig. 5: Trend of minimum TOC (DU) at Maitri for the period 1979 - 1988 and 1989-2001(Source of data: Nimbus - 7 TOMS and Earth Probe TOMS).

1979-1988. Thereafter during the period from 1989-2001, the depletion in ozone levels is observed to slow down and a decreasing trend of 3.79 DU yr^{-1} is observed. This indicates that although the ozone hole grew deeper during the period from 1979-2001, and the ozone depletion inside the vortex was increasing, a weakening in the ozone hole trend of the order +2.20 DU yr^{-1} was observed to occur during the period from 1989-2001 compared to 1979-1988. The recovery in ozone levels was small because, since the life time CFC's in the stratosphere varies from a few years to hundreds of years, although the world had reduced or eliminated the use of many of these gases, their impact from the past may be affecting the ozone layer during the period from 1989-2001. Thus, it may take several more years before we may observe a reasonable recovery in TOC at Maitri. The change in the slope observed between the two epochs indicates the possible contribution of Montreal Protocol, because apart from changes in stratospheric ozone concentration due to catalytic reactions in the stratosphere involving CFC'S, other natural factors and climatological processes such as photochemical balance, transport of ozone by winds and meteorological conditions in the stratosphere could also influence the recovery in ozone levels (Jain *et al.*, 2008; Ganguly and Iyer, 2009).

Conclusion

The trend, diurnal and seasonal variation of ozone at the Indian Antarctic station Maitri has been examined using satellite and ground based data. TOC at Maitri exhibits a clear seasonal variation, which is found to be maximum in January and minimum in September/October while the surface ozone does not exhibit any significant diurnal variation. The monthly mean TOC and the minimum TOC observed during the ozone hole period at Maitri exhibits a weakening in the ozone trend during the period from 1989- 2001 compared to 1979-1988. It is argued that it is difficult to attribute the weakening in the ozone trend entirely to the success of Montreal protocol, due to the influence of a large number of natural and anthropogenic factors on the ozone chemistry.

Acknowledgments

The authors acknowledge their gratitude to Prof. Shyam Lal, Physical Research Laboratory, Ahmedabad and Dr. S. Bhandari, visiting scientist at ICCSIR, Ahmedabad for their invaluable suggestions, India Meteorological Department for providing the ozonesonde and surface ozone data, NASA/NOAA for providing TOMS ozone data and Prof. J. N. Goswami, Director, Physical Research Laboratory, Ahmedabad, for providing library facility. The wind direction at Maitri has been obtained from NCEP/NCAR Reanalysis. The authors also thank Rev. Father Vincent Braganza, SJ. Honorable Principal of St. Xavier's College and the management of St. Xavier's college,

Ahmedabad, for financially supporting the work with a view to encourage research activity among college students at the undergraduate level.

References

Bhattacharya, B. B. (1987) Schirmacher Oasis: Environment, history and the Indian Station Maitri. Scientific report of the fourth Indian expedition, Tech. Publication no. 4, Department of Ocean development, New Delhi, pp. 171-186

Carswell A. I. Donovan, D. P. Bird, J. C. Duck, T. J. Pal, S. R. and Whiteway, J. A. (1996) Measurements at the Eureka Arctic NDSC station with a Raman DIAL system. In: *A. Ansmann, and R. Neuber, (eds.)* Advances in Atmospheric Remote Sensing with Lidar, Springer Verlag, Berlin

Crutzen, P. J. and Arnold, F. (1986) Nitric acid cloud formation in the winter Antarctic Stratosphere: a major cause for the spring time ozone hole. Nature, v. 324, pp. 651

Farman, J. C., Gardiner, B. G. and Shanklin, J. D. (1985) Large losses of total ozone in Antarctica reveals seasonal CLO_x / NO_x interactions. Nature, v. 315, pp. 207–210.

Ganguly, N. D. and Iyer, K. N. (2009) Monitoring the success of implementation of Montreal Protocol in India with satellite measurements. Int.l J. Remote Sensing, v. 30 (15–16), pp. 3927–3941.

Jain, S. L. Arya, B. C. Ghude, S. D. Kumar, A. and Gupta, V. B. (2008) Ozone hole during 2002 and 2003: A comparison. Twenty second Indian Expedition to Antarctica, Scientific report, Ministry of earth sciences, Technical publication No.20, pp. 109-116

Kalnay, E., Kanamitsu, M., Kistler, R., Collins, W., Deaven, D., Gandin, L., Iredell, M., Saha, S., While, G., Woolen, J., Zhu, Y., Chelliah, M., Ebisuzaki, W., Higgins, W., Janowiak, J. Mo, K. C., Ropelewski, C., Wang, J., Leetma, A., Reynolds, R., Roy, J., Denis, J. (1996) The NCEP/NCAR reanalysis 40-year project. Bulletin of the American Meteorological Society, v. 77, pp. 437–471.

Roelofs, G. J. and Lelieveld, J. (2000) Model analysis of stratosphere-troposphere exchange of ozone and its role in the tropospheric ozone budget. In: *C. Zerefos et al. (eds.)* Chemistry and radiation changes in the ozone layer, pp. 25-43, Kluwer Academic Publishers, Netherlands.

Schoeberl, M. R. and Hartmann, D. L. (1991) The dynamics of the stratospheric polar vortex and its relation to springtime ozone depletions. Science, v. 251(4989), pp. 46-52.

Singh, C. and Sharma, S. P. (2008) Study of meteorological parameters and ozone hole phenomena at Schirmacher Oasis, Antarctica during 22nd Antarctic expedition 2003-2004.Twenty second Indian Expedition to Antarctica, Scientific report, Ministry of Earth Sciences, Technical publication No. 20, pp. 63-93.

Tripathi, O. P. and Jain, S. L. (2000) Measurement of ozone at Maitri, Antarctica. Sixteenth Indian Expendition to Antarctica, Scientific Report, Department of Ocean Development, Technical Publication No. 14, pp. 1-36.

NASA news letter: http://toms.gsfc.nasa.gov/news/news.html#18-Nov.2004.

□□□

19

Comparison of ECCO2 and NCEP Reanalysis using TRITON and RAMA Data at the Indian Ocean Mooring Buoy Point

VIVEK KUMAR PANDEY AND SUDHIR KUMAR SINGH
Department of Atmospheric and Ocean Science, KBCAOS & MNSCOSS, IIDS, Nehru Science Centre, University of Allahabad, Allahabad-211002

Abstract

To enhance better understanding and improve the predictive capability for the ocean's role in future climate change scenarios, Estimating the Circulation and Climate of the Ocean, Phase II (ECCO2) and National Centers for Environmental Prediction (NCEP) assimilation/reanalysis datasets were compared with the TRITON and RAMA mooring buoys observational data at the point 1.5ºS, 90ºE in the Indian Ocean. Root mean square deviation (rmsd) and bias values of the GODAS/NCEP reanalysis, and TRITON and RAMA observations were used for comparison of the meridional and zonal currents and salinity at the surface, and temperature at different depth of the mixed layer. On the basis of comparison we can say that the GODAS/NCEP reanalysis is not much satisfactory to use as input for the best prediction because its performance is not better in comparison to the observational data. It may also be concluded that the performance of the ECCO2 is better than the GODAS/NCEP on the basis of the comparison of the plots but still not very clear due to unavailability of equal ECCO2 temporal data set. We suggest that there might be certainly some forcing and assimilation techniques are missing to make better performance of the assimilation/ reanalysis.

Keywords : Predictive capability, GODAS, RAMA, TRITON, Indian Ocean.

Introduction

There is great need to better represent the exchange of heat and carbon dioxide at the air-sea interface. The physical representations of upper oceanic processes need to be improved in climate models based on experimental studies of the vertical structure of temperature, salinity, and absorption of solar radiation. Improved definition of the time dependent temperature and salinity distribution in the global ocean is essential, including the air-sea fluxes of heat and fresh water. This will require full implementation of a system with the capabilities of the current and planned ocean observation systems, satellites or in situ observation. The operational network of meteorological observatories on land placed all over the world have enabled us to monitor the change of the global atmosphere. At the same time, we have less observational platforms covering the world oceans and less understanding of oceanic environmental changes. The maintenance mechanisms of the Indian Ocean warm pool are not well understood in terms of the difference of evaporation and precipitation (E-P) forcing, heat and salt balance, etc. The warm pool in the eastern Indian Ocean may have impacts on climate changes through inter-ocean interaction by shifting of the Walker circulation, by controlling MJO activities which may influence ENSO onset, and by changing the heat and fresh water transports from the Pacific to Indian Oceans via the Indonesian Throughflow. Through the TOGA (Tropical Ocean and Global Atmosphere) decade, basin wide observing networks of mooring arrays, surface drifters, XBT lines, and tide gage stations have been developed to monitor and study ENSO (El Nino/Southern Oscillation). Thus, it is important to observe intraseasonal, seasonal and interannual oceanic changes in the eastern Indian Ocean to elucidate such inter-basin interaction and the region's impact on climate. In particular, the TAO (Tropical Atmosphere Ocean) array successfully monitored the 1991-93 El Nino (McPhaden, 1993; McPhaden *et al.*, 1998) and the 1997-98 El Nino (McPhaden *et al.*, 1999), and contributed greatly to the improvement of ENSO prediction models. By this success, the oceanography and meteorology communities in Japan recognized the importance and effectiveness of such a mooring array.

The advanced understanding of the role of the tropics in forcing mid-latitude weather and climate was learned primarily through the observations of the tropical moored buoy array (TAO/TRITON) (Tropical Atmosphere Ocean /Triangle Trans-Ocean Buoy Network) in the Pacific. A similar array in the Atlantic basin (Prediction and

Research Moored Array in the Atlantic) (PIRATA) now also contributes to better understanding, model output evaluation, improved forecasts, and improved ability to discern the causes of longer-term changes in the Oceans. The TAO/TRITON array in the Pacific Ocean is operated by NOAA's National Data Buoy Center (NDBC) in cooperation with Japan. PIRATA in the Atlantic is operated jointly by NOAA's Atlantic Oceanographic and Meteorological Laboratory (AOML) and NOAA's Pacific Marine Environmental Laboratory (PMEL), in cooperation with Brazil and France. The next challenge is to advance the tropical moored array across the Indian Ocean, in cooperation with international partners. The RAMA (Research African-Asian-Australian Monsoon Array) will complete global coverage of the Earth's tropical oceans; the RAMA system design is for 43 moorings. In addition to monitoring the air-sea exchange of heat and water, the moored buoys provide platforms for supporting instrumentation to measure the air-sea exchange of carbon dioxide in the tropics. The Indian Ocean RAMA is being implemented by PMEL in cooperation with Japan, France, India, and Indonesia. Near-real-time data are distributed on the Global Telecommunications System; near-real-time data from TAO/TRITON are available from NDBC; near-real-time data from PIRATA and RAMA are available from PMEL. Historical data are archived and distributed by the NOAA National Oceanographic Data Center and the NOAA National Climatic Data Center.

To increase understanding and predictive capability for the ocean's role in future climate change scenarios, a new Global Ocean Data Assimilation System (GODAS) has been developed at the NCEP) using the Geophysical Fluid Dynamics Laboratory's Modular Ocean Model version 3 (MOM.v3) and a three-dimensional variational data assimilation scheme. The operational Global Ocean Data Assimilation System (GODAS) of NCEP generates the initial conditions and validation to the NCEP Climate Forecast System (CFS). This analysis is available from 1979 to present. In GODAS we have relied heavily on ocean profiles (XBT, TAO/TRITON, ARGO) and on SST (Reynolds analysis). The stability of the TAO/TRITON array since its establishment in the early 1990s and growth of the ARGO array mean that *in-situ* profiles retain their central importance. Although GODAS is well constrained by *in-situ* data in the tropical Pacific, satellite altimetry, available since 1992, plays an important role outside of that region. The rapid growth of the Argo array in the last few years has both increased the number of *in-situ* observations used in GODAS from

5,000 to °6,000 per month in the 1980s and 1990s to 10,000 per month today and greatly improved the global coverage of *in situ* observations. The question remains whether adding new data sets to the operational GODAS will have an impact on performance of GODAS. A study on performance of the GODAS in the near equatorial Indian Ocean is need of time before using GODAS data as initialization of the coupled model for the region. This study will give new facts to explore the region using model in future by suggesting necessary improvement in GODAS assimilation system. The GODAS fields have been used extensively in real time operation and research at NCEP. At NCEP the primary purpose of the GODAS is to provide ocean initial conditions for seasonal to interannual (S/I) forecasting with a coupled ocean-atmospheric model. Because S/I prediction has focused on the ENSO, ocean observations in the tropical upper ocean have been most important to our operations.

The specific objective of this paper is to suggest for the optimization to the reanalysis/assimilated data to reduce large-scale biases and drifts of the model relative to observations therefore we had chosen the TRITON mooring buoy which were deployed in the eastern tropical Indian Ocean in October 2001. One TRITON buoy is located at 90°E, 1.5°S to determine the upper-layer variations associated with the zonal jets along the equator. Another buoy was deployed at 95°E, 5°S, to investigate the contribution of the surface heat flux and the ocean dynamics/thermodynamics. This study will present the variability of temperature and salinity and current velocity in GODAS and ECCO2, and compare it to the data from RAMA, TRITON moorings observational data and TMI SST at the mooring buoys points 90°E, 1.5°S. To demonstrate the validity of the model fields, we are conducing model-data and *in-situ* observational data.

Data

TAO/TRITON

An approach to observing the climate system to address the research in tropical Indo-Pacific Ocean, the Global Mooring Buoy Network Program played the important role in Tropical Ocean Climate Study. Development of the Tropical Atmosphere Ocean (TAO) array of moored buoys in the tropical Pacific was motivated by the 1982 to1983 El Nino event and developed by the PMEL under NOAA's Equatorial Pacific Ocean Climate Studies program. The array has been extended to the west and into the Indian Ocean with the deployment of the TRITON array implemented by the Japan Marine Science and Technology Center (JAMSTEC). The TAO and TRITON arrays combined in January 2000 to recognize the introduction of TRITON buoys in the western Pacific by JAMSTEC. The TRITON buoys observe wind, atmospheric temperature, humidity, rainfall, sunshine, seawater temperature, salinity, and current. Water temperature and salinity recorded at regular intervals up to a 750 m depth.

The combined TAO/TRITON data are managed and distributed as a unified and integrated data set by the TAO Project. The tropical moored buoy network consists of two arrays including the TAO/TRITON (Pacific) array and the PIRATA (Atlantic) array.

RAMA

The advanced understanding of the role of the tropics in forcing mid-latitude weather and climate was learned primarily through the observations of the tropical moored buoy array Tropical Atmosphere Ocean/Triangle Trans Ocean Buoy Network (TAO/TRITON; McPhaden *et al.*, 1998; Kuroda and Amitani, 2000) in the Pacific. A similar array in the Atlantic basin Prediction and Research Moored Array in the Tropical Atlantic (PIRATA; Bourles *et al.*, 2008) now also contributes to better understanding, model output evaluation, improved forecasts, and improved ability to discern the causes of longer-term changes in the oceans. The next challenge is to advance the tropical moored array across the Indian Ocean, in cooperation with international partners. The RAMA (Research African-Asian-Australian Monsoon Array) will complete global coverage of the Earth's tropical oceans. The TAO/TRITON array in the Pacific Ocean is operated by NOAA's National Data Buoy Center (NDBC) in cooperation with Japan.

The Indian Ocean RAMA is being implemented by PMEL in cooperation with Japan, France, India, and Indonesia. RAMA is the Indian Ocean equivalent of the TAO/TRITON and the PIRATA which respectively support basin scale observing system in the tropical Pacific and Atlantic Ocean. RAMA is an observational network of moorings to monitor surface meteorology and subsurface ocean parameters related to climate in the Indian Ocean. RAMA is an element of the Indian Ocean Observing System (IndOOS) supported by multiple nations, and is a combination of complementary satellite and *in-situ* measurement platform for climate research and forecasting. Long range of time scale and quick changes that occurred in the Indian Ocean needs moored buoy array providing time series data with high temporal resolution as an essential element of IndOOS. Though RAMA is built for understanding and prediction of the east African, Asian, and Australian monsoons, it will benefit nations outside the Indian Ocean region because of teleconnections that influence the far field.

Global Ocean Data Assimilation System (GODAS)

Global Ocean Data Assimilation System (GODAS) was developed to be the replacement for the Pacific Ocean Data Assimilation System (ODAS), and to provide the oceanic initial conditions for the new NCEP coupled Climate Forecast System (CFS). The GODAS became operational in 2003 and the CFS went operational in 2004. A description of the GODAS is provided by Behringer and Xue (2004). Since 1995, the NCEP has been producing real time ocean analysis and reanalysis using ODAS (Ji *et al.*, 1995). The ODAS was conFigured for the Pacific basin and used the Geophysical Fluid Dynamics Laboratory's Modular Ocean Model version 1(MOM.v1) and a three-dimensional variational data assimilation scheme originally designed by Derber and Rosati (1989). The last version of the Pacific ODAS (referred to as RA6 hereafter) was a significant improvement over the earlier versions and has been used widely in scientific community (Behringer *et al.*, 1998). The GODAS is based a quasi-global conFiguration of the GFDL MOMv3 (Pacanowski and Griffies, 1998) and give pentad and monthly product. Model Domain is 75°S-65°N (quasi-global). Data input is temperature and synthetic salinity that is constructed from temperature and local temperature/salinity climatology (annual mean field of T/S from NODC World Ocean Data base). GODAS is forced by momentum flux, heat flux, and fresh water flux from the atmospheric NCEP reanalysis 2 (Kanamitsu *et al.*, 2002). In addition the temperature the top model level is relaxed to weekly analyses of sea surface temperature (Reynolds *et al.*, 2002), while the surface salinity is relaxed to annual salinity climatology (Conkright *et al.*, 1999). Very short relaxation periods are used (5 days for temperature and 10 days for salinity). The GODAS also produces pentad and monthly oceanic fields in real time, so they are valuable dataset to serve research, application, and operational communities.

ECCO2

The NASA Modeling, Analysis, and Prediction (MAP) program is funding a project called ECCO2: High-Resolution Global-Ocean and

Sea-Ice Data Synthesis. ECCO2 produces increasingly accurate syntheses of all available global-scale ocean and sea-ice data that start to resolve ocean eddies and other narrow current systems, which transport heat, carbon, and other properties within the ocean.

These programs aims to produce increasingly accurate syntheses of all available global-scale ocean and sea-ice data at higher resolutions that start to resolve ocean eddies and other narrow current systems.

Comparison Results

The quasi global GODAS/NCEP pentad and ECCO2 monthly data has been compared with the TRITON and RAMA daily data with the special emphasis on GODAS. RAMA and TRITON are the high frequency data of the mooring buoys. The study focus on the RAMA and TRITON mooring buoys position 90°E, 1.5°S time series data. The NCEP data was available in pentad form therefore we have changed all the available daily data on pentad time scale data. We had calculated and plotted the rmsd, bias and difference for comparison of GODAS with RAMA and TRITON and included the monthly ECCO2 general plot for data characteristics studies with these observations. In overall study we have used the word NCEP in the Figure at the place of GODAS/NCEP. RMSD is used to make a quantitative comparison between the NCEP reanalysis, and TRITON and RAMA mooring buoys observational data. The bias value of the data analysis tells about the nature of the model data with respect to the observational data. We had studied the nature of the NCEP and ECCO2 temperature time series at the different depth such as surface (0 m), 50 m, 100 m, 125 m and 200 m in both the cases TRITON and RAMA moorings. The nature of the NCEP and ECCO2 surface zonal/ meridional current is also studies with the NCEP surface salinity comparison studied with TRITON and RAMA data. The entire time series plot has been taken for the period from 2002 to 2006 except for salinity plot for NCEP versus RAMA in which comparison has been done for only two years.

In the Fig.1 the bias value is -6.27 between TRITON and NCEP which stated that the NCEP reanalysis under estimate the observational data on a wide range *i.e* time average of difference is high, and rmsd value is the 29.9 show that there is much fluctuation from mean in the meridional current at the surface. Difference value takes the limits 80 to -100 cm/s. ECCO2 reanalysis represent the variation in the plot. This implies that the surface zonal current NCEP reanalysis is poor source.

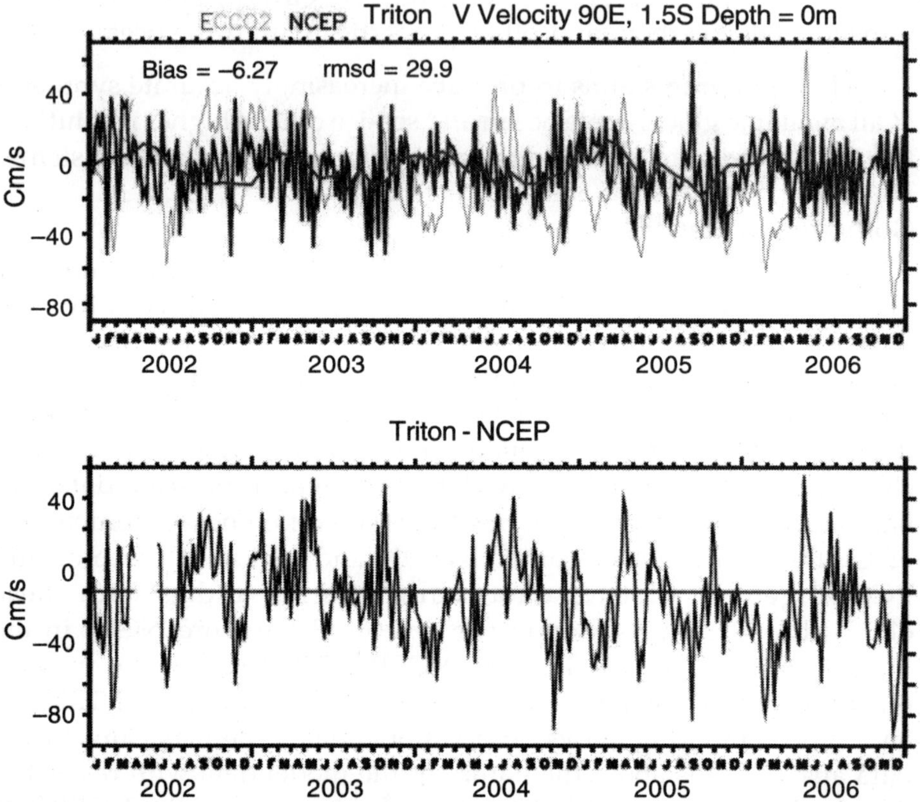

Fig.1: The upper panel represents the time series plot of the ECCO2 vs NCEP model reanalysis/assimilation vs TRITON (black) observational meridional velocity of surface currents. Bias and rmsd values show for the TRITON vs NCEP. Lower panel represents the difference plot of TRITON observational and NCEP model reanalysis at 90ºE, 1.5ºS.

According to the Fig.2 bias value is 4.09 between TRITON and NCEP which stated that the TRITON under estimate the NCEP reanalysis in the case of zonal current *i.e.* time average of difference is high while rmsd value is the 43.56 which represents more fluctuation from mean in the zonal current at the surface is very high. The difference plot takes its limit 120 to -140 cm/s. The ECCO2 is also not performing well within the plot with respect to observational data. This implies that the NCEP reanalysis surface zonal current also not perform well.

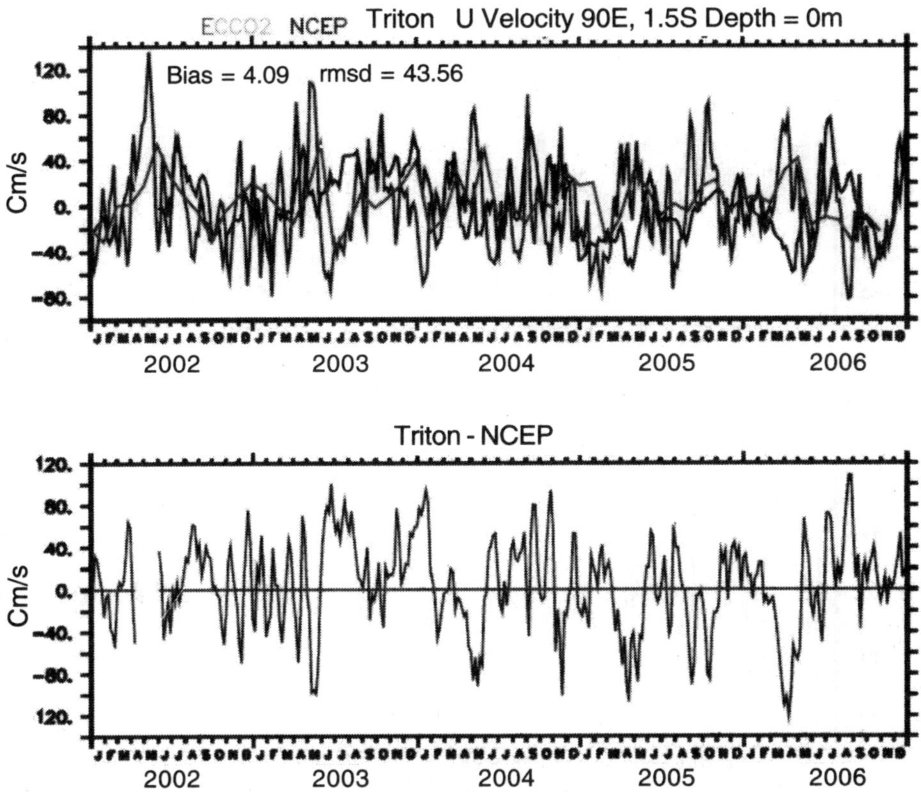

Fig. 2: The upper panel represents the time series plot of the ECCO2 vs NCEP model reanalysis/assimilation vs TRITON (black) observational zonal velocity of surface currents. Bias and rmsd values show for the TRITON vs NCEP. Lower panel represents the difference plot of TRITON observational and NCEP model reanalysis at 90ºE, 1.5ºS.

NCEP and ECCO2 is compared with the RAMA (Fig. 3). NCEP reanalysis meridional have overall same nature as with TRITON in Fig. 1 in this case bias is -6.28 very near to TRITON case -6.27 with the same rmsd and difference limits. It can say that the TRITON and RAMA both gives identical data value of the ocean variables at this mooring buoys positions.

Fig. 4 represents the comparison plot of NCEP and ECCO2 with the RAMA. NCEP reanalysis zonal current have overall same nature as with TRITON in Fig. 2. In this case bias, rmsd and difference limits have same value. ECCO2 perform the same nature as in TRITON. From Fig.1 to 4 it is clear that the current data of the NCEP reanalysis does not perform well with respect to TRITON and RAMA observation.

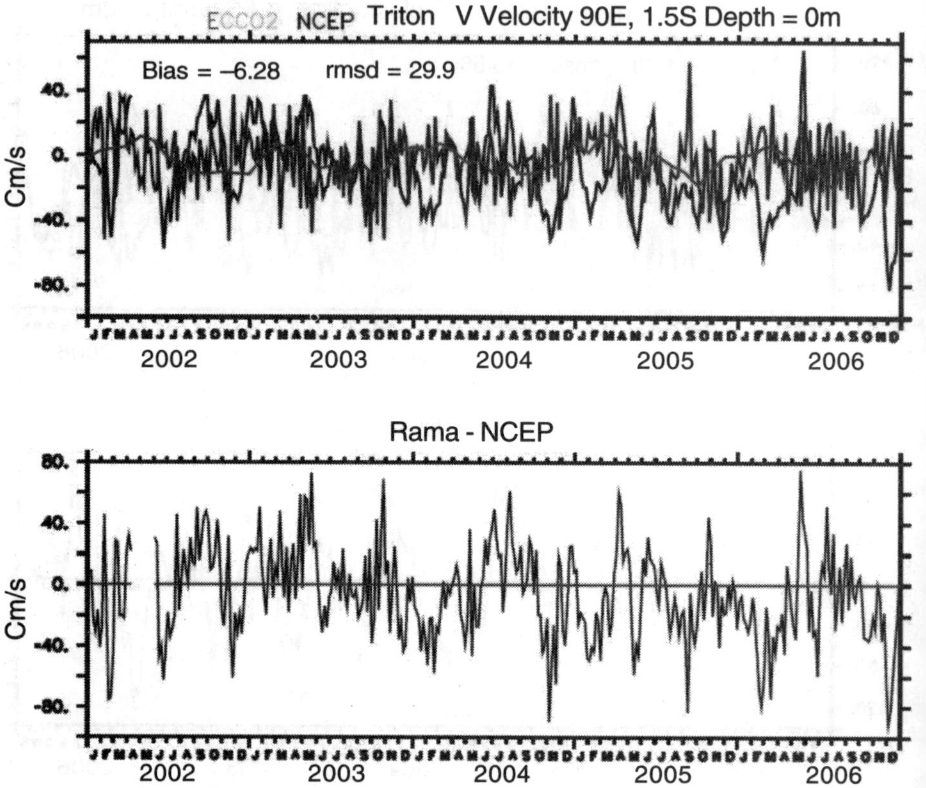

Fig. 3: The upper panel represents the time series plot of the ECCO2 vs NCEP model reanalysis/assimilation vs RAMA (black) observational meridional velocity of surface currents. Bias and rmsd values show for the RAMA vs NCEP. Lower panel represents the difference plot of RAMA observational and NCEP model reanalysis at 90ºE, 1.5ºS.

Bias value is 0.41 between TRITON vs NCEP which stated that the TRITON under estimate the NCEP reanalysis in the case of surface temperature while rmsd value is the 0.68 which represents fluctuation from mean. The difference plot takes its limit 2.5°C to -2.0°C. The ECCO2 is also deviated from observation in the plot. This implies that the NCEP reanalysis temperature comparatively perform well than current data (Fig. 5).

Fig. 6 shows that the bias value is 0.13 between TRITON vs NCEP which stated that the TRITON under estimate the NCEP reanalysis in the case of temperature at depth 50 m while rmsd value is the 0.9 which represents fluctuation from mean. The difference plot takes its limit 3.5°C to -2.0°C. The ECCO2 is also deviated from observation in the plot. This implies that the NCEP reanalysis temperature at this depth perform comparatively well than surface temperature.

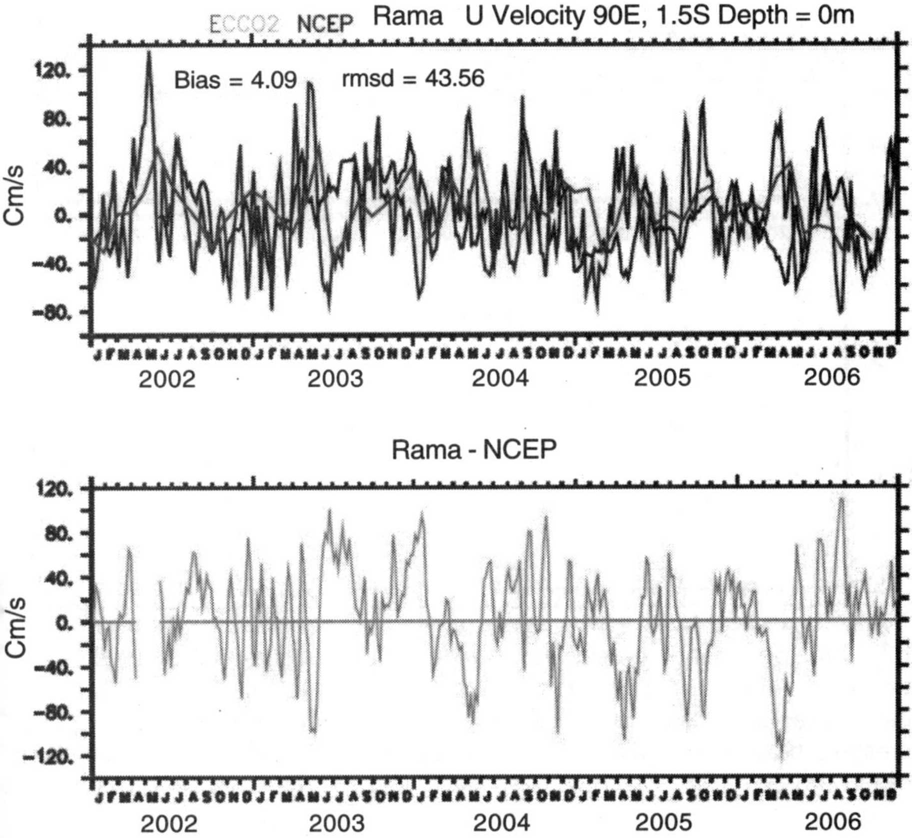

Fig. 4: The upper panel represents the time series plot of the ECCO2 vs NCEP model reanalysis/assimilation vs RAMA (black) observational zonal velocity of surface currents. Bias and rmsd values show for the RAMA vs NCEP. Lower panel represents the difference plot of RAMA observational and NCEP model reanalysis at 90ºE, 1.5ºS.

According to the Fig. 7 the bias value is -1.88 between TRITON vs NCEP which stated that the NCEP under estimate the TRITON in the case of temperature at depth 100 m while rmsd value is the 4.78 which represents fluctuation from mean is high with respect to the 50 m data. The difference plot takes its limit 10°C to -14.0°C. The ECCO2 is deviating from observation as evident by the plot. Thus the NCEP reanalysis temperature at this depth perform comparatively bad than temperature at 50 meter.

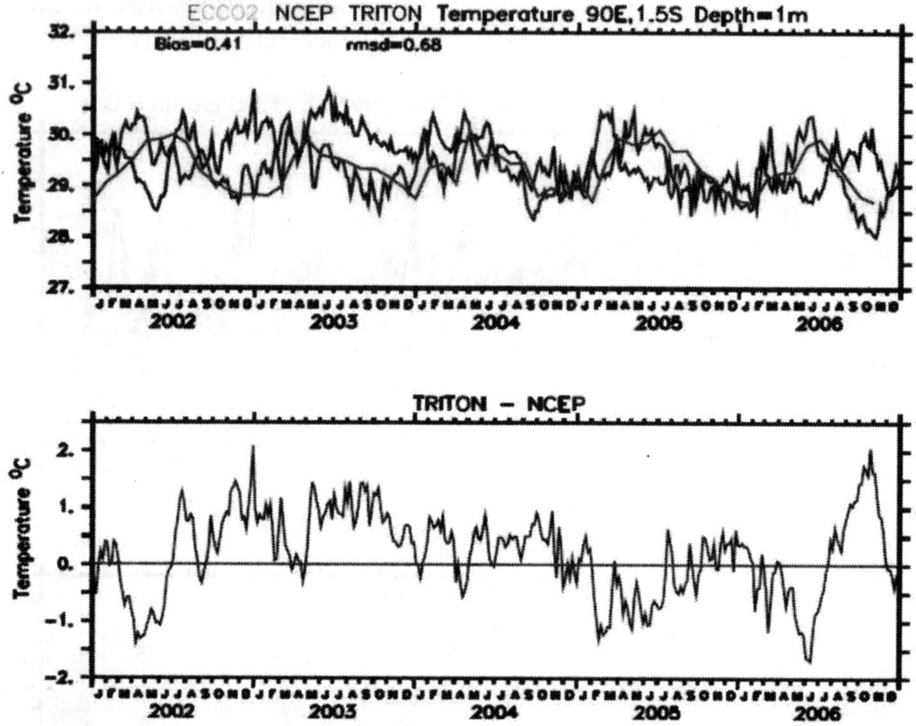

Fig.5: The upper panel represents the time series plot of the ECCO2 vs NCEP model reanalysis/assimilation vs TRITON (black) observational surface temperature. Bias and rmsd values show for the TRITON vs NCEP. Lower panel represents the difference plot of TRITON observational and NCEP model reanalysis at 90ºE, 1.5ºS.

Bias value is -0.74 between TRITON vs NCEP which stated that the NCEP under estimate the TRITON in the case of temperature at 125 m depth while rmsd value is 3.75 which represents fluctuation from mean which is less than for 100 m value (Fig. 8). The difference plot takes its limit 10 ℃ to -16.0 ℃. The ECCO2 is also match with the observational data plot. Thus performance of the NCEP reanalysis temperature at this depth also very bad.

Fig. 9 represents the comparison TRITON vs NCEP having bias value -0.27 which stated that the NCEP under estimate the TRITON in the case of temperature at 200 m depth while rmsd value is 1.11 which is more than 125 m data. The difference plot takes its limit 3.0 ℃ to -3.5 ℃. The ECCO2 is also deviated from observation in the plot. Thus, performance of the NCEP reanalysis temperature at this depth better than other depth.

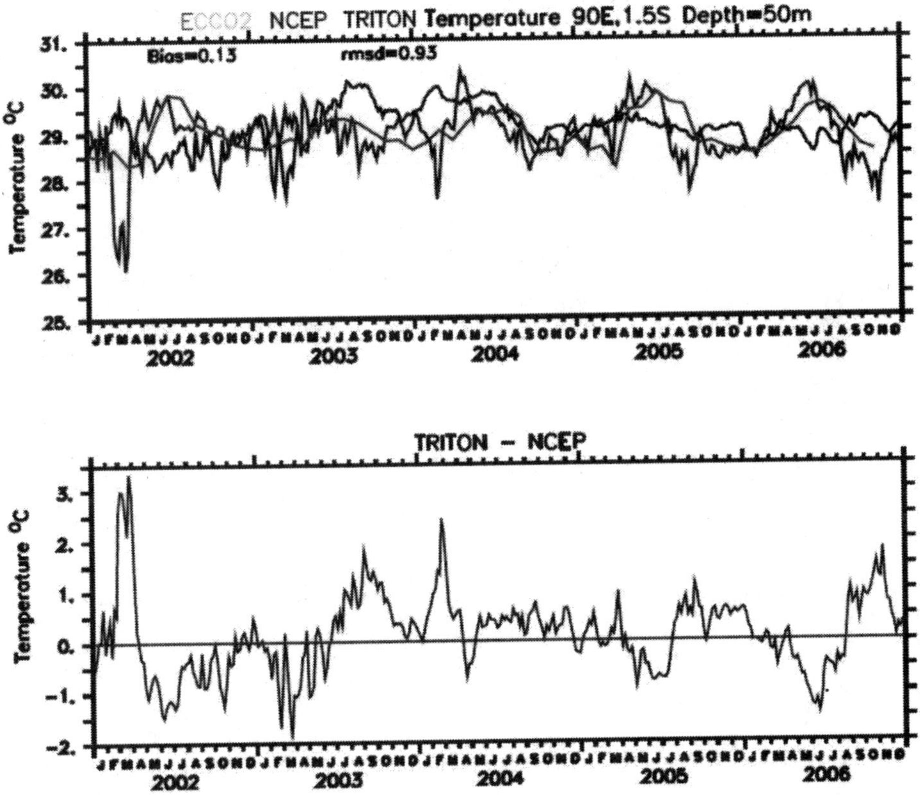

Fig. 6: The upper panel represents the time series plot of the ECCO2 vs NCEP model reanalysis/assimilation vs TRITON (black) observational 50 meter depth temperature. Bias and rmsd values show for the TRITON vs NCEP. Lower panel represents the difference plot of TRITON observational and NCEP model reanalysis at 90ºE, 1.5ºS.

According to the Fig. 10 the bias value is 0.41 between TRITON vs NCEP which stated that the TRITON under estimate the NCEP reanalysis in the case of surface temperature while rmsd value is the 0.68 which represents fluctuation from mean. The difference plot takes its limit 2.5°C to -2.0°C. The ECCO2 is also deviated in the plot from observational plot. This implies that the NCEP reanalysis temperature comparatively perform same with TRITON as well as with RAMA.

The bias value is 0.13 between RAMA vs NCEP which stated that the RAMA under estimate the NCEP reanalysis in the case of temperature at depth 50 m while rmsd value is the 0.93 which represents fluctuation from mean (Fig. 11). The difference plot takes its limit 3.5°C to -2.0°C. The ECCO2 is also deviated from observation in the plot. This implies that the NCEP reanalysis temperature at this depth perform comparatively well than surface temperature.

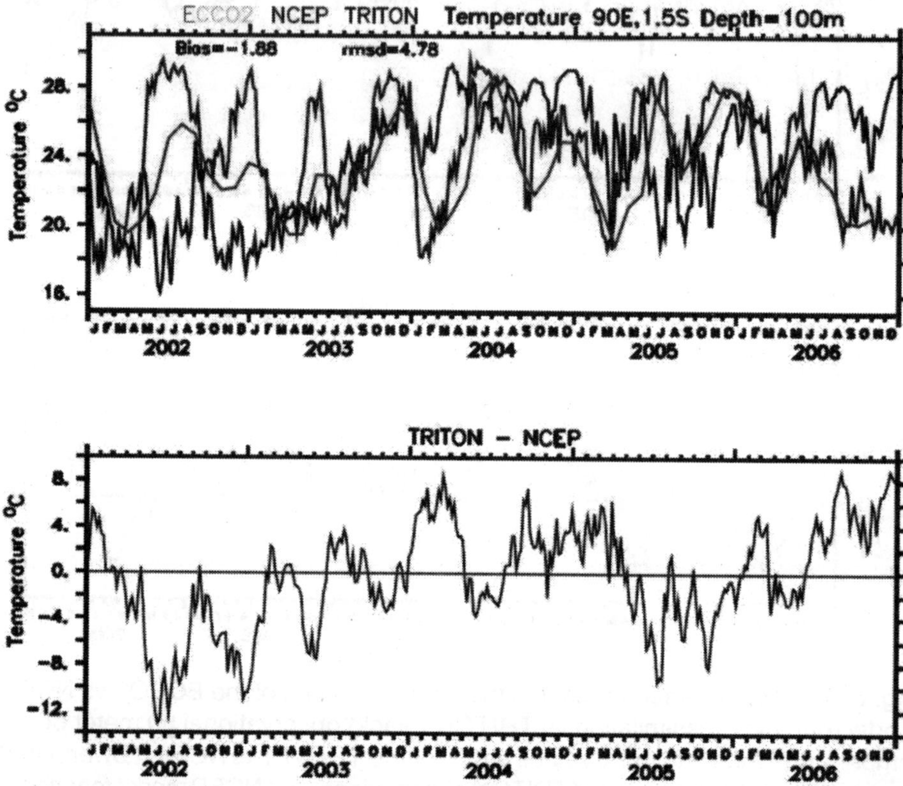

Fig. 7: The upper panel represents the time series plot of the ECCO2 vs NCEP model reanalysis/assimilation vs TRITON (black) observational 100 meter depth temperature. Bias and rmsd values show for the TRITON vs NCEP. Lower panel represents the difference plot of TRITON observational and NCEP model reanalysis at 90ºE, 1.5ºS.

Fig. 12 shows that the bias value is -1.88 between RAMA vs NCEP which stated that NCEP under estimate the RAMA for temperature at depth 100 m while rmsd value is the 4.78 which represents fluctuation from mean. The difference plot takes its limit 10°C to -14.0°C. The ECCO2 is also deviated from observation in the plot. Thus the NCEP reanalysis temperature at this depth performs comparatively bad than temperature at 50 m.

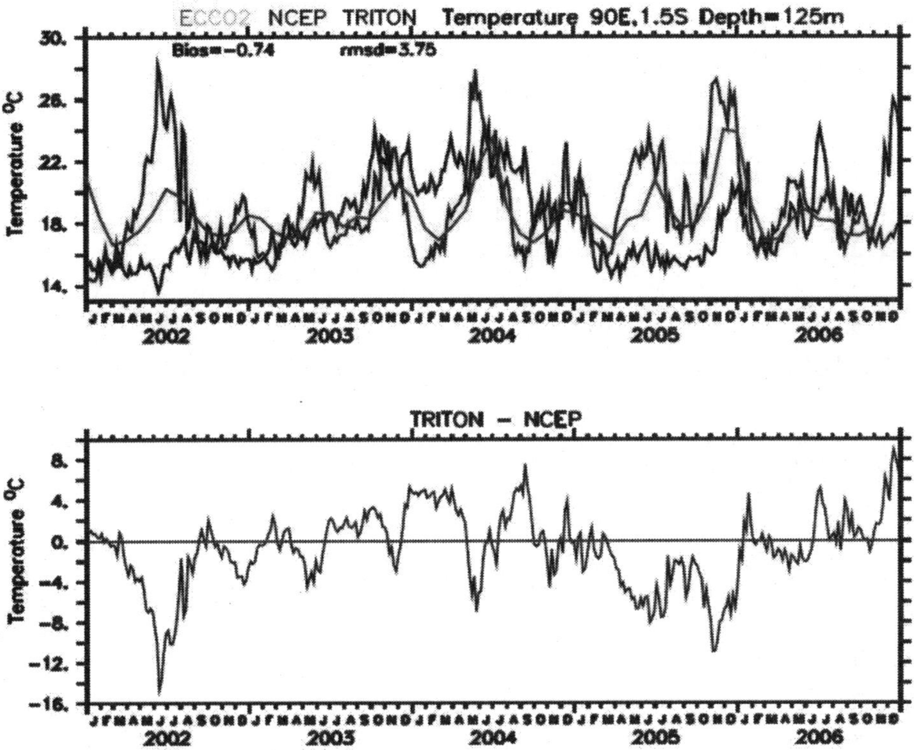

Fig. 8: The upper panel represents the time series plot of the ECCO2 vs NCEP model reanalysis/assimilation vs TRITON (black) observational 125 meter depth temperature. Bias and rmsd values show for the TRITON vs NCEP. Lower panel represents the difference plot of TRITON observational and NCEP model reanalysis at 90ºE, 1.5ºS.

Bias value is -0.74 between RAMA vs NCEP represented by the Fig. 13 which stated that the NCEP under estimate the RAMA in the case of temperature at 125 m depth while rmsd value is 3.75 which represents fluctuation from mean. The difference plot takes its limit 10°C to -16.0°C. The ECCO2 is also deviated from observation in the plot. Thus performance of the NCEP reanalysis temperature at this depth also very bad.

According the Fig. 14 the bias value is -0.27 between RAMA vs NCEP which stated that the NCEP under estimate the RAMA in the case of temperature at 200 m depth while rmsd value is 1.11 which represents fluctuation from mean. The difference plot takes its limit 3.0°C to -3.5°C. The ECCO2 is also deviated from observation in the plot. Thus, performance of the NCEP reanalysis temperature at this depth better than other depth.

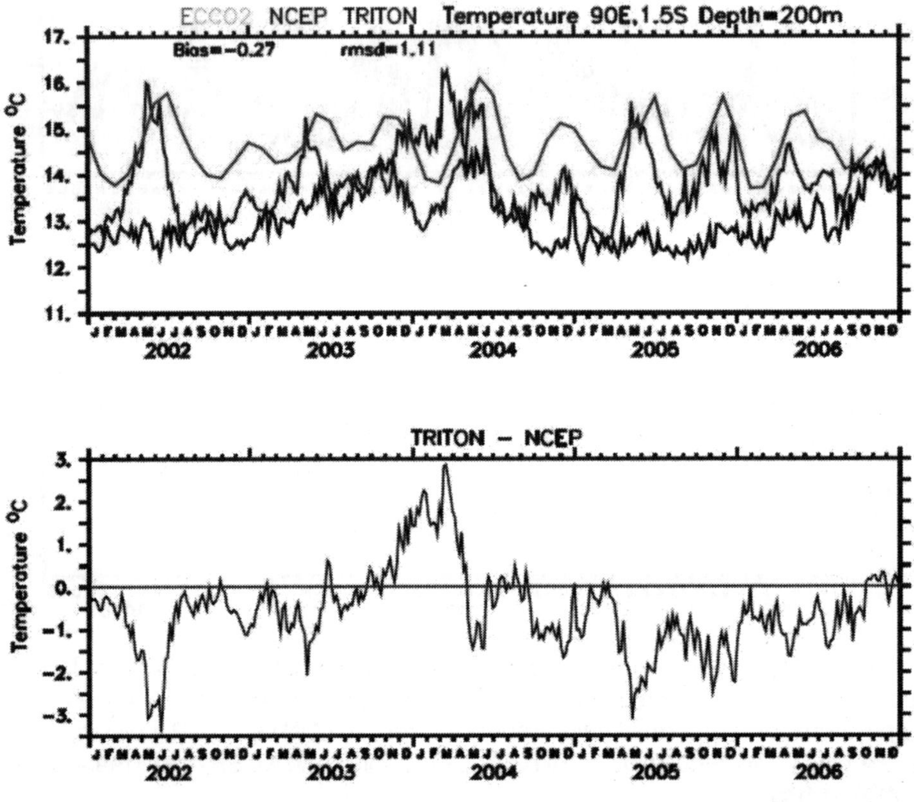

Fig. 9 : The upper panel represents the time series plot of the ECCO2 vs NCEP model reanalysis/assimilation vs TRITON (black) observational 100 meter depth temperature. Bias and rmsd values show for the TRITON vs NCEP. Lower panel represents the difference plot of TRITON observational and NCEP model reanalysis at 90ºE, 1.5ºS.

Bias value is 0.22 between TRION vs NCEP in Fig. 15 which stated that the TRITON under estimate the NCEP reanalysis in the case of surface salinity while rmsd value is 0.45 which represents fluctuation from mean. The difference plot takes its limit 1.6 psu to -1.0 psu. Thus NCEP salinity performs better than temperature.

Fig. 16 represents the time series plot of RAMA vs NCEP salinity for the period 2005 to 2006. The bias value is -0.1 between RAMA vs NCEP which stated that the RAMA under estimate the NCEP reanalysis in the case of surface salinity while rmsd value is 0.27 which represents fluctuation from mean. The difference plot takes its limit 0.8 psu to -0.8 psu. Thus NCEP salinity performs better for this particular time period.

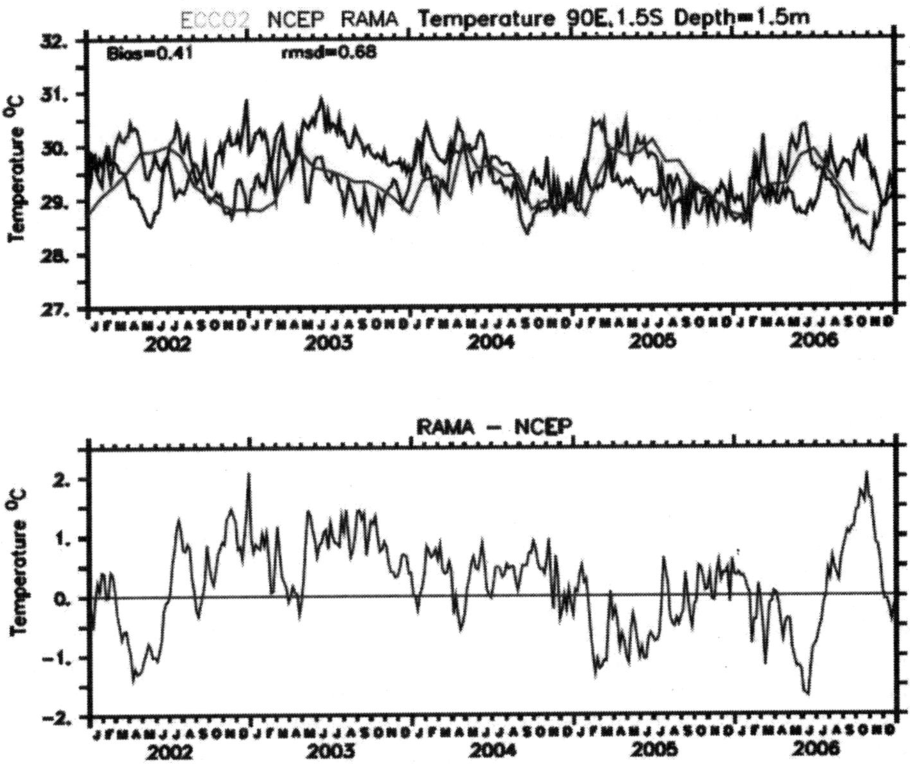

Fig. 10 : The upper panel represents the time series plot of the ECCO2 vs NCEP model reanalysis/assimilation vs RAMA (black) observational surface temperature. Bias and rmsd values show for the RAMA vs NCEP. Lower panel represents the difference plot of RAMA observational and NCEP model reanalysis at 90ºE, 1.5ºS.

Discussion

This study avails to the evaluation of performance of GODAS/NCEP using the TRITON and RAMA in the tropical equatorial Indian Ocean in similar way to Pacific Ocean (Behringer and Xue, 2004). The reanalysis/assimilation data have large biases with respect to the TRITON and RAMA. The zonal and meridional currents of the NCEP reanalysis is very poor in performance than temperature and salinity.

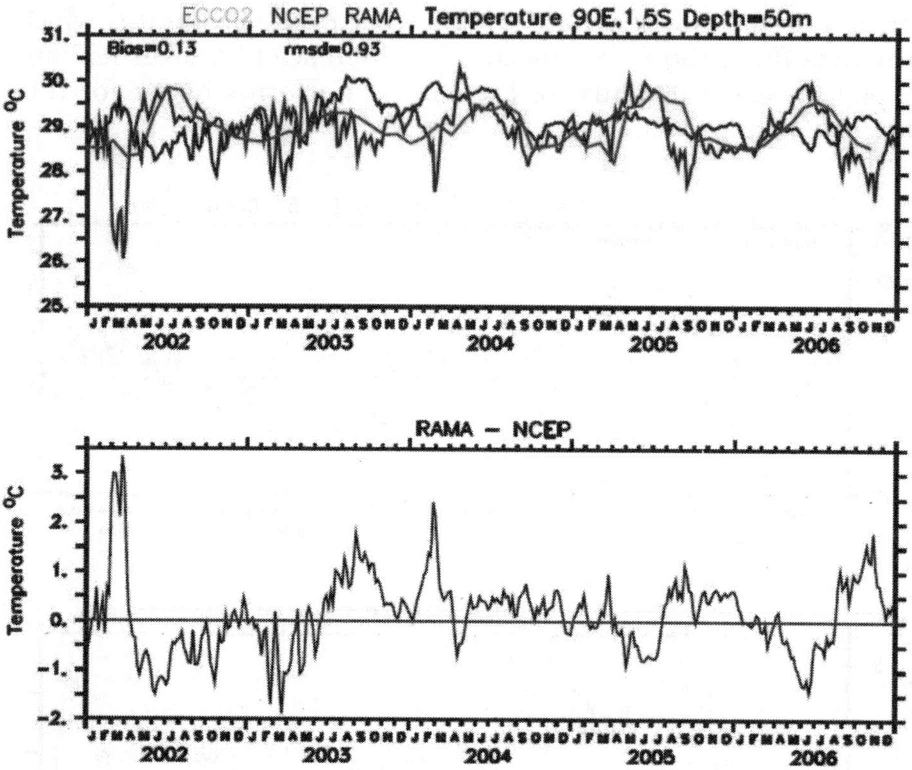

Fig. 11: The upper panel represents the time series plot of the ECCO2 vs NCEP model reanalysis/assimilation vs RAMA (black) observational 50 meter depth temperature. Bias and rmsd values show for the RAMA vs NCEP. Lower panel represents the difference plot of RAMA observational and NCEP model reanalysis at 90ºE, 1.5ºS.

The NCEP and ECCO2 reanalysis data of the temperature under estimated and over estimated the TRITON and RAMA at the different depth. A common rmsd and bias values were found for the same depth of TRITON and RAMA comparison with the NCEP reanalysis. It state that both the observational mooring buoys performing equally well at this mooring position. The study of interannual variability of Indian Ocean base on the NCEP reanalysis arise the question on the results in the presence of such biases in the data. There is problem with GODAS and ECCO2 reanalysis/assimilation data that it have very much biases in current data with enough/less biases in the temperature/ salinity that may exist due to the insufficient availability of observational data in the Indian ocean used in assimilation method for reanalysis and

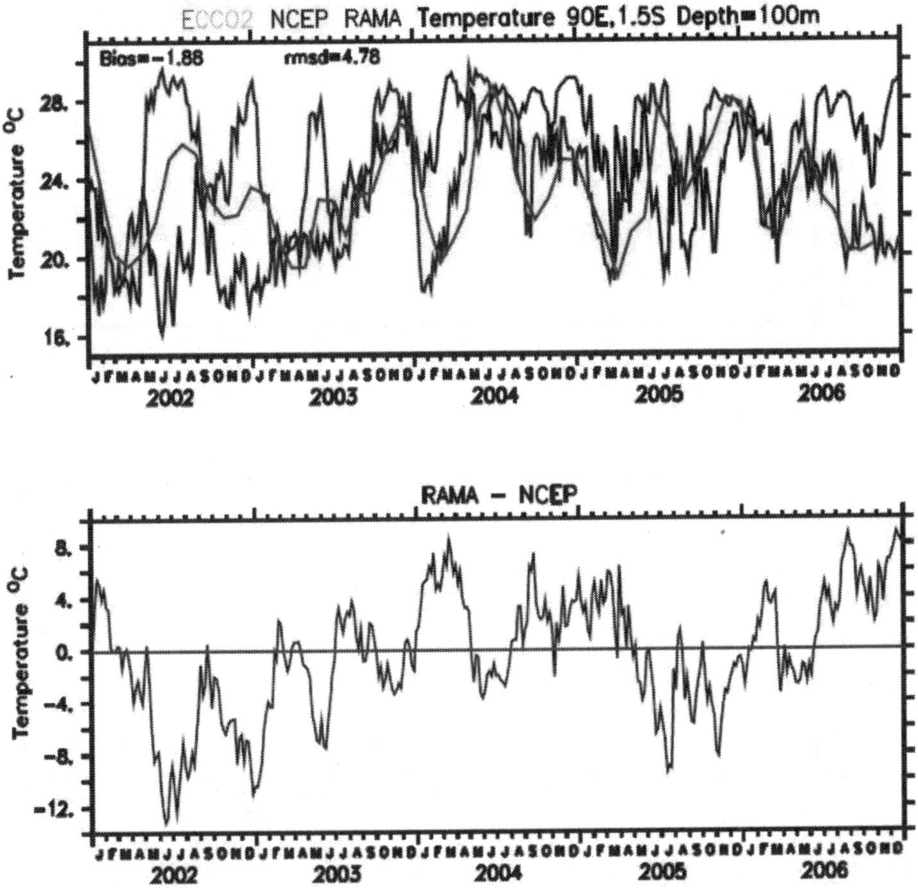

Fig. 12: The upper panel represents the time series plot of the ECCO2 vs NCEP model reanalysis/assimilation vs RAMA (black) observational 100 meter depth temperature. Bias and rmsd values show for the RAMA vs NCEP. Lower panel represents the difference plot of RAMA observational and NCEP model reanalysis at 90ºE, 1.5ºS.

inappropriate forcing used in the model simulation therefore it is very important to correct the ocean assimilation data for the climate prediction. On the basis of this study we can conclude that due to the dull performance of the GODAS data the climate forecast study results can not be very accurate to the real nature. In additional result it may also be concluded that the performance of the ECCO2 is better than the GODAS/NCEP on the basis of the comparison of the plots but still not very clear due to unavailability of equal ECCO2 temporal data set.

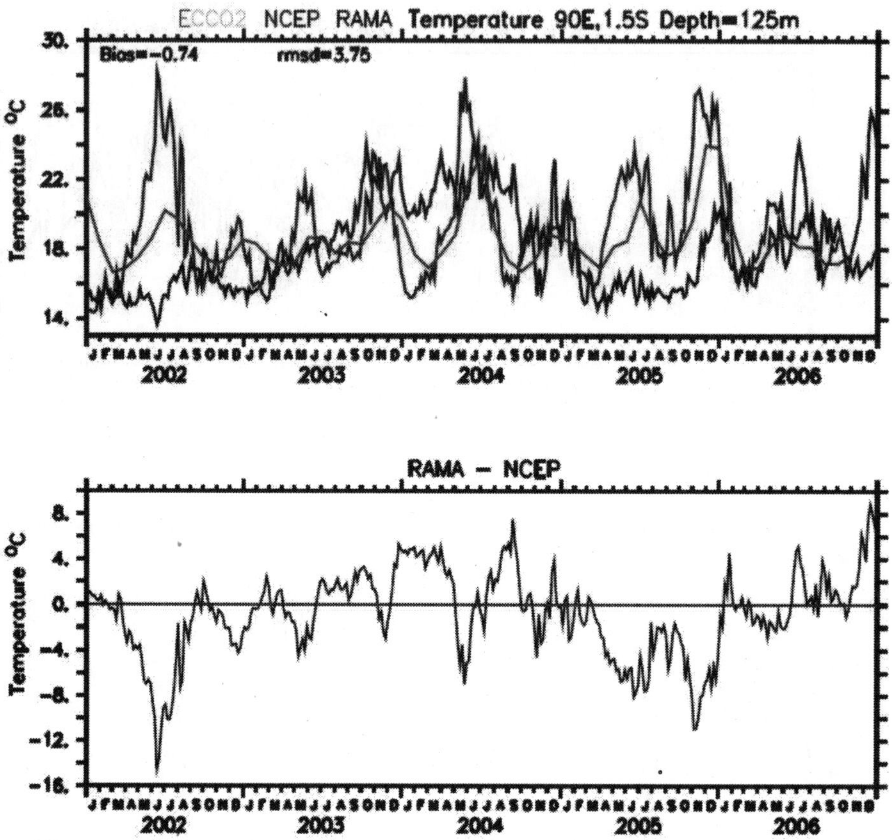

Fig. 13: The upper panel represents the time series plot of the ECCO2 vs NCEP model reanalysis/assimilation vs RAMA (black) observational 125 m depth temperature. Bias and rmsd values show for the RAMA vs NCEP. Lower panel represents the difference plot of RAMA observational and NCEP model reanalysis at 90ºE, 1.5ºS.

Acknowledgement

Authors thank to the ECCO2/MAP/NASA and GODAS/NCEP programs for providing the reanalysis/assimilation data by their web page (http://ecco2.jpl.nasa.gov/products/ and http://www.cpc.ncep.noaa.gov/products/GODAS), and also thanks to the RAMA and TAO/TRITON (PMEL /NOAA) for providing the Indian Ocean Mooring Buoys observational data by the data delivery web page (http://www.pmel.noaa.gov/tao/data_deliv/deliv-nojava-rama.html and http://www.pmel.noaa.gov/tao/data_deliv/deliv.html). Special thanks to Prof. D. Sengupta, Centre for

Atmospheric and Oceanic Sciences, IISc Bangalore for giving me the idea and suggestions, and Prof. Avinash Chand Pandey, HOD for encouragement and coordination to carry out such type of work at our new Department.

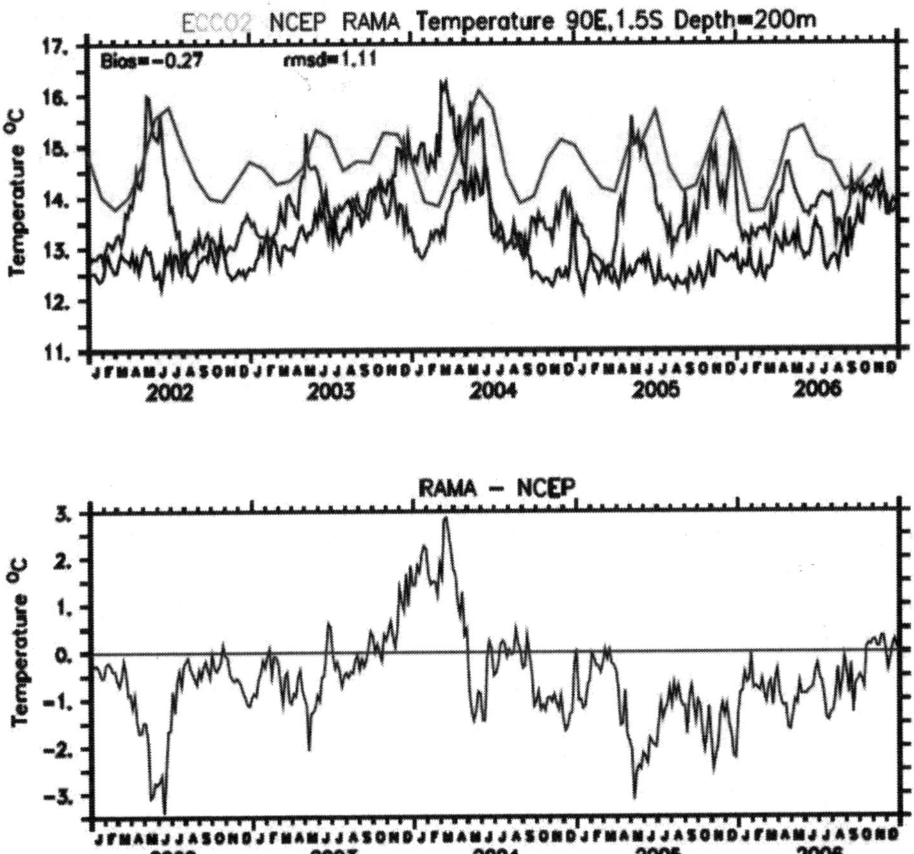

Fig. 14 : The upper panel represents the time series plot of the ECCO2 vs NCEP model reanalysis/assimilation vs RAMA (black) observational 200 m depth temperature. Bias and rmsd values show for the RAMA vs NCEP. Lower panel represents the difference plot of RAMA observational and NCEP model reanalysis at 90ºE, 1.5ºS.

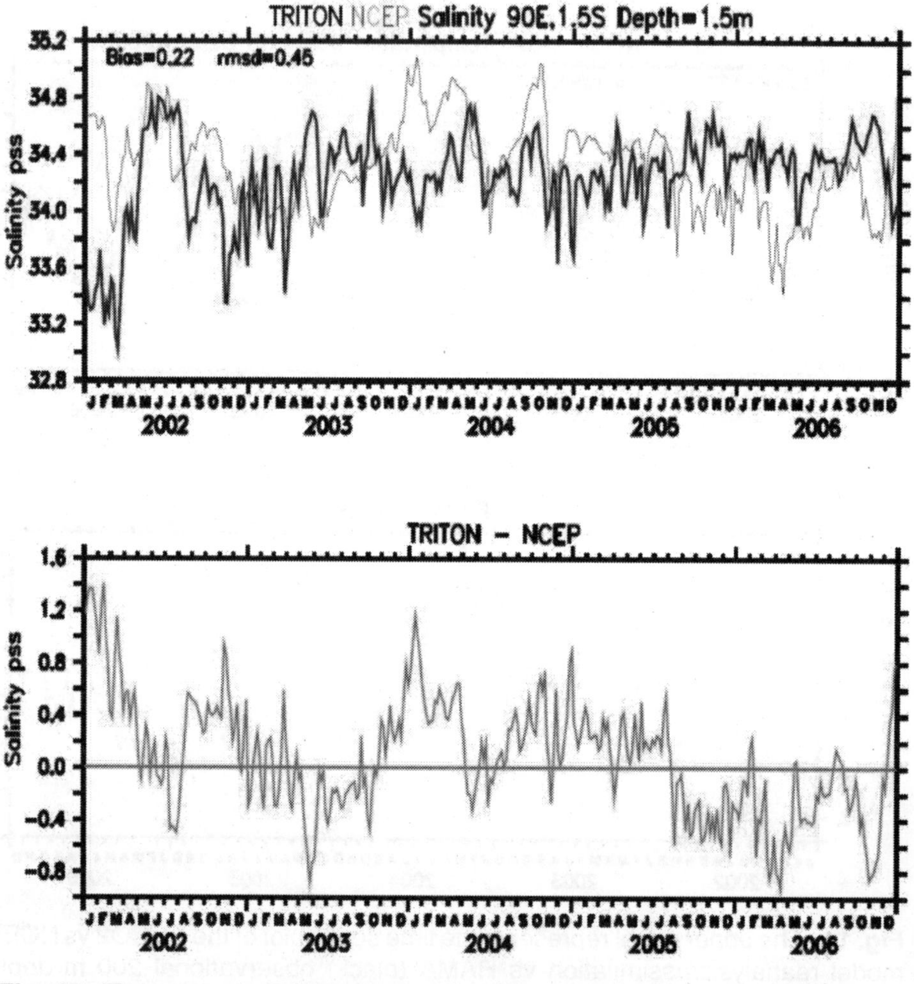

Fig. 15: The upper panel represents the time series plot of the NCEP model assimilation vs TRION (black) observational surface salinity. Bias and rmsd values show for the TRITON vs NCEP. Lower panel represents the difference plot of TRION observational and NCEP model reanalysis at 90ºE, 1.5ºS.

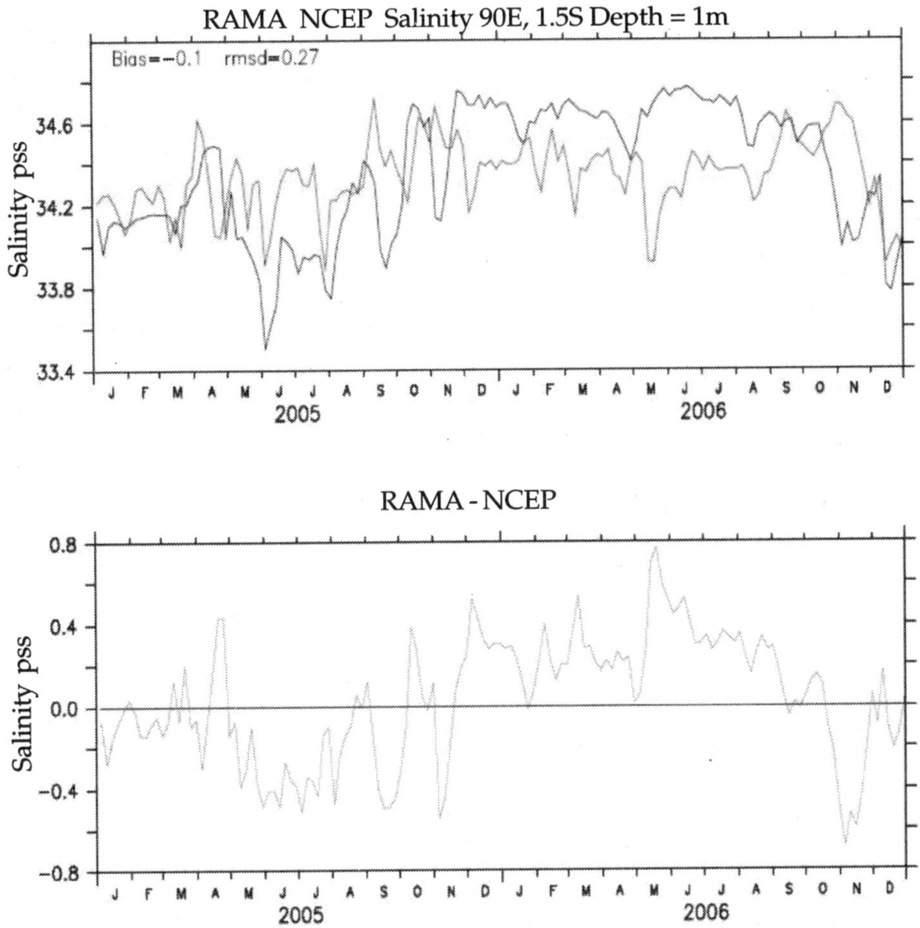

Fig. 16 : The upper panel represents the time series plot of the NCEP model assimilation vs RAMA (black) observational surface salinity from 2005-2006. Bias and rmsd values show for the RAMA vs NCEP. Lower panel represents the difference plot of RAMA observational and NCEP model reanalysis at 90ºE, 1.5ºS.

References

Behringer, D. W. and Xue, Y. (2004) Evaluation of the global ocean data assimilation system at NCEP: The Pacific Ocean, Eighth Symposium on Integrated Observing and Assimilation System for Atmosphere, Ocean, and Land Surface. AMS 84th Annual Meeting, Seattle, Washington, 11-15.

Behringer, D. W., Ji, M. and Leetmaa, A. (1998) An improved coupled model for ENSO prediction and implications for ocean initialization. Part I: The ocean data assimilation system, Mon. Wea.Rev, v. 126, pp. 1013-1021.

Conkright, M. E., Levitus, S., Brien, T. O', Boyer, T. P., Stephens, C., Johnson, D. Baranova, O., Antonov, J., Gelfeld, R., Rochester, J. and Forgy, C. (1999) World Ocean Database 1998, Documentation and Quality Control Version 2.0. National Oceanographic Data Center Internal Report 14, National Oceanographic Data Center, Silver Spring, Maryland.

McPhaden, M. J. (1993) TOGA-TAO and the 1991-93 El Nino-Southern Oscillation Event. Oceanography, v. 6, No. 2, pp. 36-44.

Ji, M., Leetmaa, A. and Derber, J. (1995) An ocean analysis system for seasonal to interannual climate studies. Mon. Wea. Rev., v. 123, pp. 460-481.

Derber, J., and A. Rosati, 1989: A global oceanic data assimilation system. J. Phys. Oceanogr., v. 19, pp. 1333-1347.

McPhaden, M.J., Busalacchi, A.J., Cheney, R., Donguy, J.R., Gage, K.S., Halpern, D., Ji, M., Julian, P., Meyers, G., Mitchum, G.T., Niiler, P.P., Picaut, J., Reynolds, R.W., Smith, N. and Takeuchi, K. (1998) The Tropical Ocean-Global Atmosphere (TOGA) observing system: A decade of progress. J. Geophys. Res., v.103, pp. 14,169-14,240.

McPhaden, M. J., Delcroix, T., Hanawa, K., Kuroda, Y., Meyers, G., Picaut, J. and Swenson, M. (1999). The ENSO observing system, Proceedings of OceanObs99 -Solicited Papers. OCEANOBS99

Conference, 18-22 October 1999, St. Raphael France. Centre National d'Etudes Spatiales, Toulouse, France.

Reynolds, R. W., Rayner, N. A., Smith, T. M., Stokes, D. C. and Wang, W. (2002) An improved in situ and satellite SST analysis for climate. J. Climate, v. 15, pp. 1609-1625.

Pacanowski, R.C. and Griffies, S. M. (1998) MOM 3.0 Manual. NOAA / Geophysical Fluid Dynamics Laboratory. Princeton, NJ, USA 08542.

❑❑❑

20

On Some Recent Applications of Gravimetry to Earth Sciences

V.M. TIWARI
National Geophysical Research Institute, Hyderabad (CSIR), 500 606

Abstract

In recent years, development of modern instruments for ground, air and satellite gravity measurements have allowed to record gravity change to the accuracy of few micro-Gals. Concurrently, there is tremendous progress in analysis methods and availability of constraining information to understand gravity changes. Therefore, gravity methods are now used for numerous geophysical applications, such as modeling of hydrological mass changes, crustal deformation, ice-mass balance in the Polar Regions, in addition to the conventional applications in exploration and geodynamics. A brief review of some recent applications is discussed here.

Keywords : Gravity methods, subsurface geology, petroleum industry, mass movement, oceanographic studies

Introduction

Gravimetry (gravity method) is a geophysical technique in which changes in the gravitational attraction of earth are measured and causes of observed changes are interpreted in terms of mass distribution over time and space. Any process that can produce mass change can be studied through Gravimetry. Thus, there are numerous applications of gravimetry in earth sciences such as mapping of subsurface geology, monitoring of snow and water mass movement, tidal studies and oceanographic studies. Gravimetry has been used since early 1920s in the petroleum industry for mapping of subsurface density changes as gravitational field strength is directly proportional to the mass, in turn density of subsurface materials. The lateral variations in the density of subsurface materials produce gravity changes, which is referred as gravity anomaly and is defined as departure of observed gravity filed from theoretically computed gravity field. In present contribution, some recent applications of gravity method are presented.

Recent Advances in Measuring and Modeling of Gravity Field

Absolute Gravity Measurements

Absolute gravity values are required for processing of gravity data and computation of gravity anomalies. In early days the absolute gravity values were determined from pendulum and further extended to other places using relative/ spring type of gravity meters. Until 1990s absolute gravity values worldwide were referenced to Potstdam base, which has accuracy of 1mGal (Torge, 1989). In mid eighties absolute gravity measurements based on free falling reflector corner cube in vacuumed chamber has began, which has accuracy of 0.01 mGal (Faller, 2002). Now FG5 the only commercially available absolute gravity meter measures gravity field with the accuracy of 1 uGal. In India first absolute gravity measurement with FG5 is made in 2003 (Tiwari *et al.,* 2006).

Airborne Gravity and Gradiometery

Airborne gravity and gradiometery are used often for reconnaissance survey of mineral exploration due to the fast coverage of large region. Development of new instrumentation capabilities and GPS positioning of moving platform have improved the accuracy of measurements. Gradiometery is now commonly being used for mineral exploration due to its small spatial resolution. However, there are no commercial gravity gradiometer available though some companies like Fugro, BHP provide their services for airborne gravity gradient measurements.

Satellite Gravimetry

Satellite gravity measurements have started in the eighties when Seasat satellite used radar altimeter to provide marine geoid globally (Nerem *et al.,* 1995; Fu and Cazenave, 2001). Over the period several satellite altimetry missions have improved the spatial and amplitude resolutions of marine geoid and gravity field and they are now available in public domain (Sandwell and Smith, 1997). The greatest advantages of these data are uniform global coverage. Satellite Laser Ranging has provided long wavelength information of shape and datum (Nerem *et al.*, 1995). In recent years CHAllenging Minisatellite Payload (CHAMP) and Gravity Recovery and Climate Experiment (GRACE) gravity missions started measuring global terrestrial and marine gravity field directly (Reigber *et al.*, 1996; Tapley *et al.*, 2004). GRACE has also provided temporal changes in the earths gravity filed with unprecedented accuracy and now being used for several hydrological, oceanographic and climate studies. A new satellite Gravity Field and Steady-State Ocean Circulation Explorer (GOCE) has been launched in March, 2009 and is the most advanced gravity space mission to date. GOCE is low-orbiting satellite, flying just 250 km above the earth. GOCE has three pairs of sensitive accelerometers set in three dimensions which will experience the gravitational acceleration of the Earth and allow the simultaneous measurement of six independent components of the gravity field.

Processing and Modelling

Methods of processing of gravity data are developed to suit the new kind of acquisition for example processing of data on moving platform, corrections for terrain, tidal etc. Identification of sources of gravity anomalies and separation of regional and residual are addressed using different methods. Power spectrum of gravity anomalies are conventionally being used to the identification of sources and also to separate long and short wavelength anomalies (Spectral and Grants, 1970 Hahn *et al.*, 1976). Regional residual separation has been effective done using a new technique based on finite element method (Mallick and Sharma, 1999). Inversion and deconvolution filters are used to obtain causative bodies and density distributions (Dimri, 1992). Based on scaling properties of power spectrum new approach of estimating subsurface source distribution have been proposed (Maus and Dimri, 1995). Continuous wavelet transform are now being used to locate the boundaries of causative sources (Marlet *et al.*, 2001). Computation of gravity responses of polygons used for 2D bodies (Talwani, 1959) are modified for computations of 3D bodies (Pedersen, 1979; Gotze and Lahmeyer, 1988) as well for fast computations (Singh and Guptasharma, 2001). GIS based software like IGMAS has provided a much simpler method of 3D modeling of gravity data (Schmidt, and Götze, 1998). New modifications to gravity computations are made to suit the inversion of gravity gradient data as well (Li, 2001; Zhdanov *et al.*, 2004).

Recent Applications

Lithospheric and Geodynamics

Gravity anomalies are used to decipher subsurface density structure caused by geodynamic processes traditionally (*e.g.* Mishra *et al.*, 2008). However, inherent problem of non-uniqueness of gravity modeling has limited its application in deriving density model up to the lithospheric depth. Now, the availability of many seismic and seismological data are enabling to construct density model up to lithospheric depth. One of such example is presented in Fig. 1, which

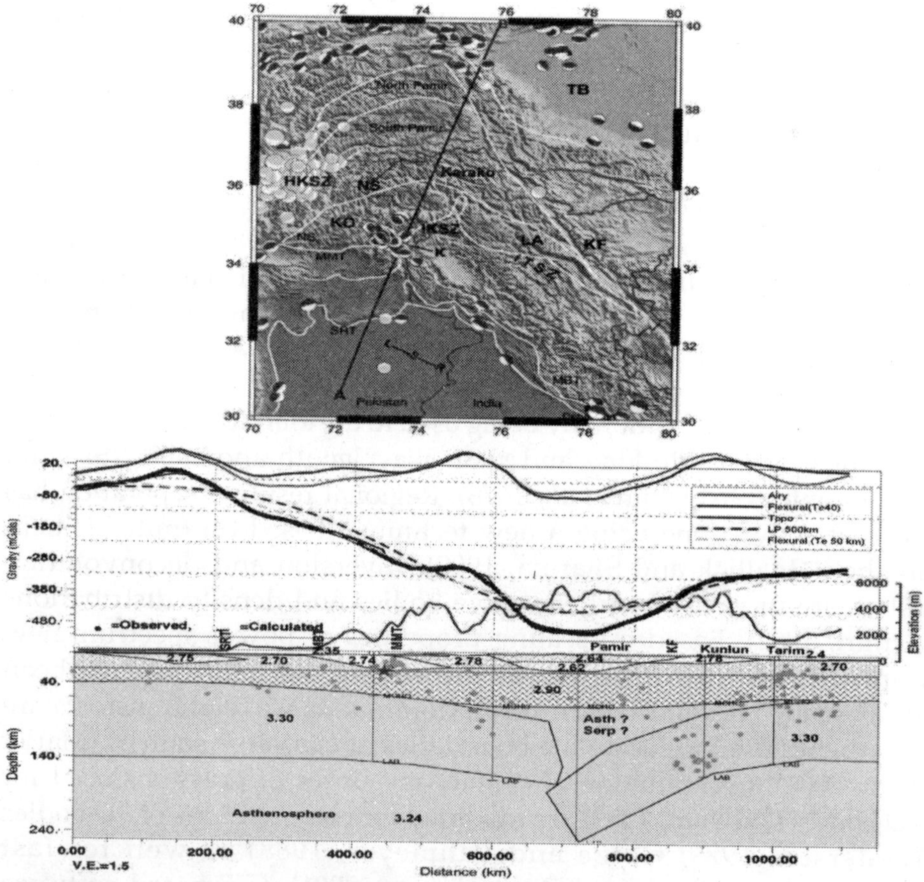

Fig. 1: Lithospheric density model in the NW Himalaya is derived from modeling of gravity anomalies with constraint from seismological results (After Tiwari *et al.*, 2009). Location of profile is also marked in the left panel.

shows density structure of Himalayan collision zone and suggests underthrusting of Indian and Eurasian plates from either side.

The 26 December 2004 Sumatra–Andaman earthquake has once again reminded the vulnerability of seismic hazard from subduction zone earthquakes. Recent study of gravity anomalies in the subduction zones (Wells *et al.*, 2003) suggest that gravity anomalies can provide information on Seismogenic behavior of the subduction zone. It has been noticed that epicenters of large magnitude earthquakes fall in the relative low region of gravity anomalies. Sources of these anomalies are interpreted in the different way. Nevertheless exposition of this has been discussed for Andaman Sumatra subduction zone suggesting that width of Seismogenic zone based on structural model governing the magnitude of the large earthquakes (Gravemeyer and Tiwari, 2006). Fig. 2 shows an example how gravity anomalies are visually correlated to large earthquakes in the Java Andaman subduction zone.

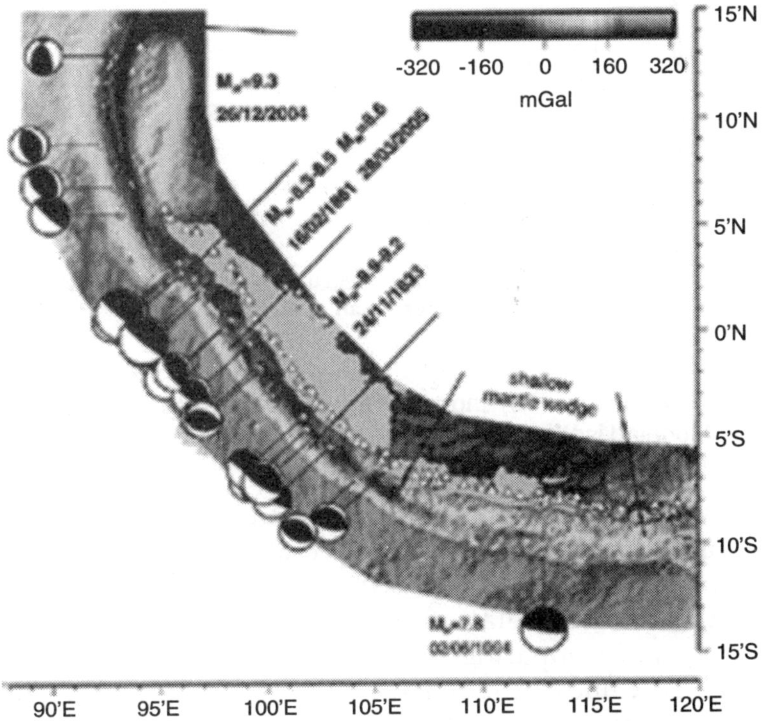

Fig. 2 : Gravity anomalies corrected for bathymetry and large earthquakes in Java-Andaman subduction zone. Most of the earthquakes are located in the relatively low gravity anomalies shown in blue color (after Gravemeyer and Tiwari, 2006).

Isostasy and Strength of Continental Lithosphere

Long term behavior of lithospheric strength is estimated from isostatic studies of topography in the ocean and continent both (Watts, 2001). Effective strength of lithosphere *i.e.* part of the lithosphere that behaves elastically for long term loading is termed as Effective Elastic Thickness (Te). It has been observed that Te in the oceanic regions generally coincides with the 450°C ± 150°C isotherm (Fig. 3). Whereas over continent it shows a large spectrum that leads a global debate about rheology of continental lithosphere (Jackson, 2002). To understand this better a large number of Te estimates using different methods are needed from different geological settings.

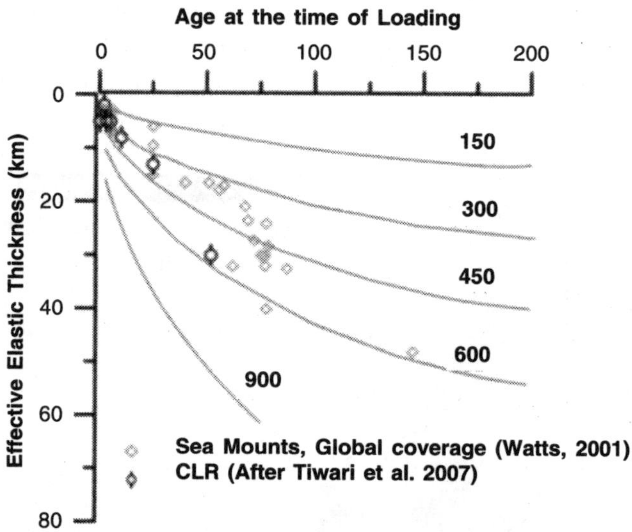

Fig. 3: Plot between EET (Te) and age of the lithosphere at the time of topographic formation. Modeled isotherms are also shown (After Tiwari *et al.*, 2007).

A numerous studies are being conducted over different continental regions using spectral as well as flexural modeling and numerical modeling approaches. Spectral approaches are based on finding the admittance and correlation between gravity anomalies and topography and comparing them with different assumed model to derive the Te values. A schematic diagram of showing admittance and coherence methods are illustrated in the Fig. 4.

It is interesting to note that Te derived over Indian peninsular region, over Deccan trap area shows very low values of about 10-15 km (Tiwari and Mishra 1999). It suggests that Te in this part of continent would certainly be less than 450°C and might be well within in the crust. Following the similar approach the Te estimated from western Himalayan region is as much as 55 km. The range of Te values (Fig. 5) found over India is intriguing and provides information for further discussion in understanding the continental elastic behavior.

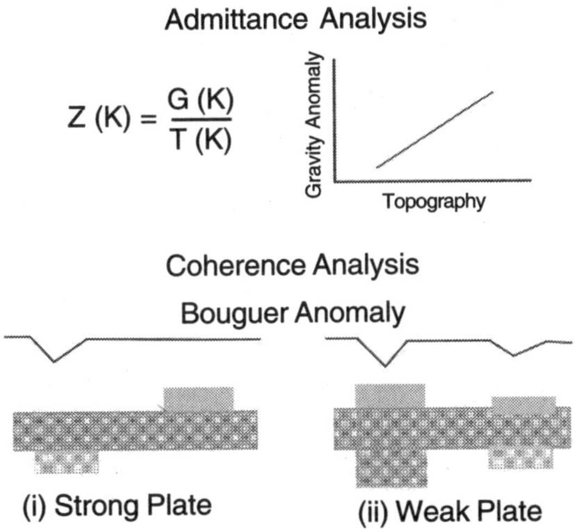

Fig. 4 : Two cross spectral methods used in estimating EET (Te) of lithosphere. Admittance is based on modeling of linear transfer function between gravity anomalies and topography. Coherence analysis uses the correlation between gravity anomalies and topography. Strong plate will show no correlation whereas weak plate will correlate and the wavelength on which correlation increase is the measure of strength.

Time-lapse Gravity for Oil Gas Reservoirs Monitoring

In recent years petroleum industry started looking for secondary recovery from oil and gas reservoirs by injecting water and gases like CO_2. Monitoring of the injected water and gas in the reservoir can be made through measuring and modeling of time lapse gravity (Hare *et.al.*, 1999; Alnes *et al.*, 2008). The amplitude of gravity changes is quite appreciable in the region where injections are large and reservoir depth is less. It is more suitable when injected material has more density difference with host. For example if water is injected to a gas field or CO_2 is injected to an oil field. This has opened a new dimension to the reservoir studies in the cost effective manner compare to the time lapse seismic, which is much expansive and cumbersome than repeat gravity measurements.

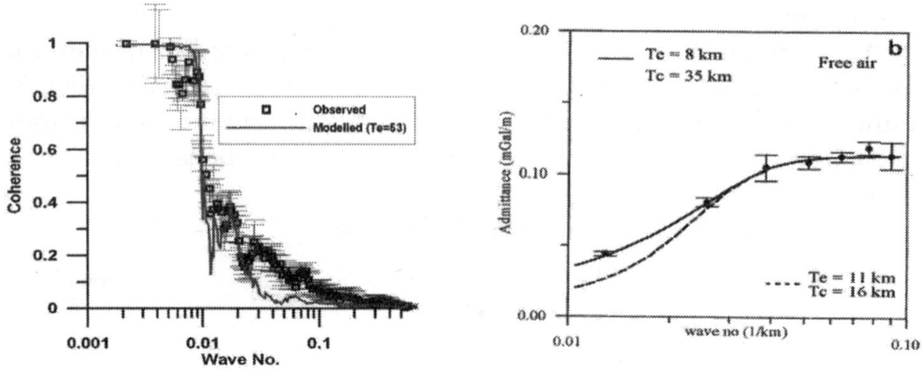

Fig. 5 : Estimate of EET (Te) over continental region (a) based Coherence in gravity anomalies and topography over NW Himalaya (after Tiwari *et al.,* 2009a). (b) estimating Te over western part of Deccan trap region (after Tiwari and Mishra, 1999).

Mapping Ice Mass Changes in the Polar Regions

Direct measurements of mass changes in polar region have been made possible through GRACE satellite gravimetry (Tapley *et al.,* 2004). This has also provided the temporal and spatial distribution of ice mass changes (Velicogna and Wahr, 2006). It has been observed that ice mass are melting more rapidly in the eastern part of Antarctica than the western part (Chen *et al.,*2008). Fig. 6 shows the mass changes computed from GRACE gravity mission over Antarctica. It has been estimated that present rate of ice mass loss from Antarctic region is 81 ± 17 km^3/yr.

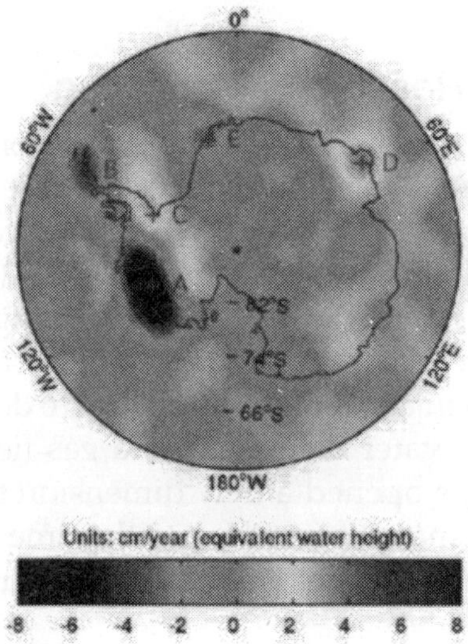

Fig. 6: Ice mass changes over Antarctica from GRACE (after Chen *et al.,* 2008)

Hydrological Applications

Since March 2002 GRACE satellite, is providing temporal gravity changes every month with an accuracy of few microGal and is expected to be in orbit till 2013 (Tapley *et al.*, 2004). The observed gravity changes can now be effectively utilized for hydrological changes (Wahr *et al.*, 1998, Wahr *et al.*, 2009). The observations from GRACE suggested a large amount of water mass loss from northern Indian region (Tiwari *et al.*, 2009). Fig. 7a shows the trend of water mass changes over Indian region. It is evident from the figure that low is centered over New Delhi and extending over entire Indus and Ganga plains. It partly encompasses Himalayan regions as well; therefore part of the decreasing trend of mass is related to melting of Himalayan glaciers. To make it more obvious a time series of mass changes derived from GRACE gravity data is plotted in Fig. 7 b. It shows a clear cut decreasing trend and also a minima during 2004-2005, which might be related to less replenishment of ground water in that year. After careful examination and utilizing new technique of mass estimation, it has been proposed that northern Indian region is losing water mass of 50 km^3/yr. Out of this large hydrological mass loss, about 30 ± 5 km^3/yr is being lost from Indian territorial region itself (Tiwari *et al.*, 2009).

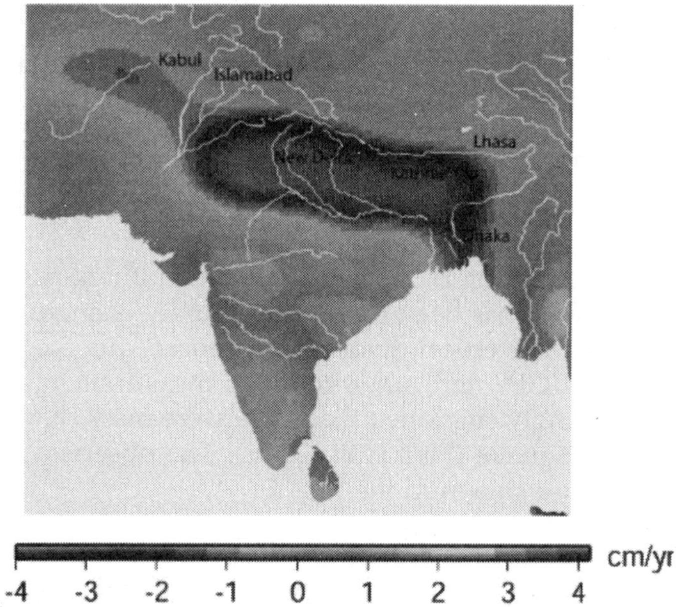

Fig. 7a: Trend of mass changes derived from GRACE gravity satellite over Indian region (after Tiwari *et al.*, 2009)

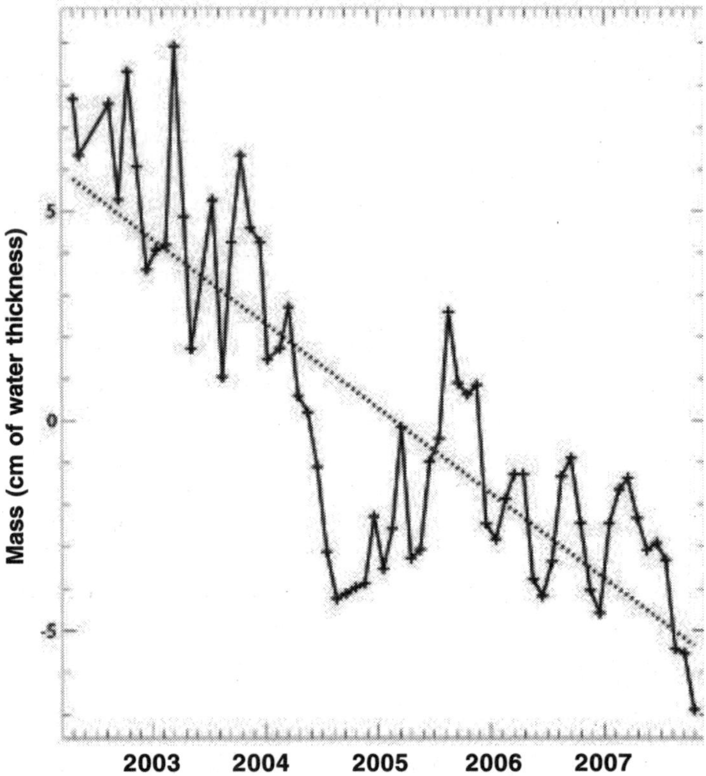

Fig. 7b: Time series of mass changes at central point (Delhi) of largest loss.

Crustal Deformation and Earthquake Studies

One of the important observation that GRACE gravity satellite mission has provide was detecting co-seismic gravity changes of 2004 Sumatra earthquake (Han *et al.,* 2006). After examining and modeling of gravity changes it has been found that gravity changes can not be entirely attributed to vertical deformation caused due the earthquake and measured by GPS and other geodetic measurements. And it is first time convincingly suggested that there were mass changes during the Sumatra earthquake (Han *et al.,* 2006). The observed changes in the gravity filed are shown in the Fig. 8.

Summary

Over last few decades, new class of gravity measurements on ground, sea and from space and air are offering a large amount of data of considerable value. Development of new techniques of modeling, interpretation, including integration with other geophysical and geological data allowed providing improved information of mass distribution and mass movements. Gravimetry thus continues to play an imperative role in geophysical studies though its inherent limitation trims down the application, yet it remains vital component of geophysical studies.

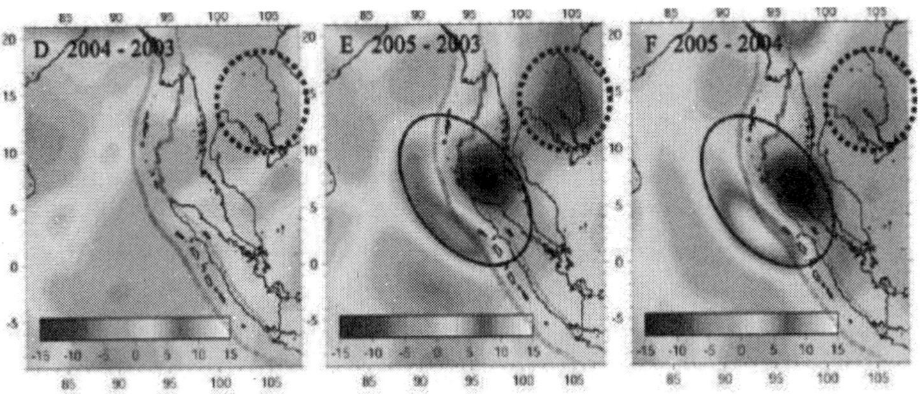

Fig. 8: Gravity changes before and after Sumatra earthquake as observed by GRACE satellite (After Han *et al.*, 2006)

Acknowledgement

Author is thankful to Dr. V.P. Dimri, Director, NGRI, Hyderabad for permission to publish the paper.

References

Alnes, H., Eiken, O. and Stenvold, T. (2008). Monitoring gas production and CO2 injection at the Sleipner field using time-lapse gravimetry. Geophysics, v. 73, pp.WA155.

Chen, J.L., Wilson, C.R., Tapley, B.D., Blankenship, D. and Young D. (2008) Antarctic regional ice loss rates from GRACE. Earth and Planet. Sci. Lett., v. 266 , pp. 140-148.

Dimri,V.P. (1992) Deconvolution and inverse theory. Elsevier Science, Amsterdam.230 p.

Faller, J. E. (2002) Thirty years of progress in absolute gravimetry: A scientific capability implemented by technological advances. Metrologia, v. 39, no. 5, pp. 425–428.

Fu, L.L. and Cazenave, A. (2001) Satellite Altimetry and earth sciences: A handbook of techniques and applications, Academy Press. 463 p

Götze, H. J. and Lahmeyer, B. (1988) Application of 3-D interactive modelling in gravity an magnetic. Geophysics, v. 53, pp. 1096-1108.

Grevemeyer, I. and Tiwari, V.M. (2006) Overriding plate controls spatial distribution of large thrust earthquakes in the Sunda-Andaman Subduction zone. Earth Planet. Sci. Lett, doi:10.1016/j.epsl.2006.08.021,

Hahn, A., Kind, E.G. and Mishra, D.C. (1976) Depth estimation of magnetic sources by means of Fourier amplitude spectra. Geoph. Prosp., v.24, pp.287-308.

Han, S. C., Shum, C. K., Bevis, M. and Kuo, C. Y. (2006) Crustal dilatation observed by GRACE after the 2004 Sumatra-Andaman earthquake. Science, v. 313, pp. 658–662, doi:10.1126/science.1128661.

Hare, J.L., Ferguson, J.F., Carlos, L. V. and Aikenand Brady, J.L. (1999) The 4-D Microgravity method for waterflood surveillance: A model study for the Prudhoe Bay reservoir, Alaska. Geophysics, v. 64, pp.78-87.

Jackson, J. (2002) Strength of the continental lithosphere: Time to abandon the jelly sandwich? GSA Today, v. 12 , pp.4 –10.

Li, X., (2001). Vertical resolution: Gravity versus vertical gravity gradient: The Leading Edge, v. 20, pp. 901–904.

Mallick, K. and Sharma, K.K. (1999) A finite element method for computation of the regional gravity anomaly. Geophysics, v.64, pp.461-469.

Marlet, G., Sailhac, P., Moreau, F. and Diament, M. (2001) Characterization of geological boundaries using 1-D wavelet transform on gravity data: Theory and application to the Himalayas. Geophysics, v. 66, pp. 1116–1129.

Maus, S. and Dimri, V.P. (1995) Potential field power spectrum inversion for scaling geology. J. Geophy. Res., v. 100, pp. 12605-12616.

Mishra, D.C., Mishra, D.C., Tiwari, V.M. and Singh, B. (2008) Geological Significance of Gravity Studies in India, Golden Jubilee Memoir, Geol. Soc. Ind., v.66, pp.329-372.

Nerem, R. S., Jekeli, C. and Kaula, W. M. (1995) Gravity field determination and characteristics: Retrospective and perspective. J. Geoph. Res., v. 100, pp. 15053,15074.

National Research Council (1997) Satellite Gravity and the Geosphere: Contributions to the study of solid earth and its fluid envelop, National Academy Press, Washinton DC. 126 p.

Pedersen, L.L (1979) Constrained inversion of potential field data. Geophysical Prospecting, v. 27, pp. 726–748.

Reigber, C., King, Z., Konig, R. and Schwintzer, P. (1996) CHAMP, a minisatellite mission for geopotential and atmospheric research. Spring Meeting, American Geophysical Union, Abstracts, S40.

Sandwell, D. T. and Smith, W. H. F. (1997) Marine gravity anomaly from Geosat and ERS-1 satellite altimetry. J. Geophy. Res., v. 102, pp. 10039–10054.

Schmidt, S. and Götze, H.-J. (1998) Interactive visualization and modification of 3D models using GIS functions. Physics and Chemistry of the Earth, v. 23, no. 3, pp. 289-295.

Singh, B. and Guptasarma, D. (2001) New method for fast computation of gravity and magnetic anomalies from arbitrary polyhedral. Geophysics, v. 66, 521 doi:10.1190/1.1444942,

Spector, A. and Grant, F. S. (1970) Statistical models for interpreting aeromagnetic data. Geophysics, v. 35, pp. 293–302 (discussion and reply — v. 36, pp. 617 and v. 39, pp. 112; errata—v. 36, pp. 448 and v. 42, pp. 1563).

Talwani, M., Worzel, J. L. and Landisman, M. (1959). Rapid gravity computations for two-dimensional bodies with application to the Mendocino submarine fracture zone. J. Geoph. Res., v. 64, pp. 49–59.

Tapley, B. D., Bettadpur S., Ries J. C., Thompson, P.F. and Watkins M. M. (2004). GRACE measurements of mass variability in the Earth system. Science, v. 305, pp. 503– 505, doi:10.1126/science.1099192.

Tiwari, V. M., Rajasekhar, R.P. and Mishra, D.C. (2009a) Gravity anomaly, lithospheric structure and seismicity of Western Himalayan syntaxis. J. of Seismology, DOI 10.1007/s10950-008-9102-6

Tiwari, V. M., Wahr. J.M. and Swenson, S. (2009b) Dwindling groundwater resources in northern Indian region, from satellite gravity observations. Geoph. Res. Lett., v. 36, L18401, doi:10.1029/2009GL039401

Tiwari, V. M. and Mishra D.C. (2008) Isostatic Compensation of Continental and Oceanic Topographies of Indian Lithosphere. Geol. Soc. Ind., Memoir 68, pp. 173-190.

Tiwari, V.M., Grevemeyer, I., Singh, B. and Morgan, J. P. (2007) Variation of Effective Elastic thickness and melt production along the Deccan-Reunion hotspot track. Earth Planet. Sci. Lett, v. 264, pp. 9-21. doi:10.1016/j.epsl.2007.08.023

Tiwari, V. M. and Mishra, D.C. (1999) Estimation of effective elastic thickness from gravity and topography data under the Deccan Volcanic Province, India. Earth and Planetary Science Letters, v.171, pp.289-299.

Tiwari, V. M., Singh, B., Rao, M.B.S.V. and Mishra, D.C. (2006) Absolute gravity measurements in India and Antarctica. Current Science, v. 91, pp.686-689.

Torge, W., (1989), Gravimetry: Walter de Gruyter, 465 p.

Velicogna, I. and Wahr, J. (2006) Measurements of Time-Variable Gravity Show Mass Loss in Antarctica, Science, v. 311, pp. 1754-1756.

Wahr, J., Tiwari V M. and Swenson S. (2009) Monitoring groundwater variability from space: the GRACE satellite gravity mission. Red Book, IAHS Publ. pp. 263-270.

Wahr, J., Molenaar, M. and Bryan, F. (1998) Time variability of the Earth's gravity field: Hydrological and oceanic effects and their possible detection using GRACE. J. Geophys. Res., v. 103, 30,205– 30,229, doi:10.1029/98JB02844.

Watts, A.B. (2001) Isostasy and flexure of the lithosphere, Cambridge Univ. Press, pp. 458.

Wells, R.E., Blakely Sugiyama, Y., Scholl, D.W. and Dinterman, P.A. (2003) Basin-centered asperities in great subduction zone earthquakes: a link between slip, subsidence, and subduction erosion? J. Geophys. Res., v. 108 doi:10.1029/2002JB002072.

Zhdanov, M. S., Ellis, R. and Mukherjee, S. (2004) Three-dimensional regularized focusing inversion of gravity gradient tensor component data. Geophysics, v. 69, pp. 925–937.

□□□

21

Indian Monsoon Cycles through the Last Twelve Million Years

SHYAM MURTI GUPTA
National Institute of Oceanography, Dona Paula-Goa, 403 004

Abstract

It is an overview for general understanding of the nature, extent, causes and effects of the Indian monsoon cycles in the modern time and in the geological past. An attempt is made to present monsoon, an interdisciplinary subject, in simple ways by briefly introducing the atmospheric, oceanic, solar, planetary, and orbital forcing on monsoon, based on established phenomenon. It limits to encompass the statistically significant and reconstructed cyclic patterns in the Indian summer monsoon, primarily reported due to changes in the proxies for the Earth-Ocean-Atmosphere interactions (resulting into land-sea thermal contrast, inter tropical convection zone, winds, clouds, rains, El-Nino, sea surface temperature), solar (Sunspots, flares, and irradiance), lunar (tidal force), planetary (gravitational perturbations), and the Earth's orbital forces (the eccentricity, precession of equinoxes, axial tilt and orbital plane inclination), changing the amount of incoming solar radiation (insolation) during the last ~12 million years (Ma).

Keywords : Indian Monsoon, El-Nino, Southern Oscillation

Introduction

In India, the word "Monsoon" is synonymous to "Mausam" which stands for the 4-months season of rain and wind accounting for ~90% rains from June to September, which is known as "Chaumasa" – starting with the plowing, sowing and harvesting of Kharif crop by the farmers. Another view is that, it originates from word "Mausim" used by Arab sailors for the southwest sea breeze from the Arabian Sea towards India. Meteorologically, monsoon is the seasonal southwest sea breeze that brings rains over India. Historical accounts narrate that generally monsoon sets in southern India in the first week of June ("आसाढ़स्य प्रथम दिवसे मेघमाल्लिब्ट्सानु") Meghadutam, Kalidasa, ~434 A.D.), and within 4-6 weeks it spreads all over India. Monsoon played an important role in ancient India's prosperity ("Artha-Shastra", Kautilya, ~321-296 B.C). Prolonged droughts are believed to force agrarians of the Indus civilization to migrate elsewhere, and greener pastures of the Indian fields drew central Asians (Greek, Turks, Afghans, Arab, Persians, Mongols, French, Portuguese, British etc.) to invade, plunder, trade, and eventually settle in India during last ~500-2500 yeas B.P. Monsoon is also known for floods/drought causing agrarian economy to suffer, and its prediction based on certain trends in physical variables with predictive possibilities (Nicholls, 1983; Meehl, 1987; Hastenrath, 1987; Sadhuram, 1997; Gadgil *et al.*, 2002) are the most sought after, but the difficult task. A brief account on monsoon, its forcing mechanism, proxy indices, and its significant cycles in time scales of years, decades, centuries to thousand of years (ka) in the last ~12 million years (Ma) are summarized. Open access digital data of the published works on modern Indian Rainfall, Sunspot numbers and total solar irradiance (TSI), and solar insolation are analyzed afresh by the Continuous Wavelet Transform and multi-taper method (MTM) for the spectral analyses, and the graphics are modified and designed in order to avoid copy right on published figures.

Indian Monsoon

Meteorologically monsoon is derived due to differential summer heating by the Sun's Incoming Solar Radiation (insolation) between the tropical Indian Ocean (causing evaporation, cloud formation and high atmospheric pressure with monsoonal SW winds), and the Asian landmass (causing extreme summer heat, low pressure and rainfall) due to atmospheric pressure and moisture difference. Thousands of years ago (ka) Indian astronomers accounted in the astronomical treatise "Surya-Siddhanta" (Burgess, 1860; Shukla, 1957; Thompson, 1997), that the Sun heats up the oceans from "Uttarayana" commencing on "Makar-Sankranti" when Sun is apparently over the tropics of Capricorn (Sishir Ritu, ~14th January) up till "Karka-Sankranti" with Sun over tropics of Cancer (Varsha Ritu,~14th July), which results into rains over India, then Sun returns back ("Dakshinyana") towards tropics of Capricorn (Fig.1A).

The Sun is apparently over the equator twice in a year on ~21st March and ~21st September due to Earth's axial tilt (23.5°) resulting in equal length of the day and night (equinoxes), and thereafter moving towards summer and winter solstices on ~21st June (longest day) and ~21st December (longest night), respectively. Twice crossing over of equator by Sun heats up the tropical ocean, evaporates surface waters, makes clouds, which get transported by the monsoonal winds blowing towards India and resulting into rains. It is known as southwest Indian Summer Monsoon (ISM), and accounts for ~90% of total annual rain fall over India. A modern analog to above ancient description is the "Hadley cell" (Halley, 1686; Hadley, 1735), that brings moist air from oceans heated by the Sun, producing gigantic and towering clouds over enormous region south of equator (Fig. 1B). Moisture in the cloudy air condenses releasing latent heat and resulting in cold dry air at a height of ~15 km, spreading out over the tropic, and then descending at ~25-30° latitudes, which is the reason that most of the deserts are located between 20-30°. Dry air again returns towards equator, evaporating the sea, picking up the moisture on the way, making towers of clouds over the tropical ocean, and thereby completing Hadley cycle, which is widely accepted (Webster *et al.*, 1998).

Earth-Ocean-Atmosphere Interaction

Physical and Topographic Constrains

Indian monsoon is bounded by the unique continental topography (Fig. 2A) that is characterized by the low lands in South-East Asia and Australia facilitating atmospheric interaction between the Western Pacific and Southern Indian Ocean trade wind (easterlies) domain unhindered. Trade winds during northern summer (May-July) are stronger than northern winter (Dec.-Feb.) in the southern Indian Ocean. While blowing east to west they encounter the high lands in Madagascar, East African and Arabian mountain ranges (>1-2 km), which act as a physical barrier and turn them into monsoonal winds heading towards Indian subcontinent all along the Somalian and Arabian Margins in the Arabian Sea. The paleo-Tethyan mountain chain extending from Middle-East Asia, Hindukush, Himalaya, Tibet and Arakan-Yoma ranges (>3-5 km) restrict the monsoon winds to cross over the Indian subcontinent and to reach central Asia. Consequently, most of monsoon rains are confined over the Indian

A

भचक्रनाभौ विषुवद्द्वितयं समसूत्रगम्
अयनद्वितयं चैव चतस्रः प्रथितास्तु ताः ७
तदन्तरेषु संक्रान्तिद्वितयं द्वितयं पुनः
नैरन्तर्यात् तु संक्रान्तेर्ज्ञेयं विष्णुपदीद्वयम् ८
भानोर्मकरसङ्क्रान्तेः षण्मासा उत्तरायणम्
कर्कादेस्तु तथैव स्यात् षण्मासा दक्षिणायनम् ९
द्विराशिनाथा ऋतवस्ततोऽपि शिशिरादयः
मेषादयो द्वादशैते मासास्तैरेव वत्सरः १०
53 (५३)

वैशाखादिषु कृष्णे च योगः पञ्चदशे तिथौ
कार्तिकादीनि वर्षाणि गुरोरस्तोदयात् तथा १७
सुर्य-सिद्धान्त, खन्ड १४, सुत्र ७-१०,१७. 54 (५४)

B

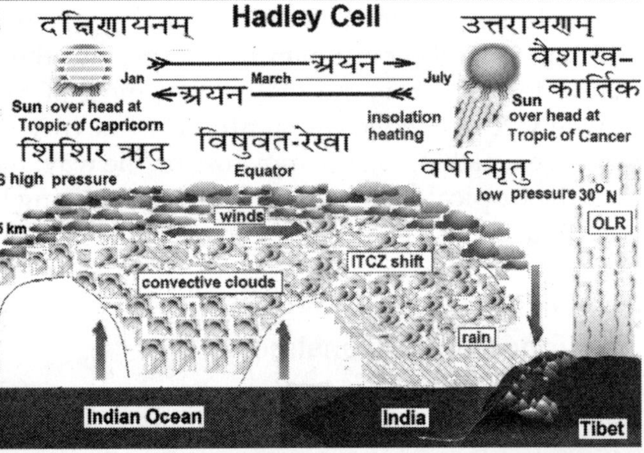

Fig. 1.(A) Description of apparent crossing over of the Sun from tropics of Capricorn (December/January) across the equator in its northward movement (Uttarayana), en-route heating of the sea during summer, which results into rain when Sun is over head on the tropics of Cancer during Varsha-Ritu (June/July) as described in the "Surya-Siddhanta" (~400 years A.D.?). (B) Somewhat analogous, what is described in (A), is the modern monsoon – a combination of Hadley-Cell, ITCZ, differential solar heating, pressure, outgoing long wave radiation (OLR) due to albedo effect from Tibet-Himalayan ice and forest cover, convective clouds, winds, and effect of all these parameters resulting into rains over the India. Sun during January is shown with white bands depicting its position during winter (Shisir Ritu), as the graphic is modified and designed for summer to emphasize the solar heating, convective clouds, winds, and the ITCZ position during July, when Sun's Uttarayana journey culminates at the tropic of Cancer resulting in torrential rains (Varsa-Ritu).

subcontinent and Southeast Asia, and the central Asia is almost deprived of monsoon rains due to physical barrier of the paleo Tethyan Himalayan and Tibetan mountain chains. Winds blowing over Australian, African and Arabian deserts (Sirocko *et al.*, 1991) towards sea (Fig. 2A) add enormous dust (deMenocal, 2004) into wind-induced strong coastal and open-ocean upwelling system with high biological productivity during monsoon (Clemens, 1998).

Inter-Tropical Convergence Zone

Gadgil (1988) opined that monsoon, covering a wide geographic region in Afro-Asian region (Fig. 2B), is manifestation of atmospheric flow of moisture laden air causing equatorial cloudiness, which is termed as the Inter Tropical Convergence Zone (ITCZ) (Fig. 1B, Fig. 2B-D). Chen *et al.* (1988) found that Indian monsoon is characterized by annual cycle of divergent circulation of giant sea-breeze, and it is modulated by planetary-scale ~30–50 day low-frequency mode to establish an onset-active, break, and then revival-retreat of rains. This process is on round the year, and seasonal position of ITCZ boundary reflects seasonal shift (north/south of equator) with the Sun always overhead on it (Fig. 1A, Fig. 2B-D). ITCZ cloudiness is also more or less cyclic at ~30-50 days (Sikka and Gadgil, 1980), and pack of clouds get fragmented by varying wind speed during northward journey towards India, resulting in spells (7-10 days) of heavy showers with intermittent period of no rains, which gives enough time to farmers to plough, sow Kharif crop. Rain fall and prevalent winds during July (mid-monsoon, Fig. 2E) gives an idea on the extent of monsoon rains over Southeast Asia including Indian Ocean (http://irid.ldeo.columbia.edu). Indian monsoon often turns aperiodic and is a major hurdle in predictive meteorology. Empirical seasonal forecast of monsoon have moderate success for over 100 years, but modeling efforts have not been so successful in predicting monsoon accurately, as the simulation of mean structure of monsoon has been very elusive, and observed variability of relationships with forcing responsible has been often difficult to replicate (Webster *et al.*, 1998).

Tropical Warm Pools and Thermo-Haline Circulation

Monsoon is affected by variations in numerous phenomenons in the Earth-Ocean-Atmosphere system. Some of them are (i) tropical Indo-Pacific reservoir of heat and salt as the warm pool (Fig. 2A-B), (ii) Indonesian through flow and the global thermohaline circulation by the conveyor belt distributing tropical heat and salt to polar regions (Fig. 2C), (iii) volcanic and cosmic dust/aerosols in atmosphere scattering away considerable amount of solar radiation and thereby cooling effect, (iv) atmospheric contents of greenhouse gases (CO_2, CH_4, NO_2, CFCs, O_3) that absorb outgoing long-wave radiation (OLR) and heat up the atmosphere, (v) vapours and low lying clouds with a feedback mechanism (more the warmth, more water vapours absorbing more outgoing radiation), and (vi) ice and vegetation cover (albedo, ratio of reflected to incident radiation acting as another feedback mechanism). All these factors and their individual/cumulative effects are still not fully understood to the predictable extent, and so the monsoon often deceives millions to suffer.

India Meteorology Department (IMD) takes the Arabian Sea and southern Indian Ocean sea surface temperature (SST) during January-March which is correlated with rains (r = ~0.55), as one of the input in forecasting Indian monsoon. Gadgil *et al.* (1984) found strong correlation with summer SSTs and cloudiness (rains). Besides, winter snow cover at Eurasian mountain causing albedo (ice reflecting radiation back to space), and east Pacific El-Nino-Southern-Oscillation (ENSO, Philander, 1983) effects, both are negatively correlated (r = -0.46) to Indian monsoon rains (Webster *et al.*, 1998). Monsoon is result of seasonal migration (Fig. 2B-D) of the ITCZ (Gadgil, 2003), and accumulation of clouds and rains due to southwest winds (Fig. 2E). Monsoon has large error (~15%) in forecast based 8-10 parameters significantly correlated with rains (Fig. 2B-C) used in predictive models. Besides, there are parameters like (i) Sunspots and radiation, (ii) Earth's orbital forcing and insolation, and (iii) planetary synod effect that are although known to affect climate on short scale, but their relationships with rains are yet to be established in empirical equations to be used in the predictive models of monsoon.

The ITCZ (Fig. 1B, Fig. 2B-D) and the tropical Indio-Pacific warm pool due to equatorial solar heating result in higher SSTs (Fig. 3A), which enhances evaporation at sea and supplies the vapors and clouds from ocean to atmosphere (Chao and Chen, 2001). Clark *et al.* (2000) found that SST during fall-winter before monsoon in tropical Indian Ocean correlates positively with subsequent monsoon rainfall, and they are negatively correlated with SSTs and the All India Rainfall Index (AIRI, Pathasarathy *et al.*, 1994) in the subsequent autumn in the northern Indian Ocean. They reported highest correlation (0.87) for the years following 1977 between the AIRI and the central Indian

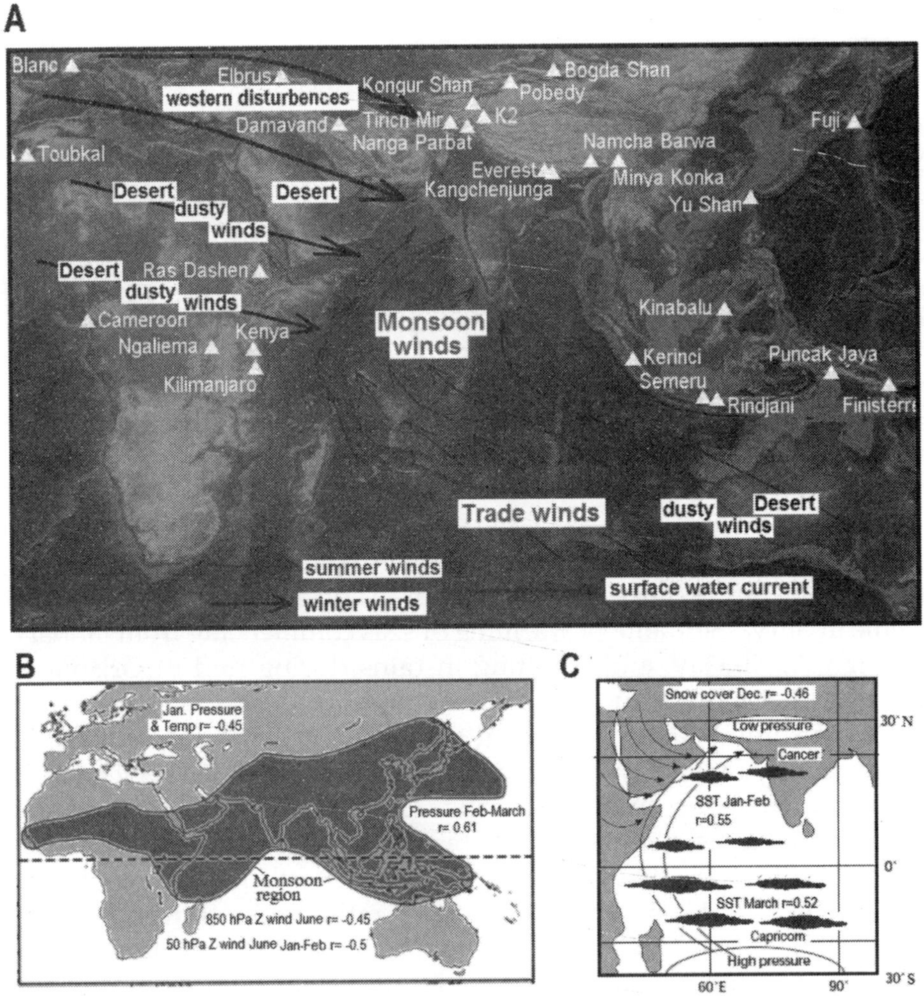

Fig. 2: (A) High mountains (>1-2 km) of the Madagascar, East Africa, Arabia act as a orographic wall to divert trade winds blowing between Australia and Madagascar towards Indian subcontinent along African Margin over Arabian Sea, which culminates its northerly journey over India due to lofty mid-Asian high lands (>3-5 km) including Himalaya and Tibet during monsoon season. Mountain peaks >3 km are shown as triangle. Occasionally, low pressure over central Europe invades the monsoon systems as the western disturbances, and is shown as arrows from Mediterranean towards India across the lofty mountains in Afghanistan. Winter winds over African and Arabian and trade winds over Australian deserts bring eolian dust into Arabian Sea and central Indian Ocean (Clemens, 1998). (B) Monsoon regime over the Afro-Asian region. (C) Atmospheric circulation during Indian summer showing low pressure on India and high pressure in southern ocean, along with monsoonal wind directions from tropics of Capricorn and Cancer.

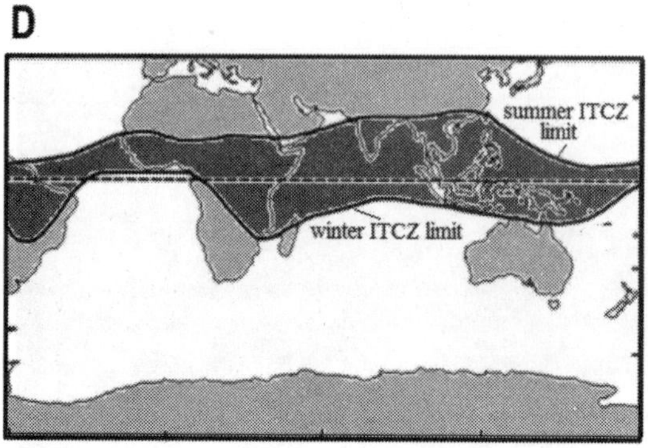

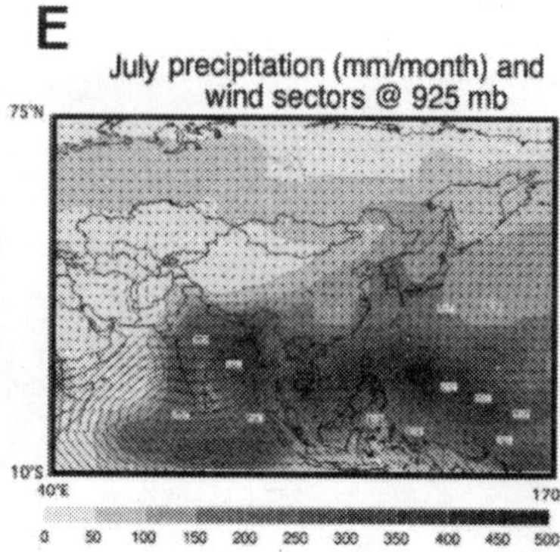

(D) Seasonal changes in northern limit of the ITCZ. (E) Mean rainfall and winds averaged for July (http://irid.ldeo.columbia.edu). Correlation (r=) with monsoon rain and forcing parameters are depicted in B and C, which are used by IMD in monsoon forecasting.

Ocean SST in the preceding September–November. These modern studies, one or the other way, directly or indirectly converge to ancient theme in Surya-Siddanta of warming of seas commencing from January through April-May, and resulting in rains during mid of Grisma to the end of Varsha Ritu from June-September. Monsoon rain water drains out into the Bay of Bengal and South China Sea resulting in lower salinity (<35 psu) in that region than non-monsoon regions (Fig. 3B) of the world oceans. Indo-Pacific tropical hot and less salty water leaks through passage between Antarctica and South Africa, becomes saltier due to en-route evaporation towards North Atlantic, and then get cooled due to lower SSTs at the Arctic circle, where it starts descending deep into water column to form a conveyor belt that returns through the deeper level back to the North Pacific along the Australia and New Zealand. Such re-distribution of heat and salt from the tropics to the pole has a great effect on regional and global climate including monsoon, and it is known as conveyor belt effect (Fig. 3C).

El-Nino and Southern Oscillation

Besides, un-usual warming of eastern tropical Pacific Ocean known as El-Nino (Philander, 1983, Fig. 3D) also play role in the strength of the monsoon (Ashok *et al.*, 2001). El-Nino years are associated with poor monsoon (El-Nino 3+4 SST trend and monsoon rain correlation r= -0.46), while La-Nina eastern Pacific cooling events relate to good monsoon, which means good Kharif rice production. Variability in Indian rice production with time between 1960 and 1996 and the all-

India rainfall index, the total summer rainfall over India, suggested that rice production has increased linearly (Gadgil, 1996). Webster *et al.* (1996) suggested that the years of deficit Indian rice production were associated with El-Nino years while rice abundant years were associated with La-Nina cold events in the eastern Pacific Ocean (Fig. 3 H). Strength and duration of the eastern Pacific warming due to the El-Nino and Southern Ocean Oscillation (ENSO), or cooling due to La-Nina, affects the domains and extents of the ocean, atmosphere, and tropical heat and salt reservoir - commonly known as the Indo-Pacific warm pool (Fig. 3A-B). El-Nino is somewhat predictable as the continuous wavelet transform (CWT) spectral analysis by method of Torrence and Compo (1998) reveals ~2.7-3.4- quasi-biennial oscillation (QBO), and ~11-years (Torrence and Webster, 1999) cycles in the modern time (Fig. 3E-d). There are several studies, which suggest that the ENSO phenomenon is cyclic in nature and its index is somewhat predictable. A sedimentary record of ENSO for the last ~2300 years in the form of lithological grey scale, Ti variation, and other productivity proxies from Peru Margin revealed highly significant ENSO cycles, mostly of solar origin, at ~250, 154, 83 (Gleissberg, 1958), 56–38, 31, 22–24 (cycles by reversal of magnetic polarity in Sun, Hale, 1924), ~11–9.4 (Sunspot), and typical ~7.4–5.7 years ENSO cycles (Agnihotri *et al.*, 2008, cf. fig. 8).

Indian Rainfall

All India homogenized mean rainfall data from 1871 (update on Parthasarathy *et al.*, 1994, IITM, Pune) has been extensively analyzed by various researchers suggesting better rainfall during La-Nina and drought like conditions during El-Nino events (Fig. 3F). The CWT spectral analysis suggests cycles at ~2.7 (similar to ENSO), 11-14 (Schwabe's Sunspot), and 80-years (Gleissberg, 1958) cycles (Fig. 3G-d). Spectral analysis (Schulz and Stattegger, 1997) of this data from 1871-2007 suggests cycles at ~54, 17, 11, 7.5, 3.2, 2.7, 2.3 and 2 years (Fig. 3I). Harmonic analysis by Multi-Taper Method (MTM - Thompson, 1982; Ghil *et al.*, 2001; www.atmos.ucla.edu/tcd/ssa) as used in Gupta (2009), results in cycles at ~80 (Gleissberg), 22 (Hale), 7.5, 5.1, and 2.7 years (Fig. 4A), which are reconstructed and compared with rainfall time series (Figs. 4B-F). Varâha-Mihira (Brahat-samhitâ, Ch. 8, v. 24-25) had accounted the ~3, 5, 7, and 11 year rain cycles.

The 3-year rain cycle was attributed by Parâsara's saying "arka-varsa-nigraham" for influence of Venus on Sun induced rains, and it was associated with visibility of Venus in India. The 5-year rain cycle was described having 1st year of even rains (sama-vatsar), 2nd year early and good rains (pari-vatsar), 3rd year excessive rains (eda-vatsar), 4th year delayed rains (anu-vatsar), and 5th year deficient rains (id-vatsar), and ~7, 18 (lunar eclipse) and 60 years (cross-product of 5 and 12 years Jupiter cycle) cycles in rains as per Iyengar (2009). Similar cycles are reported in the Sun's magnetic field at ~22 (Hale, 1924), 4–7 (ENSO) and 2–3 (QBO) cycles, and also in the historical US rainfall data (Rao and Hameed, 2003).

Indian Ocean Dipole

Indian Ocean Dipole (IOD) is a quasi-periodic oscillation of SSTs between the positive, normal and negative phases of thermocline thickness (Saji *et al.*, 1999). A positive IOD phase having higher than average SSTs and higher rainfall in western Indian Ocean region corresponds with cool waters in the eastern Indian Ocean - which results into droughts in Indonesia and Australia and vice versa during negative IOD phase. IOD is reported to affect strength of monsoon in India. On an average of 4 each positive/negative IOD events occur during every ~30 year (4x7.5) period with each event lasting ~6 months. It is worthy to note that El-Nino (Agnihotri *et al.*, 2008, cf. fig. 8b) and IOD-events both have cyclic changes at ~30-31 years and ~7.5 years indicating influence of the Saturn's period.

Solar Variability and Monsoon

Agnihotri *et al.* (2002) found less intense Indian monsoon during periods of solar minima in the last millennium, and reported cycles at ~200±20, 105±15 and 60±10 years. The 60-yr cycle observed in the instrumental rainfall data appears to be of solar origin and supports the hypothesis of solar control on the Indian monsoon on a multi-decadal time scale. Agnihotri and Dutta (2003) suggested that century scale monsoon rain fall patterns in the Indian and Chinese rain is due to manifestation of solar changes by relating them to variability in Sun's radiation. Kodera (2004) studied Sun's influence on monsoon and found that it originates from the stratosphere through modulation within the equatorial troposphere, and produces a north-south seesaw of convective activity during summer over the Indian Ocean. Higher precipitation over southeast Asia and India are reported to occur during high solar activity. Therefore, variability in solar activity and its effect on Indian monsoon are presented below.

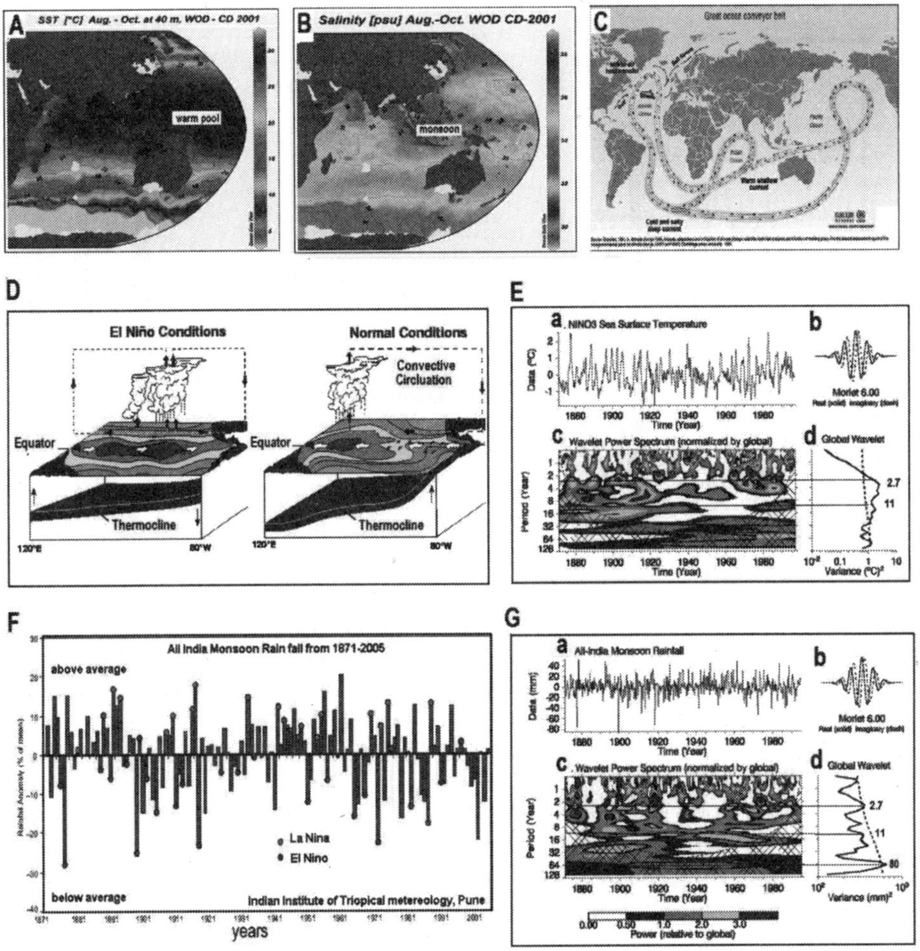

Fig. 3: (A) SST isotherms in the Indo-Pacific Ocean showing tropical warm pool and (B) surface salinity showing less saltier water in the Bay of Bengal and South China Sea due to monsoonal (Aug.-Oct.) fresh water flux from the Indian and Chinese rivers, respectively (WOD–CD-2001,ODV packag– Schlitzer, 2001). (C) Global thermohaline conveyor belt path distributing the tropical heat and salt to the north Atlantic and then reverting back to North Pacific Ocean (http://irid.ldeo.columbia.edu). (D) El-Nino and normal tropical Pacific Ocean conditions showing changes in position of convective circulation, thermocline depth and isotherms. (E) CWT spectral analysis of Nino-3 data showing 2.7 and 11-16 years cycle (Torrence and Webster, 1999). (F) Indian monsoon rainfall relative anomaly data from 1871-2005 (IITM, Pune) showing correspondence with drought and El-Nino, and good rains with La-Nina years. (G) CWT spectral analysis of Indian rainfall anomaly (Torrence and Compo, 1998) shows 2.7 (QBO), 11 (Schwab) and 80 (Gleissberg) years cycles.

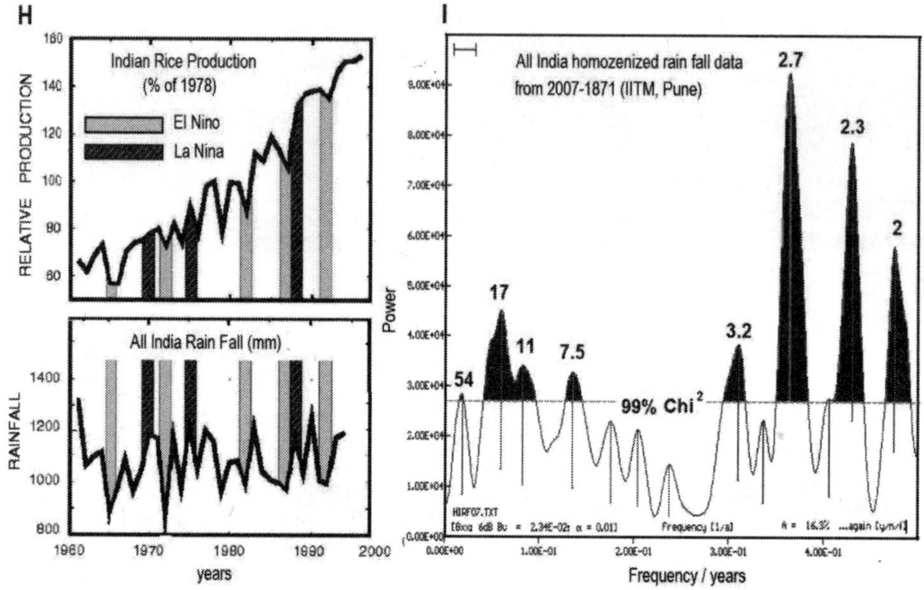

(H) Relative rice production compared to 1978 base year with upward trend due to green revolution, and All India Rain Fall during El-Nino and La-Nina years from 1960-1996 (Webster *et al.*, 1998). (I) Spectral analysis of All Indian Rain Fall from 1871-2007 showing cycles at ~54, 17, 11, 7.5, 3.2, 2.7, 2.3 and 2 years.

Variability in Sun's diameter

At present Sun's diameter is ~1 million miles (MM), and yearly small changes in it are generally not noticeable.With advent of research satellites, our understanding on Sun, Earth, Moon and other planets witnessed a phenomenal change. King (1973) reported that Sun's radius and its luminosity (solar constant) is not the same over the time. Eddy (1977), Eddy and Boornazian (1979) opined that Sun's radius is reducing at the rate of ~0.1%/century (~4 feet/hour). Eddy *et al.* (1982) reported ~0.05% change in Sun's diameter from 1863-1953 affecting its luminosity, which noticeably affected Earth's climate. Gilliland (1981) suggested large changes in Sun's luminosity during the last 265 years due changes in its radius, while Sofia *et al.* (1983) analyzed duration of total solar eclipses from 1925 to 1979 records, and found that during 1925 Sun's radius was ~375 km larger than at 1979. Sun's radiation changes due to variation in its diameter and its magnetic field, affect amount of the Total Solar Irradiance (TSI) received at the surface of Earth. Variability of TSI depends on changes in Sun's diameter, heat transport through the convective and radiative zone,

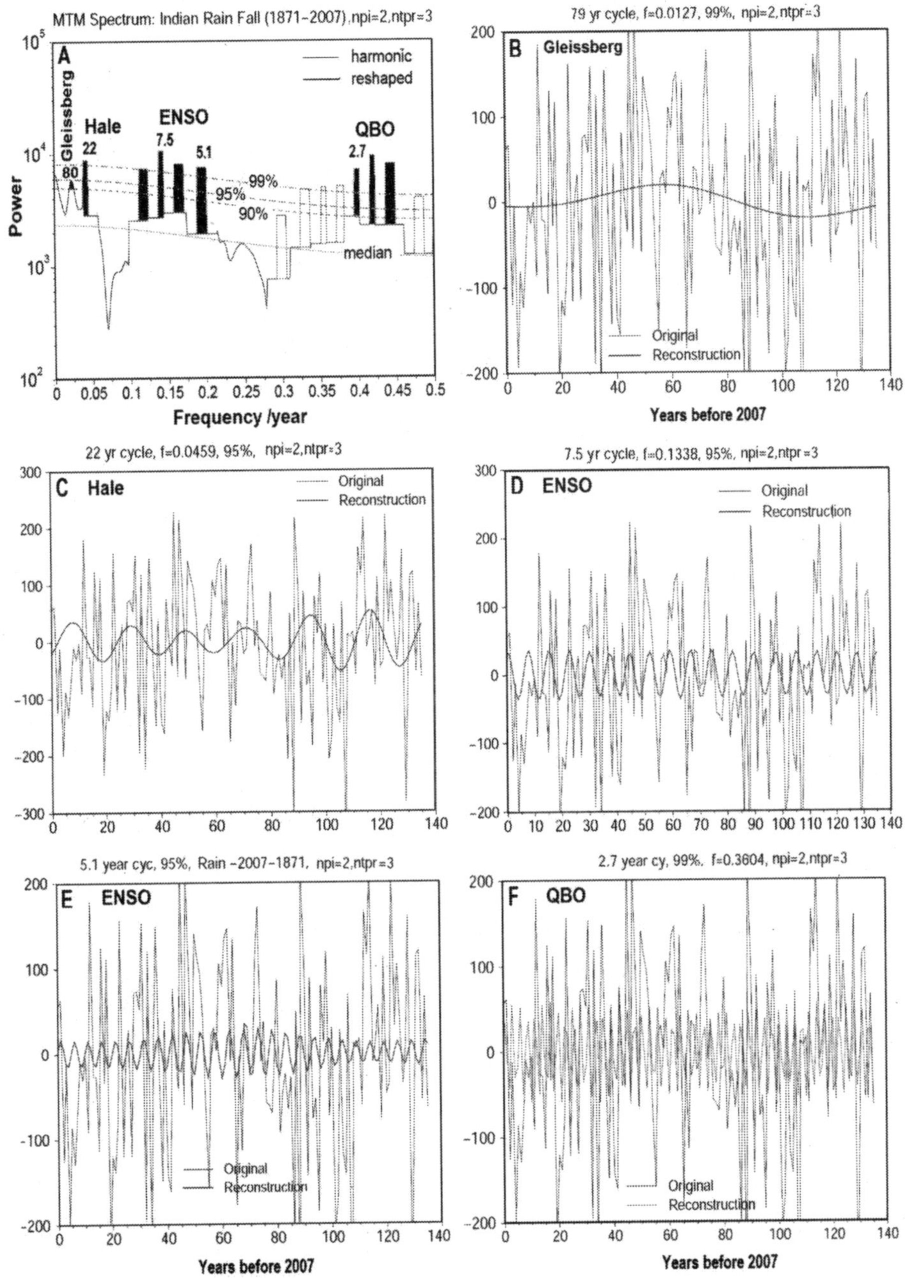

Fig. 4: MTM harmonic analysis of All India homogenized rainfall data (IITM, Pune) from 2007-1871 (A) along with reconstructions of significant cyclic trends at ~79- (B, Gleissberg cycle), 22- (C, Hale), 7 and 5 (D-E, ENSO), and 2.7 year (F, quasi-biennial oscillation) cycles.

temperature of photosphere, amount of magnetic flux on the Sun's surface, which is the perturbation in Sun's magnetic flux due to Sunspots, solar flares and prominent eruptions also known as solar tsunami (Fig. 5A). Cosmic solar rays produce cosmogenic ^{10}Be in the atmosphere due to spallation of nitrogen and oxygen atoms, while ^{14}C is formed in the atmosphere by the interaction of cosmic ray produced neutrons with nitrogen nuclei (Stuiver and Quay, 1980), and both are very good proxies for solar activity. Based on relationship between TSI and solar magnetic field, obtained from the cosmogenic radionuclide ^{10}Be measured in ice cores, TSI is reconstructed in the past. Reconstruction of TSI is based on quantitative estimate of changes in ^{14}C and ^{10}Be cosmogenic radionuclid production rates (Bard *et al.*, 2000; Bond *et al.*, 2001). Reconstruction of TSI for the last 9300 years by cosmogenic radionuclide ^{10}Be measured in ice cores was done by Steinhilber *et al.* (2009), who reported increase in TSI solar-cycles averaged from the Maunder Minimum to the present amounting to 0.9 (± 0.4) Wm^{-2}. Multi-taper method (MTM) based spectral analysis of TSI time series of Steinhilber (*et al.*, 2009) revealed Suess, Gleissberg and several other cycles (Fig. 7A), similar to those found in reconstructed Sunspot numbers of Solanki *et al.* (2004) as shown in Fig.7 (B-I). Chakraborty and Bondyopadhyay (1986) described effect of changes in solar variability influencing rainfall pattern, Chowdhury *et al.* (1981) considered it as a potential factor for droughts in India, whereas, Reid (1987) described the influence of Sun's radiation changes on global SSTs and its potentials in weather changes.

Sunspots and Changes in Sun's Magnetic Field

Hale (1924) demonstrated that the magnetic polarity in southern solar hemisphere Sunspot groups is the opposite to that in the northern solar hemisphere. Nicholson (1937) reported that at the end of each 11-year Sunspot cycle of rising and falling numbers, the polarity of spot group reverses, and at the end of 22 years, Sunspot polarities on both hemispheres of the Sun had alternated through one complete full wave of positive and negative polarity. The Sun's surface is highly dynamic with a variability of 11-year solar activity cycle in Sunspots (+ flares and eruptive prominence). During a solar active maximum, number of Sunspots increase on the surface on the Sun. Sunspots (Fig. 5A, a-b) are relatively cooler (umbra, ~4000^{0}C, 4-k^{0}C) areas of ~100,000 km in diameter encircled by warmer regions (penumbra, ~5-k^{0}C) on Sun's surface (~6-k^{0}C), and hence appear as black spots (Hoyt, 1979).

Solar flares/tsunamis are caused by coronal mass ejection consisting of the geomagnetic storm of super-hot gases and charged particles travelling at speeds of ~1 million miles/hour thrown out into the space. Usually, the deeper explosions within Sunspots are the areas from where solar flares erupt out into space. The last solar maximum was in 2001 (Fig. 5A-c), and next is awaited in 2012.

Heat received by the Earth from the Sun's radiation is affected by number of Sunspots, solar flares and TSI (Fig. 5 A, a-c). Variation in solar flares (Fig. 5A) with time suggested flares cycles at ~9-11-, ~2.25 years, and 3 months with potential in forecasts due to being in phases in the Sun's motion around the center of mass of solar system, and with a cyclic pattern in change in angular acceleration of tidal forces exerted by planets Venus, Earth, and Jupiter (Wood, 1972; Okal and Anderson, 1975) and other planets (Landscheidt, 1984) (Fig. 5F). The TSI is value of solar energy flux, over an area of 1 m^2 oriented towards the Sun, which arrives at top of Earth atmosphere (Fig. 5A,d) at a distance of 1 astronomical unit (AU, mean distance from Sun to Earth). The TSI, Sunspot, solar flare, and magnetic flux of charged particles/ radio-flux are more or less synoptic in nature (Fig. 5B). Noisy daily, monthly and yearly Sunspots data, once smoothened with mean values, derives the predictive trends (Fig. 5C-D). A comparison between annual mean Sunspots as an index of solar activity (changes in magnetic field, diameter, and solar flares), and homogenized All India rainfall data from 1871-2007 (IITM, Pune) revealed a very good visual correspondence between the two time series (Fig. 5, E). Bell (1977) observed that the August (monsoon) sea level pressure over India flipped off from positive to negative sign with annual Sunspot numbers at ~1910, in contrast to the close correspondence in the late 1940s. The switch over in their relationship at ~1920-25 was marked by a change from zonal to meridional circulations at the mid-latitudes (~30°), which reflected the reversal of Sun's polarity 22-year Hale cycle. Hiremath (2006) studied correlation coefficients for January–February, March–May, June–September and October–December and, annual mean data of Sunspots in different Sunspots cycles over last 130 years, and found that Indian rainfall is correlated with Sunspot activity with a trend of higher rainfall during low Sunspot activity. He suggested a possible physical connection between Indian rainfall and Sunspot activities and flux of cosmic rays. Sunspot activity affects annual mean temperature over India (Joseph and Amatya, 1986), which modulates landmass heating affecting the monsoon. These results concur with

Sen Gupta (1957), who reported Sunspot cycles in regional rain-fall and suggested that it negatively affects Indian monsoon system, *i.e.*, more Sunspots result in failure of the monsoon rains.

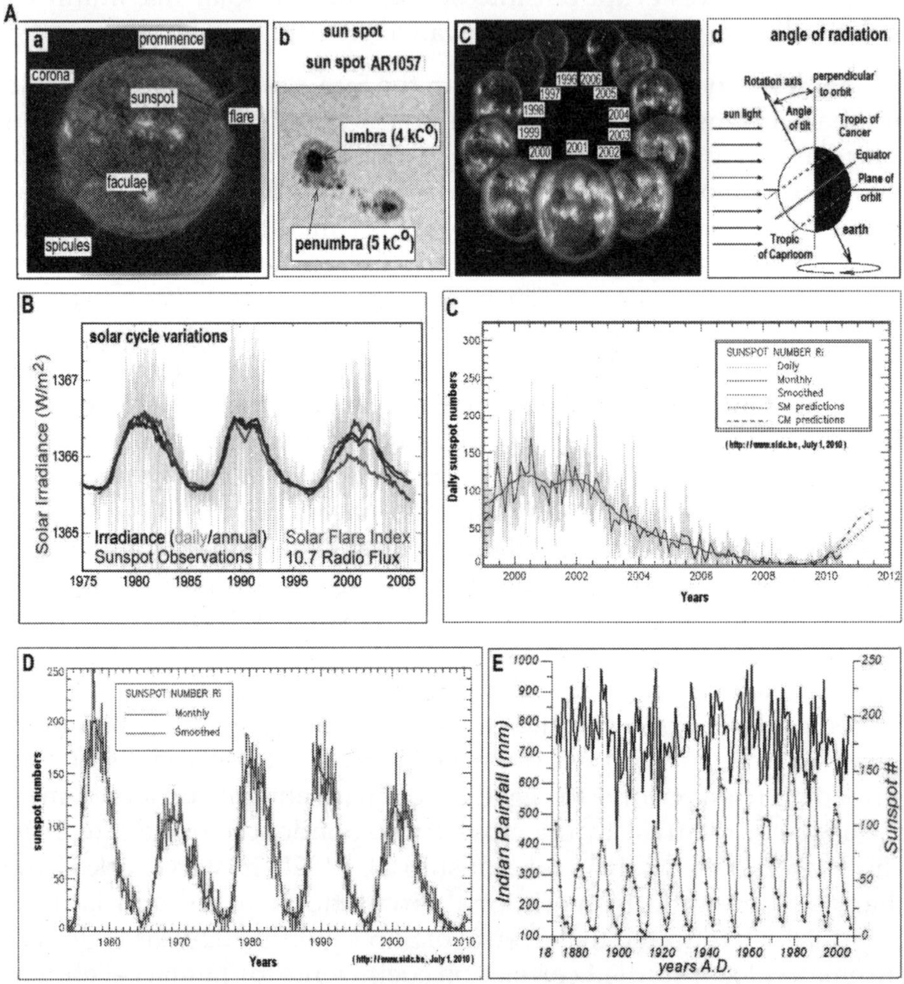

Fig. 5: (A) Surface of the Sun with corona and photosphere showing Sunspots, faculae, spicules, flares, and prominence (a), a typical Sunspot with umbra and penumbra (b), last Sunspot cycle from 1996-2006, with activity maximum at 2001 (c) and angle of radiation received at Earth over the tropics and poles (d). (B) Synoptic variation in solar irradiance, Sunspots, solar flares, and radio flux from 1875-2005 showing similar origin. (C) Daily and monthly smoothed Sunspots and their predictable trends. (D) Monthly Sunspots raw and smoothed data from 1950-2009. (E) Correspondence between All India Rainfall from 1871-2007 (IITM, Pune) and numbers of Sunspots.

(F) Edge of solar system showing a hypothetical conjunction of all the planets in one side of the Sun. Arrows from the Earth's orbit towards the planets depict forward drag (backward pull) effect of conjoined planets on Earth's speed due to synod phenomenon that increases/decreases the length of summer/winter depending on the season of conjunction event. Also shown are the Kuiper meteoritic belt and Oort clouds of comets that occasionally de-established due to galactic tides (Napier, 2001, 2006) and enters in solar system as distant visitors along with meteoritic showers and cometary-dust. Graphics in A-D and F are from www.wikipedia.com.

Sunspot Cycles in Monsoon

MTM harmonic analysis of Sunspot numbers (www.sidc.be) for the 257 years (1750-2007) revealed cycles at ~111 years, 28.5 and 17.5 years lunar tidal cycles, 9-11 (Schwabe), 4.5-7 years sub-decadal, and 3-2.5 years QBO cycles (Fig. 6A). Although, the 111 year cycle is 90% significant, it could very well be considered as a long term trend than a predictable cycle, but the century scale variability in paleo-climate data often reported. A conspicuous increase in long term trend from 0-70 to years in the time series (Fig. 6B) is analogous to global warming (GW) trend, and raises concern that GW could also be due to manifestation of increased Sunspot activity in the last 70 years. The 28.5 year cycle (Fig. 6C) is very close to Saturn's orbital period (29.45 years) around the Sun suggesting that it could be due to increase in Sunspot activity by the gravitational perturbation of Saturn. The ~17 year harmonic (Fig. 6D) is close to Soli-Lunar cycle of ~18 years, which is widely reported in tidal forcing of Sun-Moon-Earth system, and suggest that analogous to Earth's tide effect, Moon might be having similar effect on Sun, or more because Sun is gaseous (more fluid than Earth) in nature. The ~11 and 9 year are well known Schwab cycles. Amplitude of ~11 year cycle has remarkable increased since last 80 years suggesting that global warming could possibly be due to natural variability in the Sun (Fig. 6E-F). The ~7 year (Fig. 6G) and ~5-5.6 cycles (Fig. 6H-I) suggest intra-decadal variability recorded in several times series, whereas 4.7-3 years cycles (Fig. 6, J-L) are represented in ENSO variability and Indian (IITM, Pune), Chinese and US rain fall data (Rao and Hameed, 2003), and Antarctic ice core (Naik *et al.*, 2010). Besides, these individual cycles also produce larger cycles either by the way of amplitudes modulation or being in resonance with two are more closer cycles - for example pairs of (i) ~2.7 and 3 year cycles and (ii) ~4.7 and 5 and 5.5 years cycles as shown in Fig. 6, and they

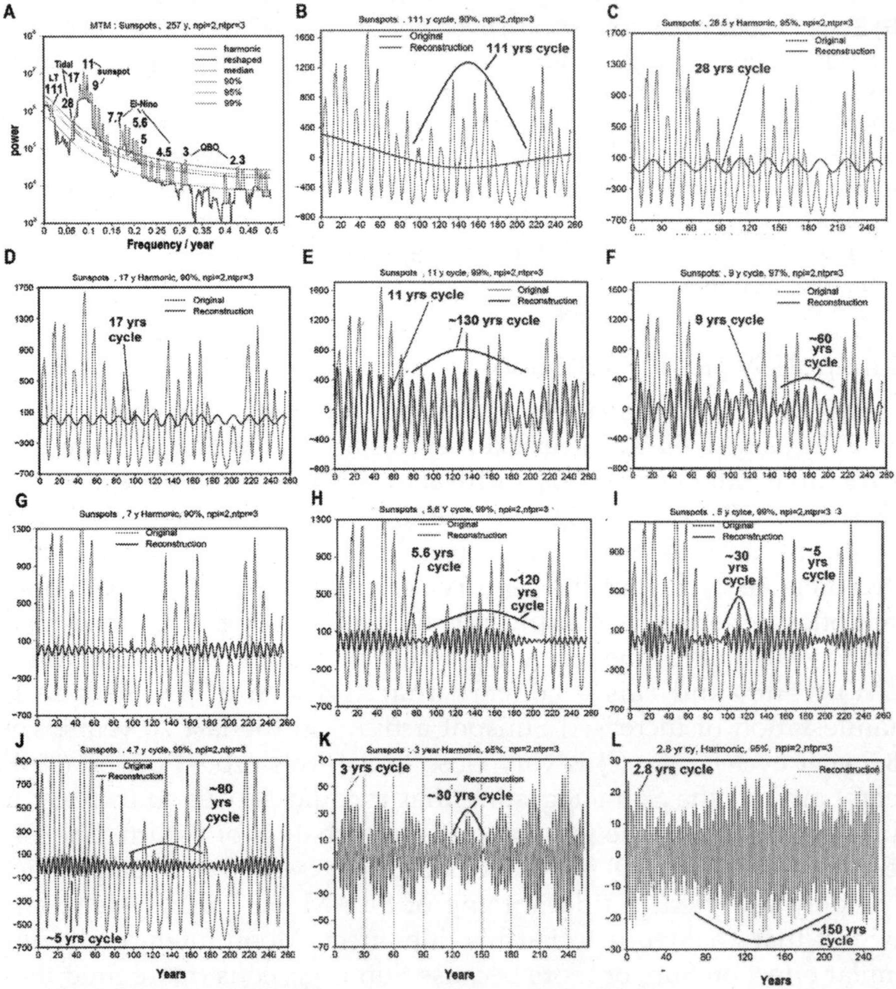

Fig. 6: MTM harmonic spectral analysis (A) of Sunspot data for last 257 years (courtesy: www.sidc.be) showing cycles at ~111- (B), 28.5- (C), 17 (D), 11- (E), 9- (F), 7- (G), 5.6- (H), 5- (I), 4.5- (J), 3- (K) and 8-years (L) cycles. Smaller cycles also produce larger cycles by amplitude modulations, shown as incomplete curved envelops of large cycles at 30, 60, 80, 120, 130, 150 years.

are shown as incomplete curved envelops as being indicative of the amplitude modulated higher cycles. The ~22- (Hale), 90- (Gleissberg, 1965; Feynman and Fougere, 1984), 208- (Suess), and 2300-year (Hallstatt) Sunspot cycles are reported from flood plain, lake, and varved sediments (Hameed *et al.*, 1983). Varved sediments deposited by Indus River off Karachi are reported to exhibit cycles ~125, 29.5, 18.7, and 12 years (von Rod *et al.* 2006, cf. fig. 11). Time series analysis of the magnetic susceptibility in a sediment core (SK 148/55) located off Ratnagiri (east Arabian Sea), resulted in precipitation cycles at ~2200, 1350, 950, 750, 470, 320, 220, 156, 126, 113, 104 and 92-years during the Holocene (Thamban *et al.*, 2007). Recently, Scafetta (2010) analyzed several global surface temperature records since 1850 and found that the records deduced from the orbits of the planets present very similar power in the spectral results. The large climate oscillations with peak-to-trough amplitude of about 0.1°C and 0.25°C, and periods of about 20 and 60 years, respectively, are reported to synchronize with the orbital periods of Jupiter and Saturn, while Schwabe and Hale solar cycles, and 9.1-year cycle synchronized to the Moon's orbital cycles are reported in these temperature records.

Sunspot in the Past

Depending on Sunspot numbers and cosmogenic nuclides ^{14}C and ^{10}Be, periods of low solar activity called as Oort minimum (980-1120 A.D.), Wolf minimum (1282-1342 A.D.), Spörer minimum (1416-1534 A.D.), Maunder minimum (1645-1715 A.D.) and Dalton minimum (1790-1820 A.D.) are reported to affect the monsoon strength. Solanki *et al.* (2004) reconstructed Sunspot number by dendrochronologically dated ^{14}C content in wood, which were positively correlated with solar irradiance in the last ~11400 years.

MTM harmonic analysis (Fig. 7B) of paleo-Sunspots data of Solanki *et al.* (2004) revealed significant cycles at ~192 (Suess), 89 (Gleissberg), 60 (2 X 30 year Saturn's period), and 42 years cycles, which correspond to planetary orbital periods of Neptune, Uranus, Saturn (Fig. 5F), respectively. All significant cyclic components in paleo-Sunspot data are reconstructed and compared with original time series, which invariably suggested changes in the amplitudes between ~4000-8000 B.P. (Fig. 7, C-I), that coincides with "Holocene Optimum" - a period of stronger monsoon than present.

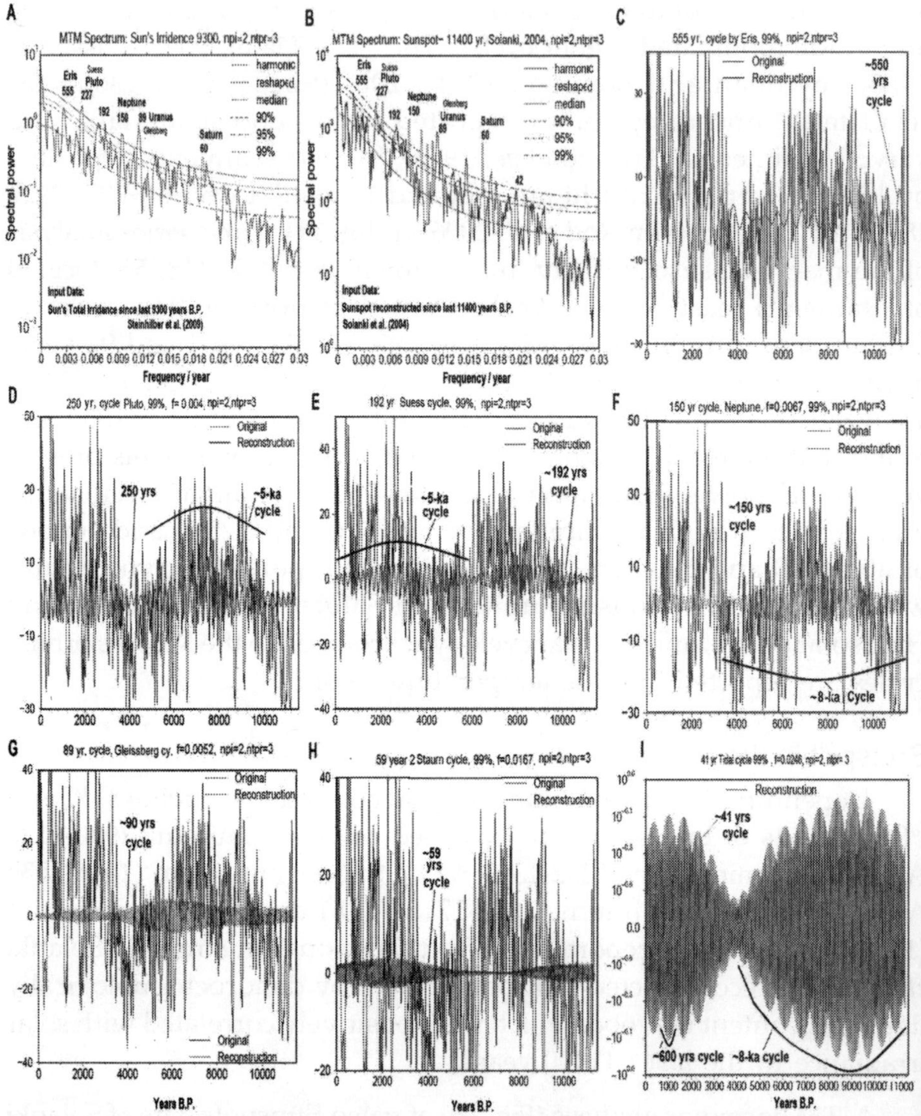

Fig. 7 : (A) MTM harmonic analysis of reconstructed TSI (Steinhilber *et al.*, 2009) for last 9.3-ka and (B) Sunspot numbers (Solanki *et al.*, 2004) for last 11.4-ka showing significant cycles at ~555 (C, Eris), 250 (D, Pluto), 192 (E, Suess), 150 (F, Neptune), 89 (G, Uranus/Gleissberg), 60 (H, 2x30 period of Saturn) and 42 (I, tidal) years cycles. Gleissberg cycle reconstructed and compared with Sunspot numbers suggesting that cycle was more pronounced during Holocene Optimum (>4-11-ka) with stronger monsoon than in last ~4-ka (G). Smaller cycles produce larger cycles by amplitude modulation shown as incomplete envelops of larger cycles at ~600-, 5000-, and 8000 years.

Foraminiferal variation of *G. bulloides* species in a sediment core revealed a ~2200 years cycle in monsoonal upwelling record from Arabian Sea (Naidu and Malmgren, 1995), which could possibly be bundle of (200 X 11) Schwab cycles. Gupta *et al.* (2005) studied variation in monsoon upwelling index *G. bulloides* percentage in Hole 723A from the Oman Margin for the last 12-ka and compared it with Sunspot numbers (Solanki *et al.* 2004) and North Atlantic Hematite percentages. They reported monsoon cycles at ~1550, 152, 137, 114, 101, 89, 83, and 79 years and compared them with Sun-spot cycles. Monsoon variation in the form of thickness in varved sediment derived from Indus river off Karachi with high resolution record for the last 5-ka, revealed variation in Indus river sediment cyclically at ~1470 (Bond cycle), and 280, 200, 125, 95, 56, 39, 30, 26, 23, 19, 14 and 12 years, due to multiples of basic ~4.4 and 9.3 years Moon's tidal cycles (von Rad *et al.*, 2006). Indian rainfall, Sunspot, TSI, El-Nino, and Indus derived varved marine sediments, all show common cycles of ~11-12 and 2.7 years and are attributed to solar radiation variability. Alternatively, these cycles could also be due to the modern micro-variations in the precession, tilt and eccentricity of the Earth's orbit. Bertrand *et al.* (2002) did multi-taper spectral analyses of the eccentricity, tilt and precession values for the last 1-ka after de-trending them, and reported similar ~11.9, 5.9, and 2.7 years cycles present in eccentricity and precession, which they attributed due to gravitational perturbation of Jupiter, its 2nd harmonic, and Venus - Earth (Ve - 2E), respectively. Jupiter completes its journey around the Sun in ~12 years, and according to Indian saying - "बारह बसरे, फूले बान्स–बरसा बिसरे, खेती नास" - every 12th year just after flowering of bamboos the crops fail due to scanty rains. Currie and Fairbanks (1985) reported drought/ floods at ~18 lunar tidal and 11 Sunspot cycles in the Chinese rains emphasizing its global implications.

Synod Effect on Earth

The gravitational pull/push of the planets like Venus, Mars, Jupiter, Saturn, Neptune, and the Trans Neptune Objects (TNO's) alters normal speed of Earth on its way around the Sun. Although, the time taken by Earth to encircle the Sun remains the same (365.25 days), Earth speeds up when it is under influence of gravitational pull of giant planets lined up in a group (within 30°) with Sun (conjunction/ synod), but when Earth travels away from them, its speed is reduced (Fig. 5F). Result is that the Earth takes less time to travel half of its journey, when it is closer to the conjunction of giant planets due to acceleration by pull effect of gravitational force of conjoined planets, whereas it spends more time in next half of journey on far side of planets conjoined with Sun. If the conjunction happens in summer half of the year and Earth is moving farther to Sun, travel time increases due to pulling backward effect resulting in few days longer summer and vice versa. For example, in January 1665 giant planets were conjoined with Sun within 45° to Earth, which resulted in increased winter, and shortening of summer by 2 days resulting in ~1% change in seasonal heating by Sun (Burroughs, 2003). The Monsoon prolonged ~5 weeks when Sun, Venus, Mercury, Mars, and Saturn are within 60° in tug of war (180°) with Jupiter in 2010 causing unprecedented rains and floods. Outer planets Jupiter, Saturn, Uranus, Neptune, and TNO bodies like Pluto, Eris and Sedna on an average take ~11.87, 29.45, 84.07, 165, 248, 557, 1250 years to go around the Sun in their orbits, respectively (en.wikipedia.org/wiki/Orbital_period). These planetary periods nearly coincides with Indian monsoon record cycles at ~11, 30 (Neff *et al.*, 2001), 90 (von Rod *et al.*, 1999) and 180-200 years (Tiwari *et al.*, 2006), 245 (von Rod *et al.*, 1999), 584 (Hong *et al.*, 2001), 1100 (Neff *et al.*, 2001; Leuchner and Sirocko, 2003), and 2200 years (2nd harmonic of 1100 years in Arabian Sea upwelling index, Naidu and Malmgren, 1995) monsoon cycles mainly due to reported Sunspot activity. Nigam *et al.* (1995) reported rain-proxy foraminifers changing at the Gleissberg cycle (80+10 years). Similarly, the Chinese rain records are reported to have cyclic variations at ~70, 80, 90, 107, 110, 123, 134, 141, 162, 198, 205, 249, 278, 324, 389, 467, 584, 834 and 1060 years (Hong *et al.*, 2001). The gravitational perturbation of planets affect in similar way in Sunspot dynamics (gaseous star) as it affects the monsoon on Earth, or vice versa. Even in modern time when changes in Earth's orbital eccentricity, precession and tilt are minimal and un-noticeable, their micro variations may be altering the monsoon at shorter biennial to decadal cycles. Such an argument is based on the MTM spectral analyses of eccentricity, tilt and precession values from 2000 to 1000 A.D, which revealed yearly and decadal cycles both due to eccentricity and precession at ~11.9 (originating by period of Jupiter), 5.9- (1/2 of Jupiter's period), 4- and 2.7- (both beat of Venus and Earth), and due to axial tilt at ~18.6 (due to Moon's orbital change) attributed to orbital/gravitational forces of Jupiter and other planets (Bertrand *et al.*, 2002).

Sun-Moon Tidal Effect

Earth orbits around the Sun and the Moon orbits around the Earth. Both have elliptical path, and inclinations in their orbital planes (ecliptics) makes a complex orbital dynamics of tidal forcing due to gravitational pull of the Sun and Moon on the Earth. Earth's rotation and Moon's orbital motion around it exerts high and low tidal stress twice daily on ocean (also in atmosphere). As Moon completes its orbit around Earth every 27.5 days, its takes 13 rounds (lunar tidal months) of Earth to accompany it around the Sun in a year, and it is basis of India's lunar calendar with provision of an "adhi-masa" and 12 lunar months. Besides, monthly lunar cycles of ~28 days, Earth-Moon-Sun system also generates two other main cycles of 8.9 years (due to advance of longitude of Moon's perigee), and 18.6 years (due to regression of the longitude of the node), in addition to ~13.6, 27.2 and 42 years tidal cycles of very complicated orbital dynamics of Earth-Sun-Moon system. All these tidal cycles (~8.9, 13.6, 18.6, 27.2, 42, 179 years) and their multiples are found in the Chinese paleo rainfall data (Currie and Fairbank, 1985). Keeling and Whorf (2000) found close similarity in mean of ~180-year tidal cycle (1/10 of the 1800 year cycle) and the 179.2 years cycle of the Sun's rotation around the center of mass of the solar system as a manifestation of planetary resonances, and reported that 90-year cycle may vary from 84 to 93 years, and its modulation produces a large cycle of ~5000 years. It is interesting that ~18, 90-95, 180, 1470, 1800, and 5000 years cycles are reported in monsoon records from stalagmites and wind strength preserved in sediments off Oman and Karachi (von Rod *et al.*, 2006). Although, Sunspot activity is within the Sun, its manifestation by gravitational and tidal forcing of planets is vital for climate changes. Sun-Moon gravitational perturbation affects Earth's axis to precess (move backward) on its orbit at the rate of 51'/year, which alters equinox-precession from ~25- to 23-ka cycle. Surya-Siddanta described precession of the Earth thousands years ago (Jha, V. jyotirvidya. wetpaint.com/page/Ayanamsha+vs+Precession, Hari, 1997).

Orbital Forcing on Monsoon

Changes in the Earth's orbital eccentricity (e), precession of equinoxes (p), axial tilt (t), and inclination (i) of orbital plane (respective to invariable plane of Sun and Jupiter) affect the climate including monsoon significantly. Orbital variables are summarized for the last 3-Ma along with MTM results with significant cycles (Fig. 8).

Earth's Orbital Eccentricity

Earth's orbit around the Sun is relatively circular at present (0.017), affecting ~6.7% change in the Sun's radiation reaching at the Earth's surface, but it was highly elliptical (0.06) in the past (Fig. 8A-a, D). Cyclic variation in shape of orbit from nearly circle (e =0.0) to ellipse (e =0.06) at every ~100- and 400-ka (Fig. 8A-a, D-E) is called eccentricity [e = (a-p)/(a+p), a= aphelion and p=perihelion distances of the Earth from Sun in the orbit, perihelion: Sun closest to the Earth, 146.1 million km; aphelion: Sun farthest away, 151.1 million km] (Berger, 1977). It changes due to variations in gravitational pull/push (perturbation) on Earth's orbit by other planets like Venus, Mars, Jupiter, Saturn etc. (Berger, 1978). High orbital eccentricity (more elliptical) produces strong seasonal contrast due to highly modulated precession values, while low change in eccentricity has a significant role on solar radiation (insolation) received (larger) by the outer atmospheric at the perihelion and aphelion. At present, it is 3.5%, and difference in insolation increases when eccentricity is greater and vice versa. At the maximum eccentricity (0.06) the difference in the insolation received by the outer atmosphere between the day at perihelion and that of the day at aphelion can be as high as 30%. MTM spectral analysis of the last 3Ma eccentricity data reveals significant cycles at ~400-, 126-, and 95-ka, which are reconstructed and compared with original eccentricity time series (Fig. 8D-E). Two minor eccentricity cycles at ~69- and 54-ka are also evident, which were mentioned by Berger (1977, p. 45) as its combined effect with tilt and precession. Normalized cumulative effect of eccentricity, tilt and precession is defined as the ETP, which causes orbital forcing on climatic changes. MTM results of the ETP for last 3-Ma exhibit triplets of the ~400, 126-, 95-ka eccentricity, ~54-, 41-, and 31-ka tilt, ~23-, 19- and 17-ka precession, and 15- and 13-ka suborbital cycles (Fig. 8G).

Earth's Wobble/Precession

While rotating on its axis on the orbital path, Earth wobbles like a spinning toy-top (gyrating motion) as a result from a torque generated by gravitational forces of the Sun and Moon together acting on the slight equatorial bulge of the Earth (Fowles, 1977). This gyrating motion of Earth's rotational axis is due to tidal forcing by Sun and Moon on the oblate-spheroid shape of Earth. Due to this torque, Earth moves slightly backwards (precess) with wobble, and it nearly takes ~25,700 years to come back at the same point. This wobbling backward motion of Earth is known as precession (p) (Fig. 8A-c, B). It is defined as p= e sin þ, were þ is Sun's longitude of perihelion with respect to moving vernal equinox. Earth's rotational axis at present points towards Polaris (North Star) during the precession maximum, but it pointed towards Vega during precession minimum at ~12.5-ka (Fig. 8B). At present northern hemisphere has winter when Earth is closest to Sun due to combined effect of tilt and precession, but at ~12.5-ka it had winter when Earth was farthest from Sun. Furthermore, the Earth's orbital ellipse itself precesses in space of arc at ~51'/year due to gravitation perturbation of Jupiter, Saturn and other larger planets. This orbital precession shortens the period of precession of equinoxes with respect to perihelion from ~25-ka to 23-ka (Berger, 1978, 1993). Variation in precession of equinoxes during the last 3 Ma is presented (Fig. 8D), and its MTM spectrum reveals the 23-, 19-, and 17-ka cycles (Fig. 8G).

Earth's Axial Tilt

Earth's rotational axis is tilted at ~23.4° at present with respect to the perpendicular on the orbital plane and it is known as axial tilt/ obliquity (t, Fig. 8A-b, C). During December axis is tilted away from the Sun, while in June it is tilted towards the Sun, resulting in northern hemisphere tilted at lesser angle and receives more radiation causing northern summer and vice versa. If the axial tilt would have been zero, there would not have been any seasons on Earth. Reason for axial tilt is not certain, but few astronomers are of opinion that ~5 billion years ago (Ga), the infant Earth was struck by a Mars size planet badly, which not only tilted its axis, but also split apart a mass which formed its satellite, the Moon. It was blessing in disguise, as seasons on Earth are due to axial tilt. At times of lesser tilt (22°) high latitudes receive lesser Sun light than during higher tilt periods, and thereby altering the climate in term of expanded ice-sheets at the poles and causing the ice-ages. In the last 3-Ma, it varied between 22°-24.5° (Fig. 8F), and its MTM results produce well known the ~54-, 41-, and 31-ka tilt cycles (Fig. 8 G).

Earth's Orbital Plane's Inclination

The 4th orbital element is inclination (i) of the Earth's orbital plane that moves up and down relative to its present position (Fig. 8C). It is termed as precession of the ecliptic or planetary precession. Up and down motion of Earth's orbital plane with respect to solar orbital plane also moves with respect to the orbits of the other planets, mainly Jupiter,

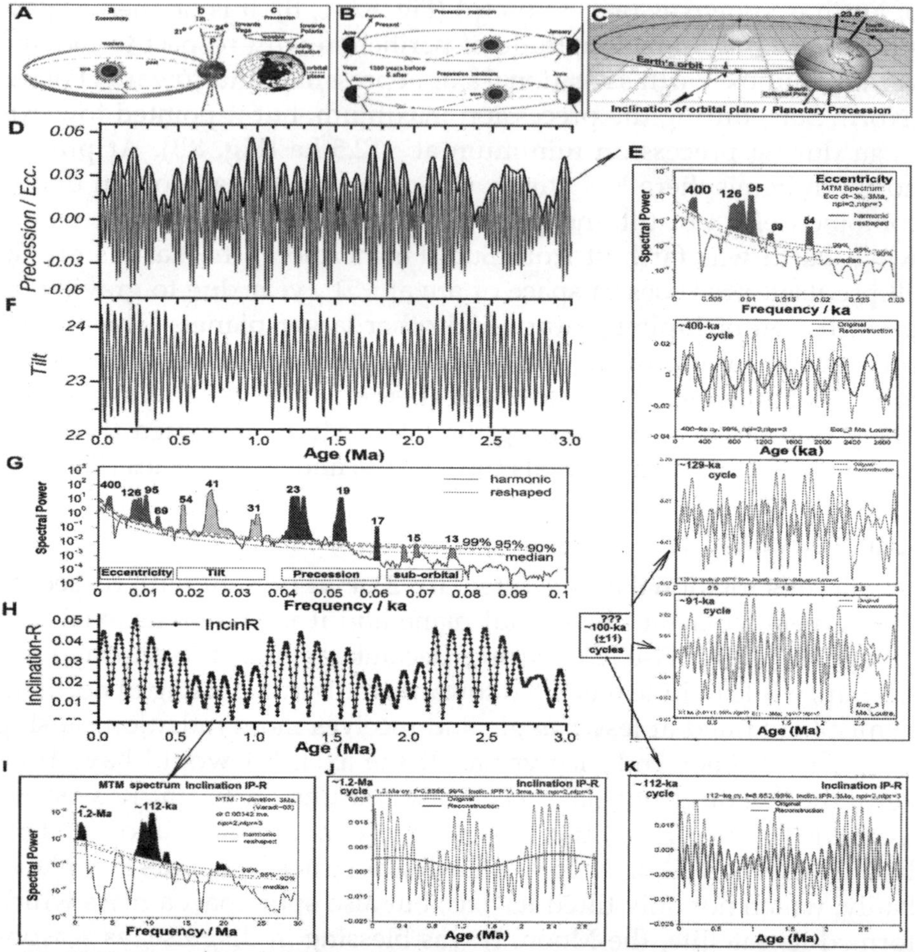

Fig. 8: Earth's orbital changes in the (A) eccentricity (a), tilt (b) and precession (c) (Berger, 1978). At present precession maximum axis points to Polaris, while it pointed to Vega at ~12.5-ka (B). Orbital inclination (C). Changes in eccentricity and precession during the last 3-Ma (D). MTM spectral power (upper) along with comparison in original and reconstructed cyclic components of ~400-, 129-, 91-ka eccentricity (lower) cycles (E). Tilt during last 3-Ma (F). MTM of combined effect of the Earth's orbital forcing "ETP" showing triplets of the ~400-, 129-, 91-ka eccentricity, 54-, 41- and 31-ka tilt, 23-, 19-, 17-ka precession, and the suborbital ~15- and 13-ka cycles (G). Orbital inclination relative to the invariable plane (Varadi *et al.*, 2003) in radians (H), and its MTM spectrum showing cycles at ~1200- and 112-ka cycles (I), and the reconstructed ~1.2-Ma (J) and 112-ka (K) cyclic components. Note the differences in amplitude changes in the ~129-ka and 91-ka cycles due to eccentricity (E), and ~112-ka cycles due to inclination (K), and all are the so called ~100-ka (+11-ka) cycles. Figures (A-C) are modified after their sources (www.wikipedia.com).

because the invariable plane (angular momentum of the solar system) is almost same to the orbital plane of Jupiter. Researchers suggested that a disk of dust and other planetary debris exists in the invariable plane, and it affects the Earth's climate in several ways. Earth moves through this invariable plane at 9th January and 9th July each year at present, when there are reports with increased radar-detected meteorites and meteoritic dust clouds. Inclination of the Earth's orbital plane was not a part in the Milankovitch's theory of climate change. Inclination has a strong (68% variance) ~100-ka cycle, which is very similar to the ~100-ka eccentricity cycles recorded in global ice volume during late Quaternary ice ages (Hays *et al.*, 1976).

Quinn *et al.* (1991) calculated inclination of Earth's orbital plane, which was used by Muller and McDonald (1997a,b) to challenge the 100-ka glacial cycles not by the eccentricity but due to inclination. Varadi *et al.* (2003) refined inclination values, which are presented for the last 3-Ma (Fig. 8H). MTM spectrum of inclination suggest cycles at ~1.2-Ma and 112-ka (Fig. 8I), and cyclic components are reconstructed and compared with original time series (Fig. 8 J-K). The eccentricity and inclination of orbital plane, both share the common forces, *i.e.*, the gravitational perturbation and torque forces exerted by Jupiter and extra Jovian planets in the Earth's orbital dynamics through out the geological past. The climate change on Earth is influenced by the gravitational pull/push (perturbation) of the planets at daily (tidal), monthly (soli-lunar), seasonal (rain), annual (ENSO), decadal (Sunspots), and millennial (amplitude modulated Sunspot cycles in Figs. 6-7) to the orbital scales (Figs. 9-10). Orbital and sub-orbital cycles in monsoon in the radiolarian records are documented by Gupta (1999, 2002, 2003, 2009).

Solar Insolation

Solar radiation (insolation) arriving at the top of surface of Earth at a given latitude and time is computed based on Earth's eccentricity (e), tilt (t), and Sun's longitude of perihelion (þ) with respect to moving vernal equinox (Berger, 1978). Influence of eccentricity and precession are mostly found everywhere, but dominant in the tropics, whereas tilt is more dominant in the polar/sub-polar regions (Berger, 1993). Insolation is straight and high during summer at the tropics when Earth is closer to perihelion (nearest point - 21st June) and low during winter when it is near aphelion (farthest point - 21st December) in its orbit around Sun at present (but not in the past), whereas polar regions receive radiation at an angle due to the axial tilt (Fig. 5A-d). When insolation is higher at 30°N over India during summer, southern Indian Ocean receives it lesser being austral winter at 30°S between Australia and Madagascar (Fig. 2C). It creates inter-hemispheric land-sea thermal contrast, generating low and high atmospheric pressure resulting into the stronger monsoon, and opposite happens with low summer insolation resulting into weak monsoon.

Researchers used insolation values of 21st June at 65°N for comparing their time series data, whereas as Berger (1977, 1993) strongly recommended monthly/seasonal and regional insolation, especially if time series are from tropic/subtropics. Leuchner and Sirocko (2003) and Gupta (2003, 2009) considered 30° insolation difference between two hemispheres (d-I=30°N minus 30°S) during northern summer as the insolation monsoon index (IMI), and they found very good relation with monsoon proxies. As differential solar heating of the mid latitudes during northern summer (Fig.1) results in the monsoonal rains, a time series of differential solar insolation (d-I, 30°N-30°S) during 21st April-21st July (90 days mean) as Indian summer monsoon index (ISM) is calculated at 1-ka for the last 10-Ma using insolation equations of Berger and Loutre (1991) in the Analyseries (Pailard *et al.*, 1996), and presented in Fig. 9A.

MTM spectral analysis of d-I revealed highly significant triplet cycles of eccentricity (~95-, 100-, 400-ka), tilt (~54-, 41-, 31-ka), precession (~23-, 19-, 17-ka), and sub-orbital cycles at ~15-, 13-, 11-, 10-, and 9-ka (Fig. 9B). MTM spectral analyses of (-I revealed highly significant triplet cycles of eccentricity (~95-, 100-, 400-ka), tilt (~54-, 41-, 31-ka), precession (~23-, 19-, 17-ka), and sub-orbital cycles at ~15-, 13-, 11-, 10-, and 9-ka (Fig. 9B). Significant cyclic components of the ~400-, 126-, and 95-ka eccentricity, 54-, 41-, 31-ka tilt, and 23-ka precession in the (-I are reconstructed and exhibited (Fig. 9C-I). Bell (2003) analyzed 800-ka climatic record comparing with precession signals and concluded that the 30-ka cycle could also be due to precession-related summer solar heat response affecting the global climates. It is in contrast to the established notion that the ~31-ka cycle (Fig. 9H) is due to tilt (Berger, 1977; Gupta, 2009), which is also evident in spectral analysis of tilt (Fig. 8-G).

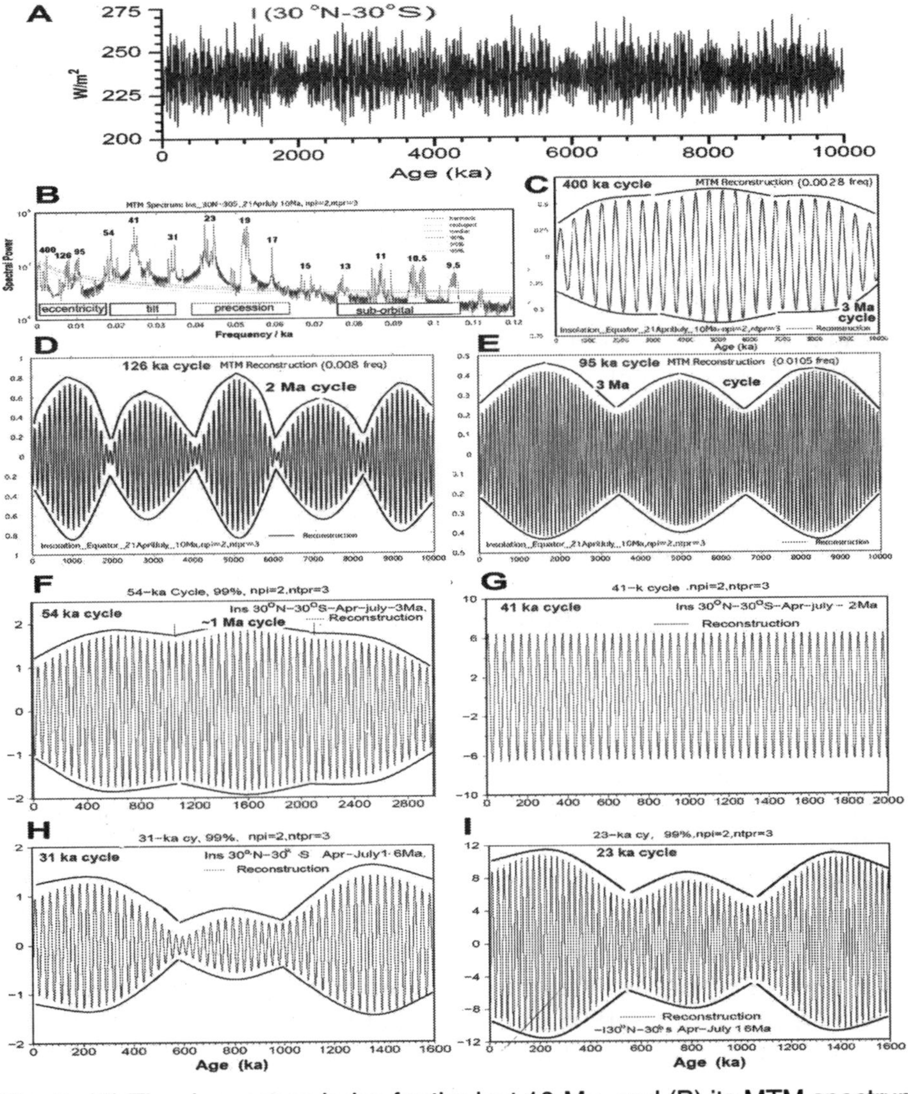

Fig. 9: (A) The -I monsoon index for the last 10-Ma, and (B) its MTM spectrum showing significant triplet cycles of the eccentricity (~95-, 126-, 400-ka), tilt (~54-, 41-, 31-ka), precession (~23-, 19-, 17-ka), and sub-orbital cycles at ~15-, 13-, 11-, 10-, and 9-ka. MTM reconstruction of equatorial summer isolation for last 10-Ma exhibiting the cyclic components of the ~400-ka (C), 95-ka (E) with amplitude modulated ~3-Ma cycle, whereas 126-ka (D) has amplitude modulated ~2-Ma eccentricity cycles. MTM reconstruction of d-I exhibits cyclic components of 54-ka with amplitude modulated cycle of ~1-Ma in the last 3 Ma (F), and (G) 41-ka cycle with uniform amplitude in last 2 the Ma. Reconstructed d-I exhibiting 31-ka tilt (H) and 23-ka precession (I) components with amplitude modulated 600-ka cycle in last 1.6 Ma.

Triplet Cycles of Earth's Eccentricity in the Monsoon

Earlier, typical ~23-, 41, and 100-ka Milankovitch (1941) cycles were reported in the Indian monsoon by Clemens *et al.* (1991), Kroon *et al.* (1991), Prell *et al.* (1992), Prell and Kutzbach (1992) and their groups. Triplet cycles of eccentricity at ~400, 126, and 95-ka (Fig. 8E) were forecasted by Berger (1977, cf. table 3), and they are evident in MTM results of the d-I monsoon index (Fig. 9B). These triplet eccentricity cycles were first reported in radiolarian transfer function based monsoonal SSTs (August) from the central tropical Indian Ocean (Gupta *et al.*, 1996). Subsequently, these were reported from the $d^{18}O$ (Clemens and Tiedemann, 1997), foraminiferal abundance (Muller and McDonald, 1997a-b) in western tropical warm pool, and in the monsoonal radiolarian indices (Gupta, 1999, 2002, 2003, 2009). Later, Gupta *et al.* (2001) have reported ~412-, 121- and 94-ka triplets cycles from benthic foraminifer *U. proboscidea* monsoon proxy from DSDP-214 site in the eastern Indian Ocean.

Milankovitch Cycles in the Monsoon

Clemens *et al.* (1991) and Prell *et al.* (1992) and their group established the role of ~23-ka precession, ~41-ka tilt and ~100-ka eccentricity cycles in the proxy monsoonal records of *G. bulloides*, $CaCO_3$, organic carbon and barium contents from the Arabian Sea upwelling zone. These three cycles are primarily known as the Milankovitch (1941) cycles in the Indian monsoon. Beaufort *et al.* (1997) studied abundance variation in the coccolithophores (*Florisphaera profunda*), an index of oceanic primary productivity, and reported the ~100-ka eccentricity and ~23-ka precession cycles, and they related it to insolation response in monsoonal productivity in the equatorial Indian Ocean. Gupta (1996) reported that modern radiolarian faunal assemblages characterized the overlying surface water SST and salinity, and clearly divide high salinity equatorial and low salinity distal Bay of Bengal waters in the central Indian Ocean. Gupta *et al.* (2002) studied radiolarian fluxes from the sediment traps moored in the southern Bay of Bengal suggested that radiolarians were invariably higher at times of higher sea surface temperature (SST >28°C) and moderately lower salinity (~33.5 psu) during the summer prior to monsoon of 1991-93. These studies suggested that higher SSTs due to summer heat and lower salinity due to rains might affect radiolarian fluxes during the monsoon. Down core studies revealed that the salinity sensitive radiolarians (Rads/g, Pyloiids, *Spongaster*, *Stylodictya-Stylochlamydium*, and *Anthocyrtidium* assemblages) changed cyclically at ~126-, 95-ka due to Earth's orbital eccentricity, ~54-, 41-, and 31-ka due to axial tilt, ~23-, 19-, and 17-ka due to changes in the precession of equinoxes, besides the ~15-, 13- and 11-ka sub-Milankovitch cycles in the last 485-ka (Gupta and Fernandes, 1997; Gupta, 1999, 2002, 2003, 2009). Besides, the reconstructed cyclic components of Rads/g also revealed a conspicuous change in the trends of the amplitudes of ~95-, 54-, 41-, 17-, 15-, and 13-ka cycles suggesting an effect of the mid-Brunhes climate shift at ~300-350-ka (Gupta, 2009). Similar mid-Brunhes climate shift is later found in the variation of *G. bulloides* from DSDP Hole 716A and 728B by Gupta *et al.* (2010).

Supra Milankovitch (>100-ka) Cycles in the Monsoon

The cycles >100-ka are termed as supra-Milankovitch cycles, and are mainly the response to the triplet eccentricity cycles, which were first reported in radiolarian transfer function based monsoonal SSTs (August) from the central tropical Indian Ocean (Gupta *et al.*, 1996). Later, Gupta *et al.* (2001) reported ~100-ka and 400-ka Earth's eccentricity from benthic foraminiferal monsoon proxies (DSDP-214 site), and Gupta and Melice (2003) found them in benthic foraminifers *U. proboscidea* and *E. exigua* from ODP-758 in the eastern equatorial Indian Ocean. Banerjee *et al.* (2010) documented the ~400-ka eccentricity cycles in the Na, Mg, Cu, Co, and Ni (Fig. 10 A-C, Ni) micronutrients carried away to sea-floor by scavenging during the monsoonal productivity, in a slowly accreting hydrogenous ferro-manganese crust dating back to ~3.5 Ma from Vityaz fracture zone in equatorial Indian Ocean. Ni is considered un-common in terrestrial sources and it is always higher in extra-terrestrial cosmic/meteorite dust (<1 ìm) flux into the ocean. They related these changes due to the monsoonal oceanic productivity, which is known to vary with insolation variance at ~100-ka eccentricity cycles in the equatorial Indian Ocean (Beaufort *et al.*, 1997). Monsoonal productivity induced fecal pellets and the particulates incorporated the micronutrients of Ni, presumably derived from cosmic (comet/meteoritic) dust, and transported it to the ocean floor on the Fe-Mn crust.

Banerjee *et al.* (2008) presented elemental variation in layers of a slow accreting hydrogenous crust having a record of the last ~12 Ma from the central Indian Ocean. Red-fit and MTM spectral analyses of Fe, Mg and Ni revealed the existence of significant (>90%) cycles at ~3, 1.5, and 1.2 Ma (Fig. 10 D-E, Ni). As the Fe/Ni ratio was higher than oceanic nature, they attributed to high Ni contents from the cosmic dust flux, which incidentally is known to have similar cycles (Johnson, 2001), supported by propositions that an estimated 5-700 small diameter meteorites are entering into Earth every Ma (Nurmi *et al.*, 2001). They suggested that the Fe, Mg, and Ni cycles at ~3, 1.5 and 1.2 Ma (main peak 1.3 Ma) could be result of geochemical response to similar 3.5-, 1.3-Ma eccentricity cycle (Berger, 1977, cf. table 3) related solar insolation changes, which are shown as the amplitude modulated envelop of bundles in the ~400-, 126-, 95-ka (eccentricity) and 54-ka (tilt) cycles (Fig. 9C-F), and at ~3, 1.5 and 1.2 Ma supra-Milankovitch cycles in the Ni contents (Fig. 10).

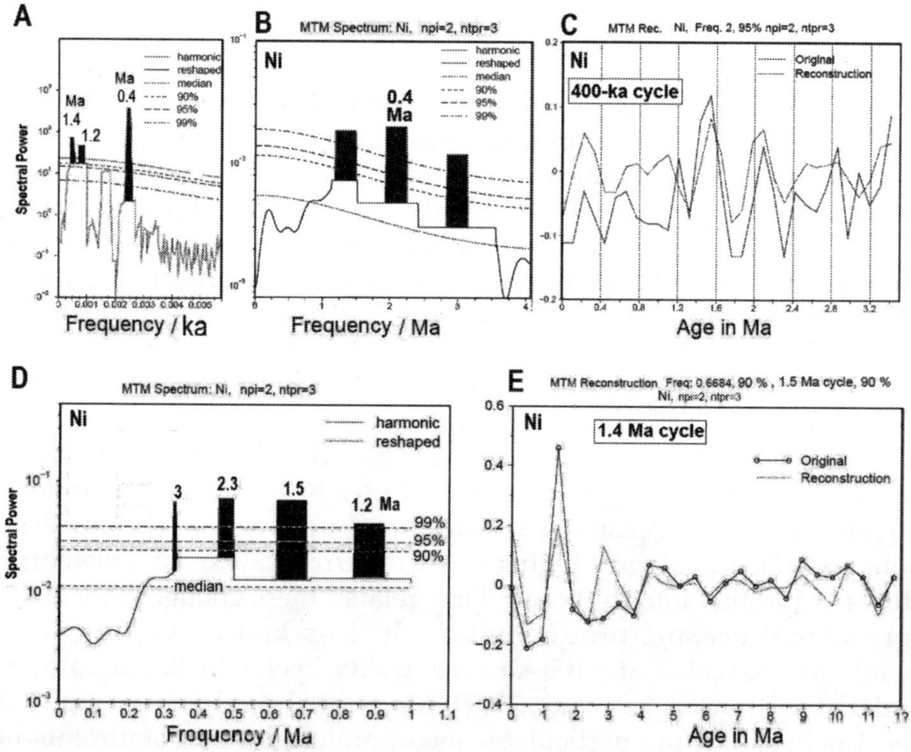

Fig. 10: (A) A part of MTM spectral analysis of d-l monsoon index sampled at ~100-ka interval showing significant cycles of ~1.4 and 0.4 Ma eccentricity cycles. (B) MTM of Nickel in Vityaz facture zone (Banerjee *et al.*, 2010) showing ~400-ka cycle, which is reconstructed and compared with original Ni time series (C). (D) MTM of Nickel from Central Indian Ocean Basin exhibiting significant ~3, 1.5, and 1.2 Ma cycles (Banerjee *et al.*, 2008) as predicted by Berger (1977, cf. table 3), and only 1.5 Ma cycle is reconstructed and compared with original Ni time series (E).

Oldest Indian Monsoon Record

Origin of the monsoon in geological past has been enigmatic. Probably, one of the earliest paleo-monsoon scenarios presented was from the radiolarian faunal assemblage changes during the late Miocene (~6-8.8 Ma) from the Sawai Bay Formation, Neill Island, Andaman (Gupta and Nigam, 1988; Guptha *et al.*, 1988) in the 1st International Conference on Radiolaria held at Marburg, Germany. A year later, Quade *et al.* (1989) reported development of Asian monsoon as revealed by the marked ecological shift from grassland to shrub vegetation (C-4) proxy variation in the molar contents of mammals during the late Miocene ~8-Ma from deposits in the northern Pakistan. Thereafter, series of papers appeared describing changes in abundance of the upwelling foraminifers like *G. bulloides,* and radiolarians like *Collosphaera Sp. A.* in the sediments from Off Oman Margin (Kroon *et al.* 1991; Prell *et al.*, 1992). Gupta (S.M.) and Srinivasan (1992) reported upwelling radiolarian fauna and the diatom blooms at ~6-8.5 Ma, and discussed the basinal shallowing due to enhanced supply of sediments derived by the Ganges and Bramhaputra rivers into Andaman Sea. Gupta and Srinivasan (1992a-b) found foraminiferal changes at ~9-Ma, which they related to monsoonal productivity in the Andaman Islands. As those foraminiferal, radiolarian and diatom species are abundance in modern monsoon dominated upwelling areas, these authors suggested that the monsoon strengthened at ~8.5 Ma in the region. Later, Gupta and Fernandes (1995) derived the long range radiolarian transfer function for the SST during August and surface salinity during July-August monsoonal peak season, and they estimated the paleo-salinity using Neill Island radiolarian data during the Late Miocene. They found that the monsoonal salinity was higher at ~8.5-Ma and lower at ~6-Ma in the Bay of Bengal, and attributed it to the lowering of the surface salinity due to enhanced river runoff by the Indian rivers (Ganges and Bramhaputra) into the Bay of Bengal. Similarly, Chen *et al.* (2003) found that changes in radiolarian flux, especially in abundance of Pyloniids, are good proxies and analogues to modern upwelling summer monsoon radiolarians. They suggested that the east Asian summer monsoon first initiated close to the middle/late Miocene boundary at ~12-11-Ma and then strengthened at ~8.24-Ma, which is in conformation with earlier studies from the Andaman Islands (Gupta (SM) and Srinivasan (1992). Since then, numerous papers have been published establishing the existence of the late Miocene monsoon in the Arabian Sea and Indian Ocean. However,

Guo *et al.* (2002) proposed that the eolian loess of central China is much older (22 Ma) than believed so far at ~8-Ma, implying that monsoon could still be older. Clift *et al.* (2004) studied variations in clay minerals in South China Sea and suggested alterations from drier to wetter climate before ~15-Ma. The timing of initial monsoon strength/origin is still not un-disputed in the global climate system.

Conclusion

Although Indian monsoon mostly sets on in early June, it often deceives casting droughts or floods causing hardship to millions in the Indian subcontinent. It varies at the intra-monsoonal scale with 40-50 days cycle due to variability in the ITCZ/Hadley cell cloud formation and its migration towards Indian subcontinent. It exhibits cycles at ~2.7, 3.5 (QBO), 5, 7, (El-Nino/La-Nina, ENSO), 9, 11, 22, 30, 45, 60, 80, 95, 150, 180, 245, 584, 1100, and 2200 years cycles, mainly reported due to Sunspot activity. Monsoon records also suggest variability due to Sun-Moon tidal effect at 8.9, 13.6, 18, 27.2, 44, 90-95, 180, 1470, 1800, and 5000 years cycles. Besides, monsoon also varies at the sub-Milankovitch/sub-orbital (~7-, 9-ka, 11-, 13- and 15-ka), Milankovitch/orbital (100-ka eccentricity, 54-, 41-, and 31-ka tilt, and 23-, 19- and 17-ka precession of equinoxes), and the supra Milankovitch (1.3 Ma and 0.4-Ma) eccentricity cycles. There are growing body of evidences implying that the changes in trade wind strength, which later transform into monsoonal winds (Fig. 2A), are also influenced by the solar heating/irradiance/insolation of the tropical/subtropical sea surface temperatures, resulting into evaporation and cloud formation. So also, there are several proxy monsoon records suggesting an important role of solar insolation (also Sunspots) in influencing the monsoon sensitivity forcing to vary it on seasonal, yearly, decadal, centennial, millennial and astronomical time scales. Several evidences exhibiting reasonable relations/rationales for monsoonal influences, like (i) Sunspots, (ii) synod effect, (iii) seasonal insolation and micro variations in the Earth's eccentricity and precession (Bertrand *et al.*, 2002), and (v) tidal forcing etc., are known to affect the monsoon strength, but to date these are not considered as the input parameters in predictive models of monsoon, either due to inconsistency in their correlation coefficients with time or difficulties in handling new variables in complex models. There seems to be a wealth of literature and historical account scattered in ancient scriptures on monsoon rains, its cyclic nature enabling pre-emptive mitigation measures adopted by rulers for benefit of the agrarian subject. For example, Kautilya used to reduce taxes on agriculture during poor monsoon, declare community feast (Sadaa-vart) during famine etc. as narrated in Artha-Sastra (Waldauer *et al.*, 1996). Due to numerous physical forcings interacting on monsoon at the same time, and their linear and non-linear response are not fully understood and assimilated in monsoon predictive models with high reliability at least at weekly level, so the farmers can plan their crops in better way. With such an advance in knowledge, when monsoon fails (flood/drought), it makes us to realize that we do not know for sure - how would be next monsoon with reasonably dependent limit of uncertainties.

Acknowledgements

Author thank to the Director, NIO, Goa, and the Director General, Council of Scientific & Industrial Research (CSIR), New Delhi for support. He thankfully acknowledges several sources for the illustrations modified after orginal graphics from www.wikipedia.com, www.google.com, www.ngdc.noaa.gov/paleo/forcing.html, Sunspot numbers (www.sidc.be), and paleo-TSI and Sunspot numbers (www.noaa.edu\paleoclimatology) and used in MTM analyses and creating the graphics used in the paper. Author thanks to Marie Loutre (Instutute of Astrophysics & Astrodynamics, Brussels) for digital data on eccentricity, precession and tilt for the last 3 Ma, and Rajesh Agnihotri for sparing All India Homogenized Rain Fall Data (www.tropmet.res.in/static_page.php?page_id=53). Article is dedicated by the author to Prof. Björn A. Malmgren (Nörttlje, Sweden) on the GURU-POORNIMA occasion as a grattitude for his teachings on quantitative paleoceanography, artificial neural networks, multivariate and the time series spectral analyses, and invoking the naturalistic views in the author. It is a part of N.I.O's on going Project GAP 2175/OLP 0011 on understanding the geology of the deep-sea oceans (N.I.O.'s contribution no./ref No. 4840).

References

Agnihotri, R. and Dutta, K. (2003) Centennial scale variations in monsoonal rainfall (Indian, east equatorial and Chinese monsoons): Manifestations of solar variability. Curr. Sci., v. 85, pp. 459–463.

Agnihotri, R., Altabet, M. A., Herbert, T. D. and Tierney, J. E. (2008) Subdecadally resolved paleoceanography of the Peru margin during the last two millennia. Geochem. Geophys. Geosyst., v. 9, pp. Q05013, doi:10.1029/2007GC001744.

Agnihotri, R., Dutta, K., Bhushan, R. and Somayajulu, B.L.K. (2002) Evidence of solar forcing of the Southwest monsoon during the last Millennium. Earth Planet. Sci. Lett., v. 198, pp. 521-527.

Ashok, K., Guan, Z. and Yamagata, T. (2001) Impact of the Indian Ocean dipole on the relationship between the Indian monsoon rain fall. Geophys. Res. Lett., v. 28, pp. 4499-4502.

Banerjee, R. Gupta, S. M., and Miura, H. (2008) A million year cycle in the Fe, Mg and Ni records of a ferromanganese crust from the equatorial Indian Ocean. AGU fall meeting, San Francisco, pp. PP43A-1511.

Banerjee, R., Gupta, S. M., Miura, H. and Borole, D. V. (2010) A ~400 ka supra-Milankovitch cycle in the Na, Mg, Pb, Ni, and Co records of a ferromanganese crust from the Vityaz fracture zone, central Indian ridge. Climate of the Past – Disc., v. 6, pp. 1311-1335, http://direct.sref.org/1814-9359/cpd/2010-6-1311

Bard, E., Raisbeck, G., Yiou, F. and Jouzel, J. (2000) Solar irradiance during the last 1200 years based on cosmogenic nuclides. Tellus, Ser. B, v. 52, pp. 985–992, doi:10.1034/j.1600-0889.2000.d01-7.x.

Beaufort, L., Lancelot, Y., Chamberlain, P., Cayre, O., Vincent, E., Bassinot, F. and Labeyrie, L. (1997) Insolation cycles as a major control of equatorial Indian Ocean primary production. Science, v. 278, pp. 1451–1454.

Beer, J., Baumgartner, S., Hannen-Dittrich, B., Hauenstein, J., Kubik, P. P., Lukasczyk, C., Mende, W., Stellmacher, R. and Suter, M. (1994) The Sun as a variable star: Solar and stellar irradiance variations. In: *Pap, J.M., et al., (eds.)*, Cambridge University Press, Cambridge, pp. 291.

Bell, G.J. (1977) Changes in sign of the relationship between Sunspots and pressure, rainfall and the monsoons. Weather, v. 32, p. 26-32.

Bell, L. G. (2003) A 30,000 yr precession-related cycle affecting climate. Theor. Appl. Climatol. v. 74, pp. 245–253, DOI 10.1007/s00704-002-0711-7.

Berger, A. L. (1977) Support for the astronomical theory of climatic change. Nature, v. 269, pp. 44-45.

Berger, A.L. (1978) Long term variations of daily insolation and Quaternary climatic changes. J. Atmos. Sci., v. 35, pp. 2362-2367.

Berger, A.L. (1993) Insolation and Earth's orbital periods. J. Geophys. Res., v. 98, D6, pp. 10341-10362.

Berger, A.L. and Loutre, M.F. (1991) Insolation values for the climate of the last 10 million years. Quaternary Sciences Review, v.10, pp. 297-317.

Bertrand, C., Loutre, M. F. and Berger, A. (2002) High frequency variations of the Earth's orbital parameters and climate change. Geophys. Res. Lett., v. 29, pp. 1893-1996, doi:10.1029/2002GL015622.

Bond, G. B., Kromer, B., Beer, J., Muscheler, R., Evans, M. N., Showers, W., Hoffmann, S., Lotti-Bond, R., Hajdas, I. and Bonani, G. (2001) Persistent solar influence on North Atlantic climate during the Holocene. Science, v. 294, pp. 2130–2136, doi:10.1126/science.1065680.

Burgess, E. (1860) The Surya Siddhanta - English Translation 1989, Gangooly, P., ed., Motilal Banarsidass, Delhi.

Burroughs, W.J. (2003) Weather cycles real or imaginary. Cambridge University Press, UK, p. 325.

Chakraborty, P.K. and Bondyopadhyay, R. (1986) Solar effect on rainfall in West Bengal. Mausam, v. 37, pp. 251-258.

Chao W. C. and Chen, B. D. (2001) The origin of monsoon. J. Atoms. Sci., v. 58, pp. 3497-3507.

Chen, M., Wang, R., Li, H-Y., Han, J. and Lu, J. (2003) Development of east Asian summer monsoon environments in the late Miocene: radiolarian evidence from Site 1143 of ODP Leg 184. Marine Geol., v. 201, pp. 169-177.

Chen, T-C, Tzeng, R-Y. and Yen, M-C. (1988) Development and life cycle of the Indian monsoon: Effect of the 30–50 day oscillation. Mont. Weather Rev., v. 116, pp.2183-2199.

Choudhury, A, Abhayankar, V.P. and Phadke, M.V. (1981) Solar activity as a potential factor for fore-shadowing drought in Indian summer monsoon regime. Mausam, v. 32, pp. 277-280.

Clark, C.O., Cole, J.E. and Webster, P.J. (2000) Indian Ocean SST and Indian summer rainfall: predictive relationships and their decadal variability. J. Climate, v. 13, pp. 2503-2519.

Clemens, S.C. (1998) Dust response to seasonal atmospheric forcing: proxy evaluation and calibration. Paleoceanography, v. 13, pp. 471-490.

Clemens, S.C. and Tiedemann, R. (1997) Eccentricity forcing of Pliocene-Pleistocene climate revealed in a marine oxygen isotope record. Nature, v. 385, pp. 301-304.

Clemens, S. C., Prell, W. L., Murray, D., Shimmield, G. and Weedon, G. (1991) Forcing mechanisms of the Indian Ocean monsoon. Nature, v. 353, pp. 720–725.

Clift, P.D., Layne, G.D. and Blusztajn, J. (2004) Marine sedimentary evidence for monsoon strengthening, Tibet uplift and drainage evolution in East Asia, In: Continent-Ocean Interactions within East Asian Marginal Seas. Geophys. Monogr., Ser. v. 149, 10.1029/149G14, pp. 255-282.

Currie, R.G. and Fairbank, R.W. (1985) Periodic 18.6 year and cyclic 11 year induced drought and flood in north eastern China and some global implications. Quat. Sci. Rev., v. 23, pp. 109-134.

deMenocal, P.B. (2004) African climate change and faunal evolution during the Pliocene-Pleistocene. Earth Planet. Sci. Lett. v. 220 pp. 3-24.

Eddy, J.A. (1977) Climate and the changing Sun. Climatic Change, v. 1, pp. 173-190.

Eddy, J.A. and Boornazian, A. A. (1979) Secular decrease in the solar diameter 1863-1953. Bull. Amer. Astron. Soc., v. 11, pp. 437.

Eddy, J.A., Gilliland, R.L. and Hoyt, D.V. (1982) Changes in the solar constant and climatic effects. Nature, v. 300, pp. 689-693.

Feynman, J. and Fougre, P.F. (1984) Eighty eight year periodicity in solar-terrestrial phenomena confirmed. J. Geophys. Res., v. 89, pp. 3023-3027.

Fowles G.R. (1977) Analytical mechanics. 3 edition. New York, Holt, Rinehart and Winston, ch. 8, p. 243.

Gadgil, S. (1988) Recent advances in monsoon research with particular reference to the Indian monsoon. Aust. Meteorol. Mag., v. 36, pp. 193–204.

Gadgil, S. (1996) Climate change and agriculture-An Indian perspective, In Climate Variability and Agriculture. (eds.), *Y. R. Abool, S. Gadgil and G. B. Pant*, pp. 1-18, Narosa, New Delhi, India.

Gadgil, S. (2003) The Indian monsoon and its variability. Ann. Rev. Earth Planet Sci., v. 31, pp. 429-467.

Gadgil, S., Joseph, P.V. and Joshi, P.V. (1984) Ocean-atmosphere coupling over monsoon regions. Nature, v. 312, pp. 141-143.

Gadgil, S. Srinivasan, J., Nanjundiah, R.S., Kumar, K.K., Munot, A.A. and Kumar, K.R. (2002) On forecasting the Indian summer monsoon: the intriguing season of 2002. Curr. Sci., v. 43, pp.394-403.

Ghil, M., Allen, M.R., Dettinger, M.D., Ide, K., Kondrashov, D., Mann, M.E., Robertson, A.W., Saunders, A., Tian, Y., Varadi, F. and Yiou, P.(2001) Advanced spectral methods for climatic time series. Rev. Geophys., v. 40, pp. 1-23.

Gilliland, R.L. (1981) Solar radius variations over the past 265 years. Astrophys. J., v. 248, pp. 1144-1155.

Gleissberg, W. (1958) The eighty years solar cycle in auroral frequency numbers. J. British Astron. Asso., v. 75, pp. 227-231.

Guo, Z. T., Ruddiman, W.F., Hao, Q. Z., Wu, H. B., Qiao, Y. S., Zhu, R. X., Peng, S. Z., Wei, J. J., Yuan, B. Y. and Liu, T. S. (2002) Onset of Asian desertification by 22 Myr ago inferred from loess deposits in China, Nature, v. 416, pp. 159-163.

Gupta, A.K. and Mélice, J-L. (2003) Orbital forcing of the Plio-Pleistocene Indian monsoons: benthic foraminiferal proxies from ODP Site 758. Curr. Sci., v. 85, pp.179-184.

Gupta, A.K. and Srinivasan, M.S. (1992a) Deep-sea benthic foraminiferal changes and terminal Miocene event at tropical Indian Ocean DSDP site 214, J. Geol. Soc. India, v. 40, pp. 262-278.

Gupta, A.K. and Srinivasan, M.S. (1992b) Uvigerina-Proboscidae abundances and Paleoceanography of the northern Indian Ocean DSDP site 214 during the late Neogene. Marine Micropaleontol., v. 19, pp. 355-367.

Gupta, A.K., Das, M. and Anderson, D.M. (2005) Solar influence on the Indian summer monsoon during the Holocene. Geophys. Res. Lett., v. 32, pp. L17703, doi:10.1029/2005GL022685.

Gupta, A. K., Dhingra, H., Mélice, J.-L. and Anderson, D.M. (2001) Earth's eccentricity cycles and Indian Summer monsoon variability over the past 2 million years: Evidence from deep sea benthic foraminifer. Geophys. Res. Lett., v. 28, pp. 413–4134.

Gupta, A.K., Sarkar, S., De, S., Clemens, S.C. and Velu, A. (2010) Mid-Brunhes strengthening of the Indian Ocean Dipole caused increased equatorial East African and decreased Australasian rainfall. Geophys. Res. Lett., v. 37, pp. L06706, doi:10.1029/ 2009GL042225, 2010.

Gupta, S.M. (1996) Quantitative radiolarian distribution in the surface sediment from the central Indian Ocean and their paleomonsoonal significances. J. Geol. Soc. India, v. 47, pp. 339-354.

Gupta, S.M. (1999) Radiolarian monsoonal index Pyloniids respond to astronomical forcing in the last 5,00,000 years in the central Indian Ocean Basin. Man & Environment, v. 24, pp. 99-105.

Gupta, S.M. (2002) Pyloniid stratigraphy – a new tool to date tropical radiolarian ooze from the central tropical Indian Ocean. Marine Geology, v. 184, pp. 85-93.

Gupta, S. M. (2003) Orbital frequencies in radiolarian assemblages of the central Indian Ocean: implications on the Indian summer monsoon. Palaeogeogr. Palaeoclimatol. Palaeoecol. v. 197, pp. 97-112.

Gupta, S.M. (2009) Radiolarian abundance - a monsoon proxy responding to the Earth's orbital forcing: Inferences on the mid-Brunhes climate shift, Earth Science India. v. 2, pp. 1-8. www.earthscienceindia.info/Gupta.htm.

Gupta, S.M. and Fernandes, A.A. (1995) Supra-generic radiolarian transfer function based late Miocene paleomonsoonal changes in the northeastern Tropical Indian Ocean. Bull. Ind. Geol. Asso., v. 28, pp. 29-51.

Gupta, S.M. and Fernandes, A.A. (1997) Quaternary radiolarian faunal changes in the central Indian Ocean: inferences to oscillating hydrographic front at 10°S of Equator. Curr. Sci., v. 72, pp. 965-972.

Gupta, S.M. and Nigam, R. (1988) Cluster analysis of radiolarian fauna from Sawai Bay Formation (Late Miocene), Neill Island, Ritchie's Archipelago, Andaman Sea, and its paleoceanographic significance. Geologica et Palaeontologica, conf. book - 1st Int. Conf. Radiolaria, (EURORAD V) Marburg, Germany, 21-24 July, 1988, p. 17.

Gupta, S.M. and Srinivasan, M.S. (1992) Late Miocene radiolarian biostratigraphy and paleoceanography of Sawai Bay Formation, Neill Island, Andaman, India. Micropaleontology, v. 38, p. 209-236.

Gupta, S.M., Fernandes, A.A. and Mohan, R. (1996) Tropical sea surface temperatures and the Earth's eccentricity cycles. Geophys. Res. Lett., v. 23, pp. 3159-3162.

Gupta, S.M., Mohan R. and Guptha, M.V.S. (2002) Radiolarian fluxes in the southern Bay of Bengal: A sediment trap study. Deep Sea Res., I, v. 49, pp. 1669-1688.

Guptha, M.V.S, Gupta, S.M. and Srinivasan, M.S. (1988) Calcareous Nannoplankton and radiolarian biostratigraphy of Sawai Bay Formation, Neill Island, Ritchie's Archipelago, Andaman Sea. Geologica et Palaeontologica, conf. book - 1st Int. Conf. Radiolaria, (EURORAD V) Marburg, Germany, 21-24 July, 1988, p. 18.

Hadley, G. (1735) On the cause of the general trade winds. Phil. Trans., R. Soc. London, v. 34, pp. 58-62.

Hale, G.E. (1924) The of Sunspot polarity. Proc. National Acad. Sci., USA, v. 10, pp. 53-55.

Halley, E. (1686) A historical account of the trade winds and monsoons observable in seas between and near tropics with an attempt to assign a physical cause of the said winds. Phil. Trans. R. Soc. London, v. 16, pp.153–68

Hameed, S., Yeh, W.M., Li, M.T., Cess, R.D. and Wang, W.C. (1983) An analysis of periodicities in the 1470-1974 Beijing precipitation record. Geophys. Res. Lett., v. 10, pp. 436-439.

Hari, K.C. (1997) True rationale of Surya Sidhhanta. Indian J. Historical Sci., v. 32, pp. 183-190.

Hastenrath, S. (1987) On prediction of India summer rainfall anomalies. J. Climate Appl. Meteor., v. 26, pp. 847–857.

Hays, J.D., Imbrie, J. and Shackleton, N. (1976) Variations in the Earth's orbit: pacemaker of the Ice-Ages. Science, v. 194, pp. 1121-1132.

Hiremath, K. M. (2006) The influence of solar activity on the rainfall over India: Cycle-to-cycle variations. J. Astrophys. Astronom., v. 27, pp. 367-372.

Hong, Y.T., Wang, Z.G., Jiang, H.B., Lin, Q.H., Hong, B., Zhu, Y.X., Wang, Y., Xu, L.S., Leng, X.T. and Li, H.D. (2001) A 6000-year record of changes in drought and precipitation in northeastern China based on a ä[13]C time series from peat cellulose. Earth Planet. Sci. Lett., v. 185, pp. 111.

Hoyt, D. (1979) Variation in Sunspot structure and climate. Climatic Change, v. 2, pp. 79-92.

Iyengar, R. N. (2009) Monsoon rainfall cycles as depicted in ancient Sanskrit texts. Curr. Sci., v. 97, pp. 444-447.

Johnson, K.S. (2001) Iron supply and demand in the upper ocean: Is extraterrestrial dust a significant source of bio available iron. Global Biogeochem. Cycles, v. 15, pp. 61- 63.

Joseph, C.P. and Amatya, B.V.S. (1986) Influence of long term changes in Sunspot activity on annual mean temperature over south India. Mausam, v. 37, pp. 276-278.

Keeling, C.D. and Whorf, T.P. (2000) The 1,800-year oceanic tidal cycle: A possible cause of rapid climate change. Proc. Nat. Acad. Sci., USA, v. 97, pp. 3814–3819.

King, J.W. (1973) Solar radius changes and the weather. Nature, v. 245, pp. 443-446.

Kodera, K. (2004) Solar influence on the Indian Ocean monsoon through dynamical processes, Geophys. Res. Lett., v. 31, pp. L24209, doi:10.1029/2004GL020928,

Kroon, D., Steins, T. and Truest, S.R. (1991) Onset of monsoonal related upwelling in the western Arabian Sea as revealed by planktonic foraminifers. In Prell, W.L. *et al.*, Proceedings of the Ocean Drilling Project, College Station, Texas (Ocean Drilling Program), Sci. Res., v. 117, pp. 257-263.

Landscheidt, T. (1984) Cycles of solar flares and weather. In: N.-A. Morner and W. Karlen eds, Climatic Changes on a Yearly to Millennial Basis, D. Reidel Publishing Company, pp. 473-481.

Leuschner, D.C. and Sirocko, F. (2003) Orbital insolation forcing of the Indian monsoon – a motor for global climate changes? Palaeogeogr. Palaeoclimatol. Palaeoecol., v. 197, pp. 83-95.

Meehl, G. A. (1987) South Asian monsoon and the tropospheric biennial oscillation. J. Climate, v. 10, pp. 1921–1943.

Milankovitch, M.M. (1941) Kanon der Erdbestrahlung. Beograd. K6ninglich Serbische akademie, (English translation: Canon of Insolation and the Ice

Age Problem, by Israel Program for Scientific Translation and published for the U.S. Department of Commerce and the National Science Foundation), pp. 484.

Muller, R.A. and MacDonald, G.J., (1997a) Simultaneous presence of orbital inclination and eccentricity in the proxy climate records from Ocean Drilling Program site 806. Geology, v. 25, pp. 3-6.

Muller, R.A. and MacDonald, G.J. (1997b) Glacial Cycles and Astronomical Forcing, Science, v. 277, pp. 215- 218.

Naidu, P.D. and Malmgren, B.A. (1995) A 2200 years periodicity in the Asian monsoon system. Geophys. Res. Lett., v. 22, pp. 2361-2364.

Naik, S.S., Thamban, M., Laluraj, M., Redkar, B. L. and Chaturvedi, A. (2010) A century of climate variability in central Dronning Maud Land, East Antarctica, and its relation to southern annular mode and El-Niño - Southern-Oscillation. J. Geophys. Res., v. 115, D16102, doi:10.1029/2009JD013268.

Napier, W. M. (2001) Temporal variation of the zodiacal dust cloud. Mont. Not. R. Astron. Soc. v. 321, pp. 463-470.

Napier, W. M. (2006) Evidence for cometary bombardment episodes. Mon. Not. R. Astron. Soc. v. 366, pp. 977–982 doi:10.1111/j.1365-2966.2005.09851.

Neff. U, Burns S.J., Mangini A, Mudelsee, M., Fleitmann, D. and Matter, A. (2001) Strong coherence between solar variability and the monsoon in Oman between 9 and 6 kyr ago. Nature, v. 411, pp. 290–293.

Nicholls, N. (1983) Predicting Indian monsoon rainfall from sea-surface temperature in the Indonesia-north Australia area. Nature, v. 306, pp. 576–577.

Nicholson, S.B. (1957) The Sunspot Cycle. American Astronomers Report, Sky and Telescope, April, 1957.

Nigam, R., Khare, N. and Nair, R.R. (1995) Foraminiferal evidences for 77-Year cycles of droughts in India and its possible modulation by the Gleissberg Solar cycle. J. Coastal Res., v. 11, pp.1099-107.

NOAA, Solar Geophysical Data (1984) Sunspot numbers for the period 1871-1984, World Data Center A, Washington D.C., U.S.A. (1975) and from a catalogue of Solar Geophysical Data published by NOAA, U.S.A. (1984).

Nurmi, P., Valtonen, M.J. and Zheng J.Q (2001) Periodic variation of Oort Cloud flux and cometary impacts on the Earth and Jupiter. Mont. Not. R. Astro. Soc. v. 327, pp. 1367–1376.

Okal, E. and Anderson, D.L. (1975) On the Planetary theory of the Sunspots. Nature, v. 253, pp.511-513.

Paillard, D., Labeyrie, L. and Yiou, P. (1996) Macintosh program performs time series analysis. EOS Trans AGU, v. 77, pp. 379.

Parthasarathy, B., Munot, A. A. and Kothawale, D. R. (1994) All-India Monthly and Seasonal Rainfall Series: 1871-1993. Theor. Appl. Climatol. v. 49, pp. 217-224.

Philander, S.G.H. (1983) El-Nino Southern Oscillation. Nature, v. 302, pp. 296-297.

Prell, W.L. and Kutzbach, J.E. (1992) Sensitivity of the Indian monsoon to forcing parameters and implications for its evolution. Nature, v. 360, pp. 647-652.

Prell, W.L., Murray, D.W., Clemens, S.C. and Anderson, D.M. (1992) Evolution and variability of the Indian Ocean summer monsoon: Evidence from the western Arabian Sea drilling program,. In: Duncan, R.A., *et al.*, (eds.), Synthesis of Results from Scientific Drilling in the Indian Ocean, Geophysical Monograph 70, America Geophysical Union, Washington, D. C., pp. 447-469.

Quade, J., Cerling, T. E. and Bowman, J. E. (1989) Development of Asian monsoon revealed by marked ecological shift during the latest Miocene in northern Pakistan. Nature, v. 342, pp. 163-166.

Quinn, T. R., Tremaine, S. and Duncan, M. (1991) A three million year integration from the Earth's orbit. Astron. J., v. 101, pp. 2287-2305.

Rao, A. R. and Hameed, K. (2003) Multi-taper method of analysis of periodicities in hydrologic data. J. Hydrology, v. 279, pp. 125–143.

Reid, G.C. (1987) Influence of solar variability on global sea surface temperatures. Nature, v. 329, pp. 142-143.

Sadhuram, Y. (1997) Predicting monsoon rainfall and pressure indices from sea surface temperature. Curr. Sci. v. 72, pp. 166–168.

Saji, N.H., Goswami, B.N., Vinayachandran, P.N. and Yamagata, T. (1999) A dipole in the tropical Indian Ocean. Nature, v. 401, pp. 360-369.

Scafetta, N. (2010) Empirical evidence for a celestial origin of the climate oscillations and its implications. J. Atmos. Solar-Terrest. Phys., v. 10 , pp. 1-18, doi:10.1016/j.jastp.2010.04.015

Schlitzer, R., (2001) Ocean Data View, http://www.awi-bremerhaven.de/GEO/ODV.

Schulz, M. and Stattegger, K., (1997) Spectral analysis of unevenly spaced paleoclimatic time series. Comput. Geosci. v. 23, pp. 929-945.

Sen Gupta, P.K. (1957) Sunspot cycle in regional rain-fall. Weather, v. 12, pp. 322-324.

Shukla, K. S. (1957) The Surya Siddhanta with the Commentary of Paramesvara. Dept. of Mathematics and Astronomy, Lucknow University Research papers.

Sikka, D.R. and Gadgil, S. (1980) On the Maximum Cloud Zone and the ITCZ over Indian, Longitudes during the Southwest monsoon. Mon. Weather. Rev., v. 108, pp. 1840-1853.

Sirocko, F. Sarnthein, M, Lange, H. and Eirlenkeuser, H. (1991) Atmospheric summer circulation and coastal upwelling in the Arabian Sea during the Holocene and the last glaciation. Quat. Res., v. 36, pp.72-93.

Sofia, S., Dunham, D.W., Dunham, J.B. and Fiala, A.D. (1983) Solar radius change between 1925 and 1979. Nature, v. 304, pp. 522-526.

Solanki, S.K., Usoskin, I. G. Kromer, B. Schüssler, M. and Beer, J. (2004) Unusual activity of the Sun during recent decades compared to the previous 11,000 years. Nature, v. 431(7012), pp. 1084–1087.

Surya Siddhanta, Maharishi University of Management, Vedic Literature Collection. PDF file - Mahoreg VedicFinal.qxd, by P.F. Reund – 29 July, 2002, pp. 1-55.

Steinhilber, F., Beer, J. and Fröhlich, C. (2009) Total solar irradiance during the Holocene. Geophys. Res. Lett., v. 36, pp. L19704, doi:10.1029/2009GL040142.

Stuiver, M. and Quay, P.D. (1980) Changes in atmospheric ^{14}C attributed to variable Sun. Science, v. 207, pp. 11-19.

Thamban, M., Kawahata, H. and Rao, V.P. (2007) Indian summer monsoon variability during the Holocene as recorded in sediments of the Arabian Sea: timing and implications. J. Oceanogr., V. 63, pp. 1009-1020,

Thompson, R. (1997) Planetary Diameters in the Surya-Siddhanta. J. Sci. Explor. v.11, pp. 193-200, 1997

Thomson, D.J. (1982). Spectrum estimation and harmonic analysis, Proceed. IEEE, v. 70, pp. 1055-1096.

Tiwari, M., Ramesh, R., Somayajulu, B.L.K., Jull, A.J.T. and Burr, G.S. (2006) Paleomonsoon precipitation deduced from a sediment core from the equatorial Indian Ocean. Geo-Marine Lett., v. 26, pp. 23-30.

Torrence, C. and Compo, G.P. (1998) A Practical Guide to Wavelet Analysis, Bull. Amer. Meteor. Soc., v.79, pp. 61-78,

Torrence, C. and Webster, P.J. (1999) Interdecadal Changes in the ENSO–monsoon System. J. Climate, v. 12, pp. 2679-2690.

Varadi, F., Runnegar, B. and Ghil, M. (2003) Successive refinements in long term integration of planetary orbits. Astrophys. J., v. 592, pp. 620–630.

von Rad U., Schaaf, M., Michels K.H., Schulz, H., Berger, W.H. and Sirocko, F. (1999) A 5000-yr record of climate change in varved sediments from the Oxygen Minimum Zone off Pakistan, northeastern Arabian Sea. Quaternary Res., v. 51, pp.9–53.

von Rad, U., Lukage, A., Berger, W.H. and Rolinski, H.D. (2006) Annual to millennial monsoonal cyclicity recorded in Holocene varved sediments from the NE Arabian Sea. J. Geol. Soc. India, v. 68, pp. 353-368.

Waldauer, C., Zahka, W.J. and Pal., S. (1996) Kautilya's Arthasastra: A neglected precursor to classical economics. Ind. Econ. Review, v. 31, pp. 101-108.

Webster P. J., Magana, V.O. and Palmer, T.N. (1998) Monsoons: Processes, predictability, and the prospects for prediction. J. Geophys. Res., v. 103(C7), pp. 14451¯14510.

Wood, K.D. (1972) Sunspots and planets. Nature, v. 240, pp. 91-93.

□□□

22

Spatial Variability of Rock Depth using Artificial Intelligence Techniques

PIJUSH SAMUI[1] AND T.G. SITHARAM[2]
[1]Centre for Disaster Mitigation and Management, VIT University, Vellore-632014
[2]Department of Civil Engineering, Indian Institute of Science, Bangalore-560012

Abstract

This study describes two Artificial Intelligence (AI) techniques for predicting spatial variability of rock depth in Bangalore. Reduced level of rock (d) at Bangalore, India is arrived from the 652 boreholes data in the area covering 220 km^2. The first AI technique uses generalized regression neural network (GRNN) that are trained with suitable spread(s) to predict d at any point in Bangalore. The second technique uses Least Square Support Vector Machine (LSSVM), is a statistical learning theory which adopts a least squares linear system as a loss function instead of the quadratic program in original support vector machine (SVM). Here, LSSVM has been used as a regression technique. A comparative study between the two developed AI techniques has been presented in this paper. The results indicate that the developed GRNN model has the ability to predict d with an acceptable degree of accuracy {Coefficient of Correlation(r) =0.885, and Root Mean Square Error (RMSE) =0.021}. Whereas, the developed LSSVM model predicts d with an acceptable degree of accuracy (r =0.967, and RMSE=0.004). This study also highlights the capability of LSSVM model over GRNN.

Keywords : Rock depth; spatial variability; artificial intelligence; least square support vector machine; artificial neural network.

Introduction

The determination of spatial variability of rock depth is an imperative task in earth science. Spatial variability of the bed/hard rock is vital for many applications in geotechnical and earthquake engineering problems such as design of deep foundations, site amplification, ground response studies, liquefaction, microzonation etc. In the literature, spatial variability of soil properties is generally studied by combining statistical analysis of site-specific data with insights from random-field theory (Yaglom, 1962; Lumb, 1975; Alonso and Krizek, 1975; Vanmarcke, 1977; Tang, 1979; Wu and Wong, 1981; Tabba and Yong, 1981; Asaoka and Grivas, 1982; Vanmarcke, 1983; Baecher, 1984; Baker, 1984; Kulatilake and Miller, 1987; Kulatilake, 1989; Kulatilake and Ghosh, 1998; Fenton, 1998; Phoon and Kulhawy, 1999; Fenton, 1999, Uzielli *et al.,* 2005). The science of prediction in the presence of correlation between samples is not at all well developed in the random field method. The interpretation of trends in the data from the random-field method as true trends in the mean or simply as large-scale fluctuations is a question which can only be answered by engineering judgement. Therefore, the statistical parameters, which have been used to model a random field, are generally uncertain. Geostatistics have been used for spatial variability modeling by many researchers (Delhomme, 1979; Soulie, 1983; Kulatilake, 1989; Soulie *et al.,* 1990; Chiasson *et al.,* 1995; DeGroot, 1996). However, Random field methods and geostatistics have been applied in spatial variability modeling with limited success (Juang *et al.,* 2001).

With an objective of predicting the spatial variability of the reduced level of the bed/hard rock (d) in Bangalore, an attempt has been made to develop models based on Generalized Regression Neural Network (GRNN) and Lease Square Support Vector Machine (LSSVM). It is also aimed at comparing the performance of these developed models for the available data in Bangalore. GRNN has been used for function approximation in this study. It has been shown that, given a sufficient number of neurons, GRNN can approximate a continuous function to an arbitrary accuracy (Specht, 1991). LSSVM is firmly based on the theory of statistical learning (Suykens *at al.,* 1999). Here LSSVM has been used as a regression technique. The paper has the following aims:

1. To examine the feasibility of GRNN and LSSVM model for predicting d at any point in Bangalore.

2. To make a comparative study between developed GRNN and LSSVM model.
3. To develop an equation based on LSSVM model for the prediction of d at any point in Bangalore.

Geotechnical Data

A large amount of geotechnical data (depth, density of the soil, fines content, Standard Penetration Test Value(N), depth of ground water table, and rock depth) consisting of 652 boreholes has been collated along with index and engineering properties of subsoil layers at different locations in Bangalore (see Fig.1). Geotechnical data were evaluated for geotechnical investigations of several major projects in Bangalore. In total, information about the 652 borelogs has been entered into the database using a GIS with ARCINFO package. The subsurface three dimensional (3D) GIS model of Bangalore has been developed with a scale of 1:20000. Fig. 1 depicts 1km×1km grid within the corporate boundary, along with other boundaries, ring roads and borehole locations. It may be noted that an average of about two boreholes are available within each grid. Geotechnical data were collated from archives of Torsteel Research Foundation (India) and the Indian Institute of Science; these data were collected as part of several major projects in Bangalore during the years 1995-2003. The data in the model are on average to a depth of 30m below the ground level. The hard rock has been identified by visual observation of the cores taken at these locations. Rock depth from ground level is the

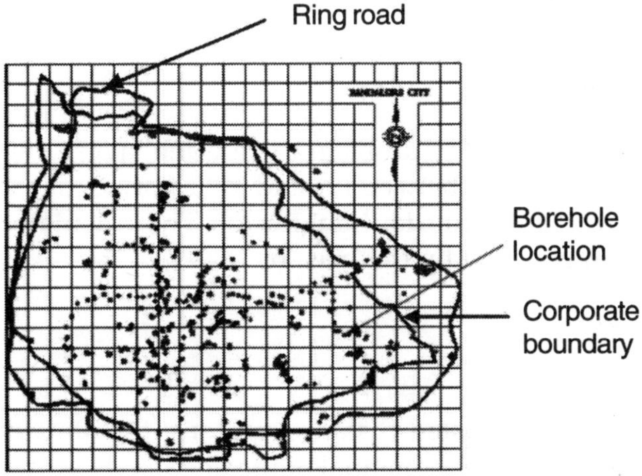

Fig. 1: Borehole location in Bangalore Map (scale: 1:20000).

difference between the ground reduced level at borehole location and reduced level of the hard rock at the same borehole location. The reduced level of the hard rock at borehole location is the difference between the ground reduced level at borehole location and depth of overburden thickness up to hard rock for the same borehole. The depth of overburden is estimated from the available borelogs.

GRNN Model

In a GRNN design (Fig. 2), hidden layer weights (W_R) are simply the transpose of input vectors from the training set. GRNN is comprised of three layers of artificial neurons (Cigizoglu and Alp, 2006). The hidden layer consists of radial basis neurons, whose transfer function is a Gaussian. A Euclidean distance is calculated between an input vector and these weights (Specht, 1991).

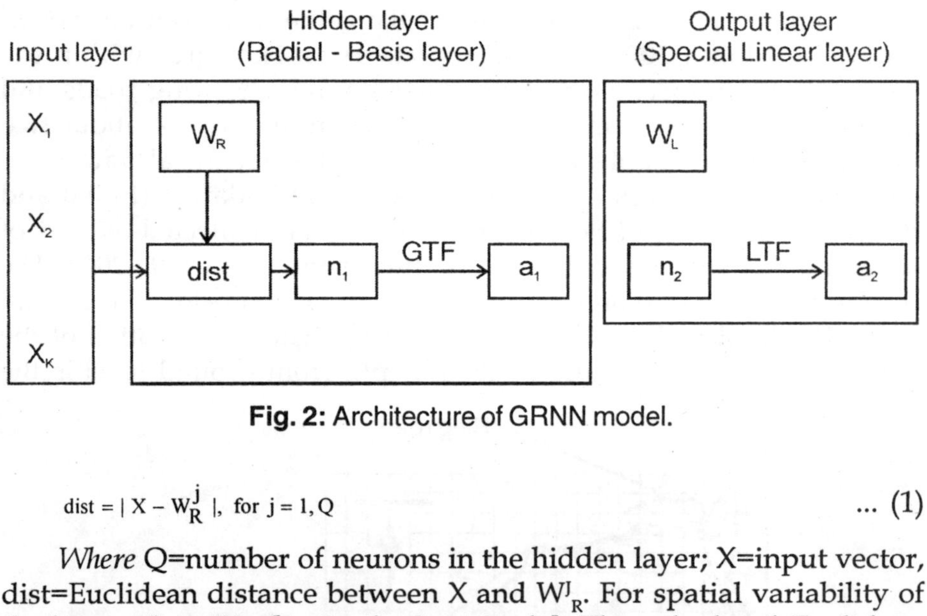

Fig. 2: Architecture of GRNN model.

$$\text{dist} = | X - W_R^j |, \text{ for } j = 1, Q \quad \dots (1)$$

Where Q=number of neurons in the hidden layer; X=input vector, dist=Euclidean distance between X and W^J_R. For spatial variability of d of Bangalore, X= [Latitude, Longitude]. The calculated Euclidean distance is then rescaled by the bias, b:

$$b = 0.8326 / s \quad \dots (2)$$

$$n_1 = dist \times b \quad \dots (3)$$

Where n_1 = the adjusted distance, and s = the spread. The radial basis output is then the exponential of the negatively adjusted distance having the form (Specht, 1991):

$$a_1 = e^{\left[-(n_1)^2\right]} \qquad \text{... (4)}$$

Therefore, if a neuron weight is equal to the input vector, distance between the two is 0 giving an output of 1. This type of neuron gives an output characterizing the closeness between input vectors and weight vectors. The weight matrix size is defined by the size of the training dataset, while the number of neurons is the number of input vectors. The output layer consists of neurons with a linear transfer function, which is (Specht, 1991):

$$n_2 = W_L \times a_1 \qquad \text{... (5)}$$

Where W_L is the weight matrix in the output layer. In designing a GRNN model, it is very important to select a proper spread(s).

The main scope of this study is to implement the above methodology for prediction of d in Bangalore. For predicting d in a given space, the two input variables (latitude and logitude of borehole in Bangalore) are used for the GRNN model in this study. There are many ways of normalizing data, but the method used in this analysis is normalizing the data against their maximum values (Sincero, 2003). In carrying out the formulation, the data has been divided into two sub-sets, such as:

(a) A *training dataset*: This is required to construct the model. In this study, 457(70% of total data) out of the 652 d values are considered as training dataset.

(b) *A testing dataset*: This is required to estimate the model performance. In this study, the remaining 195(30% of total data) data is considered as testing dataset.

's' is the most important parameter in a GRNN, has been determined using the procedure described below:

(i) Begin the training of the intended GRNN by assuming a small initial s value.

(ii) Evaluate the trained GRNN using the testing data set. This is done simply by comparing the GRNN-predicted d value with the known d value for each case in the testing data set. A root mean squared error (RMSE) of all cases is calculated:

$$\text{RMSE} = \sqrt{\sum_{i=1}^{n} \varepsilon^2 / n} \qquad \text{... (6)}$$

Where, e is the difference between the known d value and the predicted d value, and n is the number of cases in the testing data set.

(iii) Repeat steps (i) and (ii) above, a number of times using a gradually increased 's' value. With each adopted 's', a GRNN is trained and used to predict d value for each case in the testing data set, and RMSE is then calculated.

(vi) Plot the RMSE against the corresponding s value. The s that yields a minimum RMSE is considered to be the optimum 's'.

The GRNN model has been developed by using MATLAB.

LSSVM Model

LSSVM models are an alternate formulation of SVM regression (Vapnik and Lerner, 1963) proposed by Suykens *et al.* (2002). In LSSVM, Vapnik's e-insensitive loss function has been replaced by a cost function which corresponds to a form of ridge regression (Suykens *et al.*, 1999). Consider a given training set of N data points $\{x_k, y_k\}_{k=1}^{N}$ with input data $x_k \in R^N$ and output $y_k \in r$ where R^N is the N-dimensional vector space and r is the one-dimensional vector space.

The two input variables used for the LSSVM model in this study are latitude and longitude. The output of the LSSVM model is d. So, in this study, and x = [Latitude, Langitude] and y = d. In feature space LSSVM models take the form:

$$y(x) = w^T \phi(x) + b \qquad \text{... (7)}$$

Where the nonlinear mapping $\phi(.)$ maps the input data into a higher dimensional feature space; $w \in R^n$; $b \in r$ w = an adjustable weight vector; b = the scalar threshold. In LSSVM for function estimation the following optimization problem is formulated (Smola, 1998):

$$\text{Minimize: } \frac{1}{2} w^T w + \gamma \frac{1}{2} \sum_{k=1}^{N} e_k^2$$

$$\text{Subjected to: } y(x) = w^T \phi(x_k) + b + e_k,\ k = 1, ..., N. \qquad \text{... (8)}$$

The following equation for d prediction has been obtained by solving the above optimization problem (Vapnik, 1998; Smola and Scholkopf, 1998).

$$d = y(x) = \sum_{k=1}^{N} \alpha_k K(x, x_k) + b \quad \text{... (9)}$$

The radial basis function has been used in this analysis. The radial basis function is given by:

$$K(x_k, x_l) = \exp\left\{-\frac{(x_k - x_l)(x_k - x_l)^T}{2\sigma^2}\right\} \quad \text{k,l=1,...,N} \quad \text{... (10)}$$

Where σ is the width of radial basis function.

The plausibility of the above proposed models is evaluated using d in the subsurface of Bangalore. In LSVM model, the same training dataset, testing dataset, and normalization technique have been used as used in SVM model. The value of g and width of radial basis function (σ) value has been chosen by trial and error approach. In the present study, training and testing of LSSVM have been carried out by using MATLAB.

Results and Discussion

Coefficient of correlation(r) has been used as main criteria to examine the performance of developed model. Fig. 3 shows the plot between s and RMSE for GRNN model. From Fig. 3, it is clear that 0.021 is the optimum 's'. The performance of training and testing data has been determined by using the value(s=0.021) of optimum 's'. Fig. 4 shows the performance of GRNN model. Fig. 4 also confirms that the developed GRNN model has capability to predict d at any point in Bangalore. In case of LSSVM model, the design value? g and s is 50 and 0.1 respectively. Fig. 5 shows the performance of LSSVM model. Both GRNN and LSSVM have better performance in the training phase than in the testing phase.

The loss of performance with respect to the testing set addresses a machine's susceptibility to overtraining. There is a very marginal reduction in performance on the testing dataset (*i.e.*, there is a difference between machine performance on training and testing) for the GRNN as well as LSSVM model.

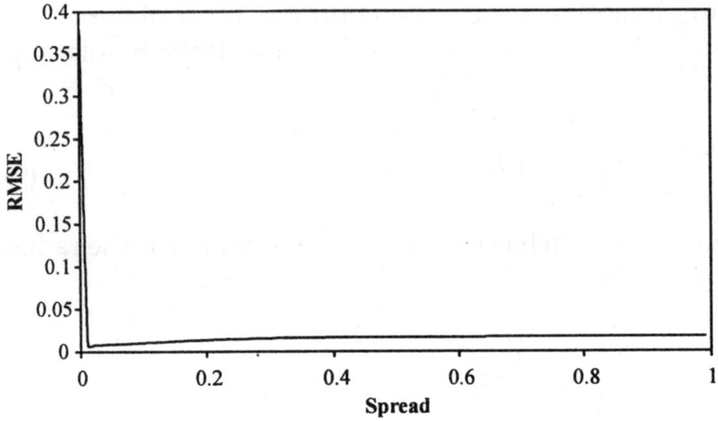

Fig. 3: Determination of optimum 's' for GRNN model.

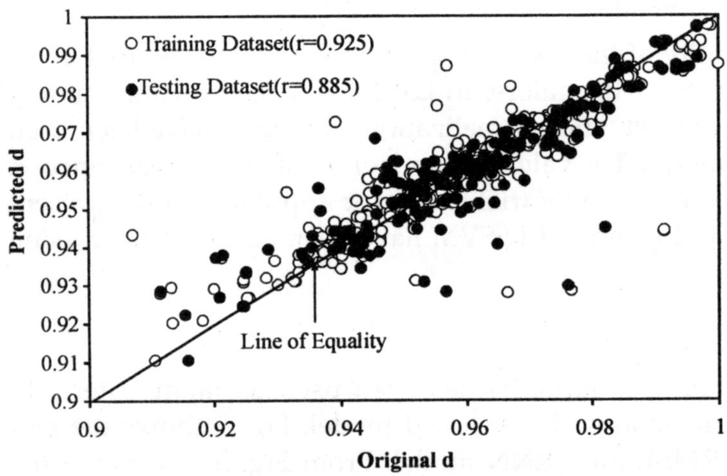

Fig. 4: Performance of GRNN model.

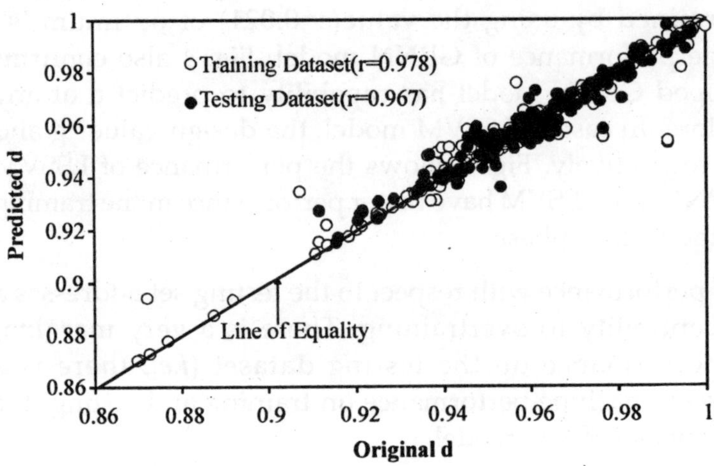

Fig. 5: Performance of LSSVM model.

This relatively small decline of performance of the LSSVM over GRNN model indicates its ability to avoid overtraining, and hence it can be expected to generalize better than GRNN. The following equation can (by putting? s = 0.1 and b=-0.075 7 in equation-9) be developed based on LSSVM for the prediction of d at any point in Bangalore.

$$d = \sum_{k=1}^{457} \alpha_k \exp\left(-\frac{(x_i - x)(x_i - x)^T}{0.02}\right) - 0.0757 \qquad \text{... (11)}$$

Where, x_i is the input of training dataset, x is input of the data whose d value has to be determined and T is transpose. Fig. 6 shows the value of a.

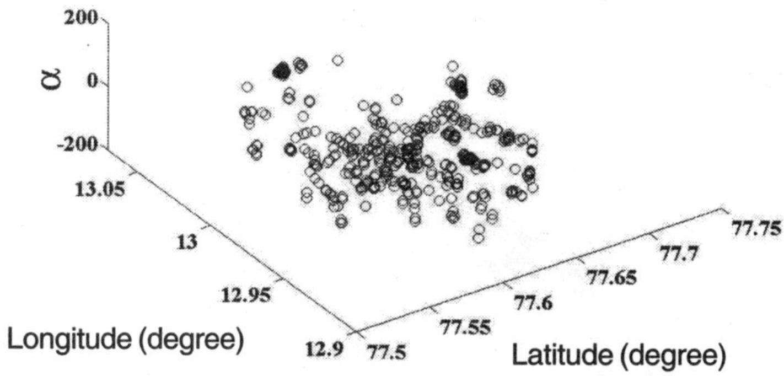

Fig. 6: Values of a.

In order to compare between the developed GRNN and LSSVM model, five points have been chosen randomly from known reduced level of rock values of 652 points in the subsurface model of Bangalore. The predicted values of these points are shown in Table-1.

Table-1 : Comparison GRNN and LSSVM model.

Bore hole no.	Longitude (degree)	Latitude (degree)	Actual reduced level of the rock (m)	Predicted reduced level of rock(m) by GRNN	Predicted reduced level of rock(m) by LSSVM
71-2	77.5765	12.9448	885.20	888.40	886.80
53-6	77.6237	12.9447	884.46	880.46	883.95
725-39B	77.6641	12.9924	893.50	897.50	892.52
51-9	77.5874	12.9331	896.60	898.60	897.56
87-4	77.5368	13.0293	900	897.45	901.35

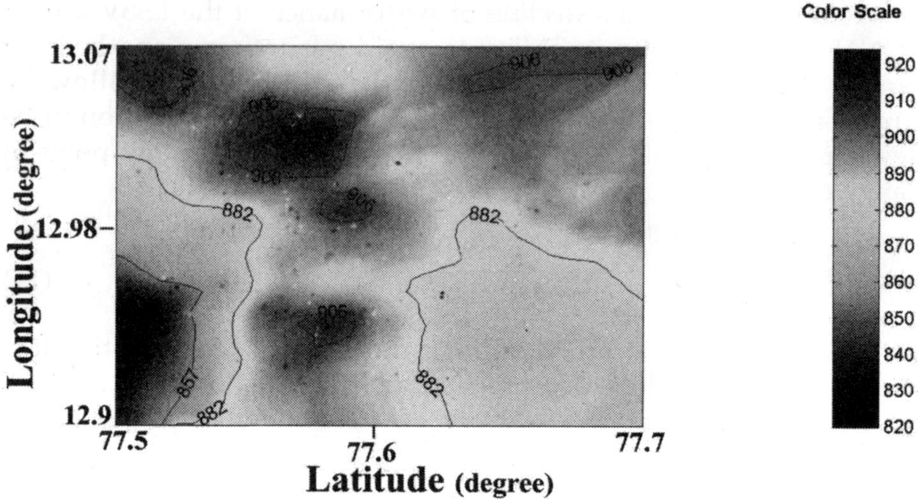

Fig. 7: Map of d for Bangalore using GRNN.

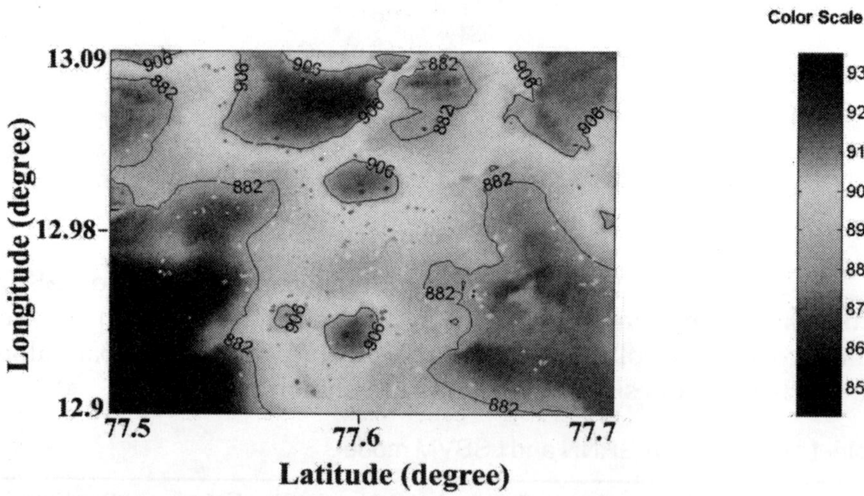

Fig. 8: Map of d for Bangalore using LSSVM.

Conclusion

This paper describes two artificial intelligence techniques (GRNN and LSSVM) for the determination of spatial variability of rock depth in Bangalore. In the development of the model discussed here, significant effort is required to build the machine architecture. However, once developed and trained, the transpired models performed the simulations in a small fraction of the time required by the physically based model. A comparative study between the developed GRNN and LSSVM model shows that LSSVM is better than GRNN. Scientists can use the developed equation based on LSSVM model for the prediction of reduced level of rock at any point in Bangalore. This paper gives a powerful model based on LSSVM for the prediction of spatial variability of reduced level of rock in Bangalore.

References

Alonso, E. E. and Krizek, R. J. (1975) Stochastic formulations of soil properties. Proc. Second Int. Conf. on Applications of Statistics and Probability in Soil and Struct. Eng., Aachen, Germany, pp. 10-32.

Asaoka, A. and Grivas, D. A. (1982) Spatial variability of the undrained strength of clays. J. Geotechnical. Eng., ASCE, v. 108(5), pp. 743-745.

Baecher, G.B. (1984) On estimating auto-covariance of soil properties. Specialty Conference on Probabilistic Mechanics and Structural Reliability, ASCE, v.110, pp.214 – 218.

Baker, R. (1984) Modeling soil variability as a random field. Math. Geol., v. 16(5), pp. 435-448.

Chiasson, P., Lafleur, J., Soulie, M. and Law, K. T. (1995) Charaterizing spatial variability of clay by geostatistics. Can. Geotech. J., v. 32, pp.1-10.

Cigizoglu, H.K. and Alp, M. (2006) Generalized regression neural network in modellingriver sediment yield. Adv. Eng. Softw., v. 37, pp. 63–68.

DeGroot, D. J. (1996) Analyzing spatial variability of in situ soil properties. ASCE 540 proceedings of uncertainty'96, uncertainty in the geologic environment: from theory to practice. ASCE Geotech. Special Pub., v. 58, pp. 210-238.

Delhomme, J. P. (1979) Spatial variability and uncertainty in groundwater flow parameters: A geostatistical approach. Water Resour. Res., v. 15(2), pp. 269-280.

Fenton, G. A. (1998) Random field characterization NGES data. Paper presented at the workshop on Probabilistic Site Characterization at NGES, Seattle, WA, v.550.

Fenton, G. A. (1999) Random field modeling of CPT data. J. Geotech. Geoenvir. Eng., ASCE, v. 125(6), pp. 486-498.

Juang, C.H., Jiang, T. and Christopher, R.A. (2001) Three-dimensional site characterisation:neural network approach. Géotechnique, v. 51(9), pp. 799-809.

Kulatilake, P. H. S. W. and Ghosh, A. (1998) An investigation into accuracy of sparial variation estimation using static cone penetrometer data. Proc. of the First Int. Symp. on Penetration Testing, Orlando, Fla., pp. 815-821.

Kulatilake, P. H. S. W. and Miller, K. M. (1987) A scheme for estimating the spatial variation of soil properties in three dimensions. Proc. of the Fifth Int. Conf. on Application of Statistics and Probabilities in Soil and Struct. Eng., Vancouver, British Columbia, Canada, pp. 669-677.

Kulatilake, P. H. S. W. (1989) Probabilistic Potentiometric surface mapping. J. Geotech. Eng., ASCE, v. 115(11), pp.1569-1587.

Lumb, P. (1975) Spatial variability of soil properties. Proc. Second Int. Conf. on Application of Statistics and Probability in Soil and Struct. Eng. Aachen, Germany, pp. 397-421.

Phoon, K. K. and Kulhawy, F. H. (1999) Characterization of geotechnical variability. Can. Geotech. J., v. 36(4), pp. 612-624.

Sincero, A.P. (2003) Predicting Mixing Power Using Artificial Neural Network. EWRI World Water and Environment, doi :10.1061/40685(2003)262.

Smola, A. J. (1998) Learning with Kernels. Ph.D. dissertation, GMD, Birlinghoven, Germany.

Smola, A. and Scholkopf, B. (1998) On a kernel based method for pattern recognition, regression, approximation and operator inversion. Algorithmica, v.22, pp. 211-231.

Soulie, M. (1983) Geostatistical applications in goetechnics. Geostatistics for natural resources characterization, Part 2, NATO ASI Series, pp.703- 615.

Soulie, M., Montes, P. and Sivestri, V. (1990) Modelling spatial variability of soil parameters. Can. Geotech. J., v. 27, pp. 617-630.

Specht, D. F. (1991) A general regression neural network. IEEE Trans. Neural Networks, v. 2, pp. 568-576.

Suykens, J.A.K., Lukas, L., Van Dooren, P., De Moor, B. and Vandewalle, J. (1999) Least squares support vector machine classifiers: a large scale algorithm. Proc. Eur. Conf. Circuit Theory and Design (ECCTD'99), Stresa, Italy, pp. 839–842.

Suykens, J.A.K., Brabanter, J.D., Lukas, L. and Vandewalle, J. (2002) Weighted least squares support vector machines: robustness and sparse approximation. Neurocomput, v. 48(1–4), pp. 85–105.

Tabba, M. M. and Yong, R. N. (1981) Mapping and predicting soil properties: Theory. J. Eng. Mech., ASCE, v. 107(5), pp.773-793.

Tang, W. H. (1979) Probabilistic evaluation of penetration resistance. J. Geotech. Eng., ASCE, v. 105(GT10), pp.117-191.

Uzielli, M., Vannucchi, G. and Phonn, K. K. (2005) Random field characterisation of stress-normalised cone penetration testing parameters. Geotechnique, v. 55(1), pp.3-20.

Vanmarcke, E. H. (1977) Probabilistic modeling of soil profiles. J. Geotech. Eng., ASCE, v. 102(11), pp. 1247-1265.

Vanmarcke, E.H. (1983) Random fields: Analysis and synthesis. The MIT Press, Cambridge, MA.

Vapnik, V.N. (1998) Statistical learning theory. Wiley, New York,

Vapnik, V.N. and Lerner, A. (1963) Pattern recognition using generalizd portait method. Automation and Remote Control, v.24, pp. 774-780.

Wu, T. H. and Wong, K. (1981) Probabilistic soil exploration: a case history. J. Geotech. Eng., ASCE, v. 107(GT12), pp.1693-1711.

Yaglom, A. M. (1962) Theory of stationary random functions. Prentice-Hall, Englewood Cliffs, NJ.

□□□

23

Effect of Depth on the Thermal Signature of Buried Metallic Object

J. A. OLOWOFELA[1], O. D. AKINYEMI[1], R. BELLO[1] AND A. A. ALABI[2]
[1]Department of Physics, University of Agriculture, Abeokuta, Nigeria
[2]Department of Physics, Lagos State University, Ojo, Nigeria

Abstract

The use of thermography for land mine detection has become a topic of great interest in recent years. The thermal properties and burial depth of the buried object also play a role in the thermal signature at the surface. The objective of this contribution is to determine the effect of burial depth on thermal signature of buried metallic object. The object used in this work is steel material buried at depth ranging from 1cm to 50cm. The two buried objects used in this work are steel of 12cm x 12cm surface area with thicknesses of 0.5cm and 3cm respectively. The soil where the objects were buried is mainly sandy. Soil above the buried objects and below it is assumed to be the same type of soil. This work was carried out in Abeokuta, Ogun State, Nigeria. There was a remarkable phase shift which increased with burial depth. A change in burial depth from 1cm to 10cm caused the maximum positive peak to shift from 46.5^{0}C to 38.0^{0}C and a change in burial depth from 40cm to 50cm, caused the maximum peak to shift from 30.0^{0}C to 29.0^{0}C. The burial depth of the buried objects has effect on the amplitude of the temperature at the surface and thus its thermal signature. It was also observed that the thickness of the buried objects has a significant effect on its thermal signature.

Keywords : Dynamic thermography, thermal capacitance, thermal signature, land mine

Introduction

The use of thermography for land mine detection has become a topic of great interest in recent years. The basic principle of all dynamic thermography based techniques is the idea that the thermal signature of the soil is altered by the presence of shallowly buried objects. This fact makes this approach very well suited for land mine (including Buried Objects and Unexploded Ordinances) detection since their different thermal properties will result in perturbations of the expected thermal pattern that can be measured by sensors (infrared and thermocouple). Moreover, this holds for every type of mine and other buried objects, despite the amount of metal content, if any, making possible the detection of small plastic antipersonnel mines. (Lopez *et al.*, 2003)

The thermal properties and burial depth of the buried object also play a role in the thermal signature at the surface (De Jong *et al.*, 1999). The situation becomes more complex as a result of the diurnal and annual heat flux cycles that drive the transport of heat to and from the surface (Remke *et al.*, Accessed 2007).

The difference in the thermal capacitance between soil and mine affects their heating/cooling rates and therefore their associated infrared emissions. Infrared cameras are used to map heat leakage patterns from the ground which, nevertheless, makes this thermography method an anomaly identification technique (Ashley, 1996). The technique essentially measures the thermal emissivity of the ground and interprets changes in emissivity as being caused by the presence of a foreign object; therefore, material characterization information is not provided. However, this technology has the advantages of being passive, can be performed remotely, by aerial search, and can cover a large area in a short time. Infrared thermography is best suited for identifying minefields (global area search), rather than searching for individual mines (local area search). It cannot, however, work when the soil and mine are in thermal equilibrium, and therefore is generally limited for use either at sunset or sunrise where a temperature gradient can be established at the ground surface (Hussein and Waller, 2000). All bodies (surfaces) with temperature greater than absolute zero emit electromagnetic radiation at all wavelengths of the spectral (0, 8) (Lienhard and Lienhard, 2004).

IR sensor systems respond to the electromagnetic radiation in the infrared spectral range. In practice, two spectral bands mostly used in

IR sensor systems are the midwave infrared (3-5 ìm) and the longwave infrared (8-12 ìm) (Jacobs, 1996). Although IR signatures of a source depend on its temperature, IR sensor systems do not measure directly the temperature of the surface but the thermal energy emitted from the surface. Theoretically, it is possible to convert from the measured IR signatures to the temperature of the surface provided the spectral emissivity of the surface and the hypothesis that the energy comes only from a gray surface is usually composed of emitted (self–emission) and reflected (environmental) radiation (Thanh, 2007).

The objective of this contribution is to determine the effect of burial depth on thermal signature of buried metallic object. The object used in this work is steel material buried at depth ranging from 1cm to 50cm. It was assumed that the soil above the buried object has the same thermal characteristics as the soil below and away from the buried object.

Materials and Method

The two buried objects used in this work were steel of 12cm x 12cm surface area with thicknesses of 0.5cm and 3cm respectively. The equipment used were four temperature sensors and data logger (Hoboware by Onset Corporation, USA) to store the data. The temperature sensor is a 12-Bit temperature smart sensor designed to work with the HOBO Data logger. All sensor parameters are stored inside the smart sensor, which automatically communicates configuration information to the logger without any programming or extensive user set up. It has measuring range of -40^0 to $+100^0C$ with an accuracy of $<0.2^0C$ from 0^0 to $+50^0C$ and a resolution of $<0.03^0C$ from 0^0 to $+50^0C$. The dimension of the sensor is 7 x 38mm with a measuring averaging option. The soil where the objects were buried is mainly sandy. The soil above and below the buried objects is assumed to be the same type. This work was carried out in Abeokuta, Ogun State, Nigeria.

One temperature sensor was placed on each of the buried objects; another temperature sensor was placed at the same depth without buried object. This is to show that heat was actually being emitted from the buried objects. The fourth temperature sensor was placed on the surface to serve as control experiment. This measures both the heat from the sun and the heat emitted from the soil surface. Fig. 1 shows the equipment used and the burial site.

Fig. 1: Equipment and the burial site.

On day one of the work, the objects were buried at a depth of 1cm with all the sensors arranged as described above. On day two, the depth was increased to 10cm with sensors arrangement the same as in day one. The depths were increased in step of 10cm in the subsequent days up to a depth of 50cm. The work was stopped at 50cm depth because at this depth, there were no longer significant difference in the thermal signatures of the two buried objects and the third sensor buried at the same depth. It should be noted that weather parameters have effect on the thermal signature at the soil surface but have little effect on the thermal signature of the buried object.

Results and Discussion

Burial depth of buried object has an effect on the temperature at the surface, and thus its thermal signature. There was a remarkable phase shift which increased with burial depth (Remke *et al.*, 2007). A change in burial depth from 1cm to 10cm caused the maximum positive peak to shift from 46.5^{0}C to 38.0^{0}C and a change in burial depth from 40cm to 50cm, caused the maximum peak to shift from 30.0^{0}C to 29.0^{0}C.

The plot showed that the temperature difference decreased with depth. A change in burial depth lead to a phase shift of the temperature curves (Fig. 2-7). The amplitude variation decreased rapidly for deeper burial depth and any significant amplitude variation was absent for deeper depth, this was in agreement with the observation of Remke *et al.*, 2007.

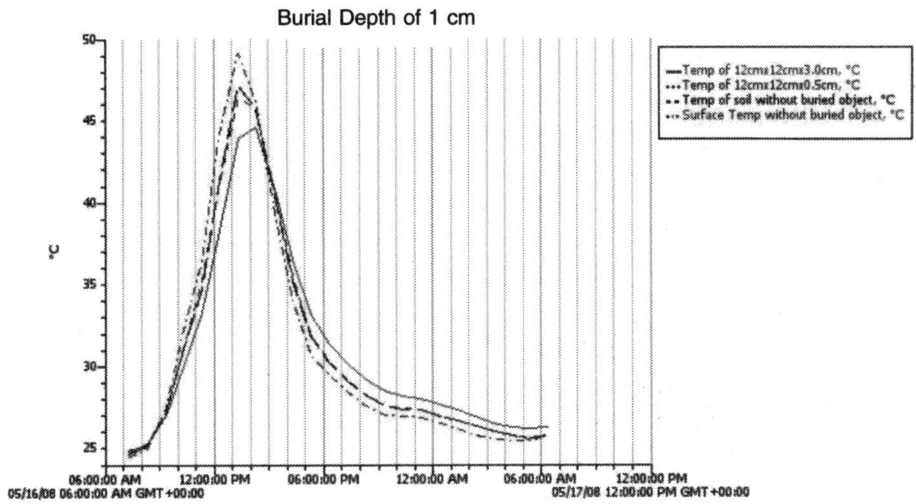

Fig. 2 : Temperature emitted by the objects at burial depth of 1cm.

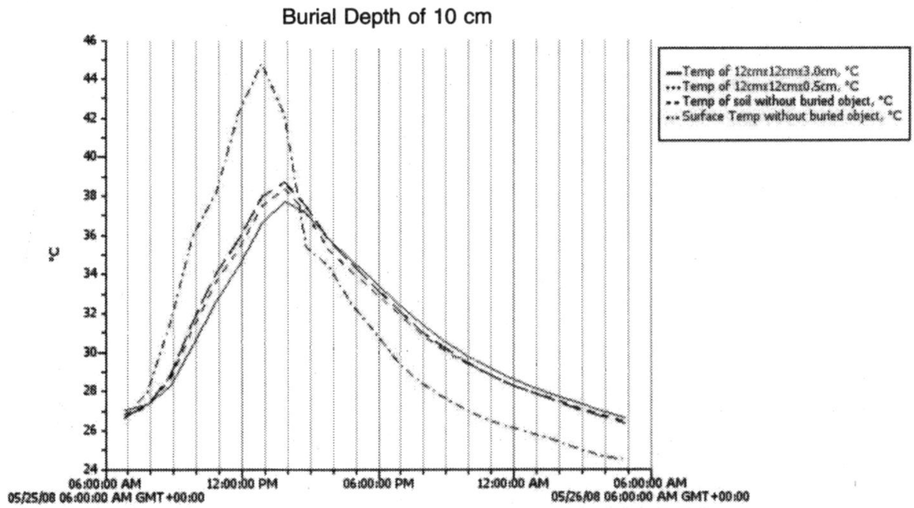

Fig. 3 : Temperature emitted by the objects at burial depth of 10cm.

Another aspect that played a role in the amplitude variation of buried objects at the surface are the thermal properties of the buried objects. The strength and phase of the thermal signature also depend on the thermal properties and burial depth of the object. Effect of thickness of buried objects on thermal signature was also observed. It was observed that the smallest one absorbed heat quickly during the day but looses its heat faster at night. This shows that heating and cooling rates of buried objects could be used to predict the size or thickness of the buried objects.

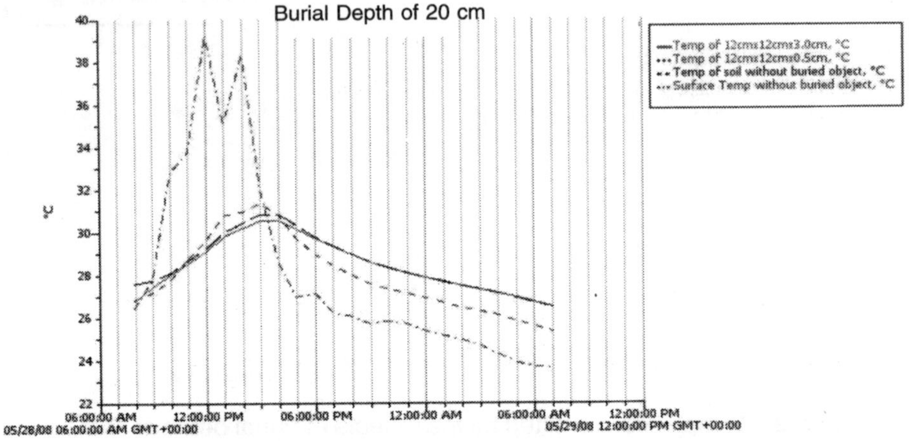

Fig. 4 : Temperature emitted by the objects at burial depth of 20cm.

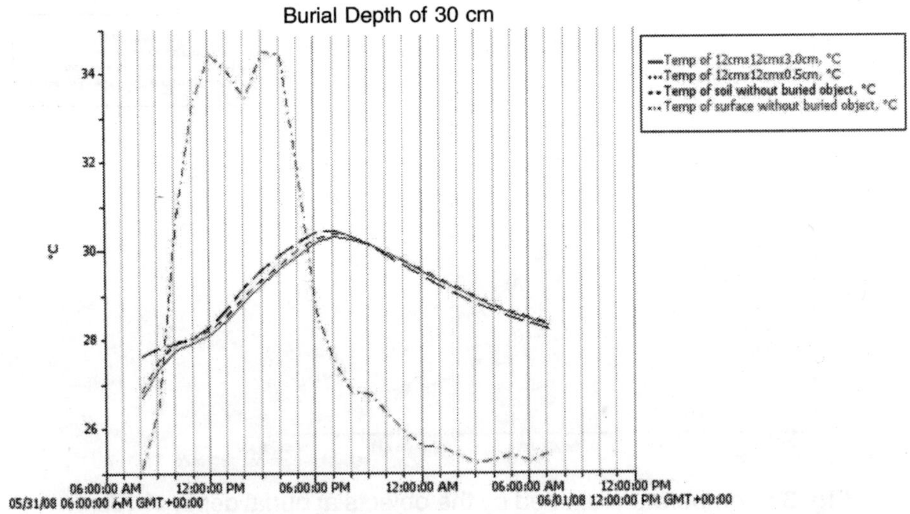

Fig. 5 : Temperature emitted by the objects at burial depth of 30cm.

The surface temperatures for all the days of the experiment reached peak between 1300 hours and 1330 hours except on the last day which reached its peak at 1430 hours. Variation observed in the last day was attributed to changes in weather conditions. The anomalies observed in the surface temperature for burial depth of 20cm and 30cm was due to light showers that occurred at that period.

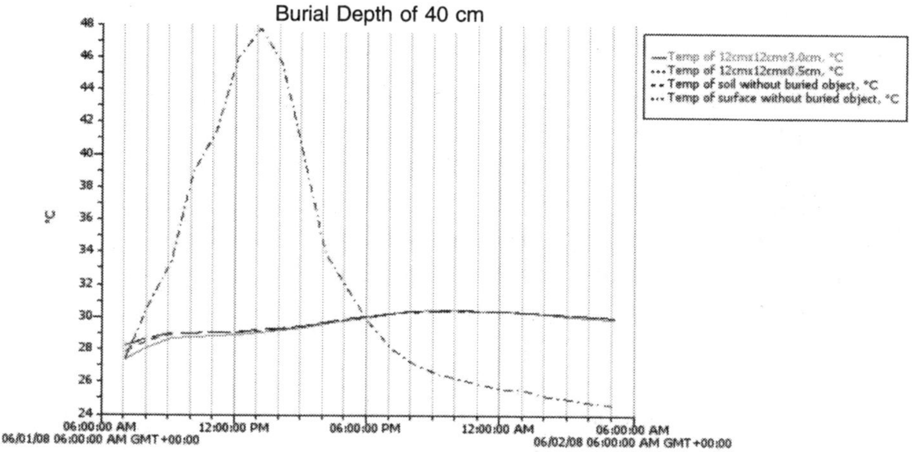

Fig. 6 : Temperature emitted by the objects at burial depth of 40cm.

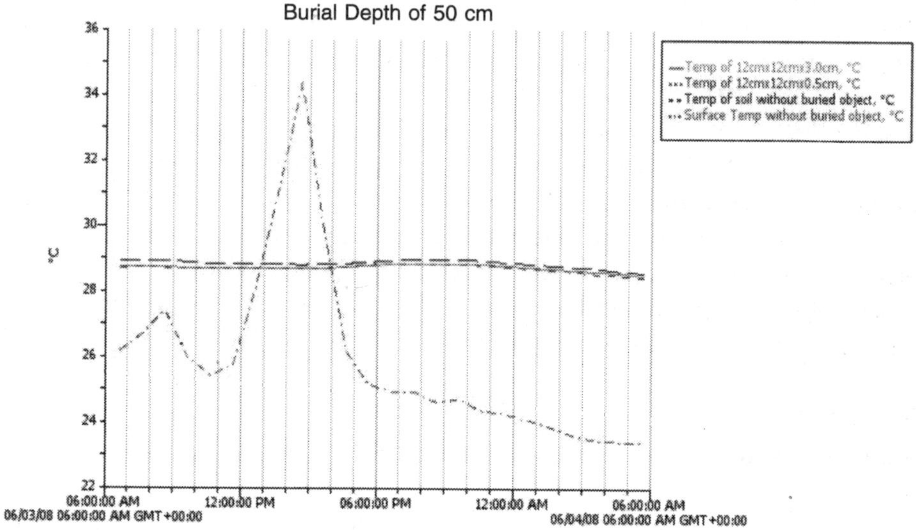

Fig. 7 : Temperature emitted by the objects at burial depth of 50cm.

The thermal signatures of the buried objects reached their peaks at times which depend on the depth of burial. For depth of 1cm, the peak temperatures occurred at time ranges from 1330 hours to 1430 hours with very close relationship with the surface temperature. For depth of 10cm, the peaks occurred at 1400 hours. The peaks of the thermal signatures for depth of 20cm occurred between 1630 and 1700 hours. For burial depths of 30 cm and 40cm, the peak of the thermal signatures occurred at 1900 hours and 21 hours respectively while the peak could not be defined for burial depth of 50cm where the thermal pattern appear almost horizontal.

With the observations above, it could be seen that considering burial depths of 10cm to 40cm, there was two hours shift for every 10cm in the time the thermal signatures of the buried objects reached their peaks. It was also observed that the temperature of the emitted heat by the buried object decreased with burial depth. For depth of 1cm the peak temperature occurred between 44^{0}C and 47^{0}C while for 40cm it occurred at 30°C.

Conclusion

Thermal sensors hold much promise for the detection of metallic, non-metallic mines and other buried objects. However, the prediction of their thermal signatures depends on a large number of factors. In this work, effect of burial depth and thickness of buried object on thermal signatures of metallic objects was investigated. The burial depth of the buried objects has effect on the amplitude of the temperature at the surface and thus its thermal signature. It was also observed that the thickness of the buried objects has a significant effect on its thermal signature, but this effect reduced gradually with increasing depth until no effect was observed at depth of 50cm. This work has shown that the deeper the burial depth of landmines and other buried objects, the lesser it will be to detect. Also, light or small landmines or other buried objects will easily be detected as lighter or small objects will emit more temperature that big or thicker objects.

References

Ashley, S. (1996) Searching for Land mines. Mechanical Engineering, v. 118(4), pp. 62.

Hussein, E. M. A and Edward, J. W. (2000) Landmine Detection: The Problem and the Challenge. Applied Radiation and Isotopes, v. 53, pp. 557 – 563.

Jacobs, P. A. M. (1996) Thermal infrared characterization of ground targets and backgrounds. SPIE Optical Engineering press, Bellingham (WA), USA, pp. 7–9.

Lienhard, J. H. IV and Lienhard, J. H. V. (2004) A heat transfer textbook. Plogiston press, Cambridge MA, third edition, pp. 28-34.

Lopez, P., Sahli, H., Vilarino, D. L. and Cabello, D. (2003) Detection of perturbations in Thermal IR signatures: An inverse problem for buried landmine detection. Proceedings of the SPIE, v. 5046, pp.242 – 252.

Remke, L. V., Borchers B, Hendrickx, J. M. H. and Sung-ho H. (2007) Soil effects on thermal signatures of buried non-metallic landmines.

Available at http://www.ees.nmt.edu/Hydro/landmines/pub/spie2003_thermal. (Accessed 2007).

Thanh, Nguyen Trung (2007) Infrared thermography for the detection and characterization of buried object. Uitgeverij VUBPRESS Brussels University Press, 23 p.

□□□

24

Modelling Electrical Conductivity of Contaminated Nigerian Riverbed Sands

OLUKAYODE D. AKINYEMI AND IYIOLA KUFORIJI
Department of Physics, College of Natural Sciences, University of Agriculture, PMB 2240, Abeokuta, Nigeria

Abstract

Investigation of electrical conductivity of contaminated earth is becoming increasingly relevant in geophysical sciences, where global environmental safety is a key issue. Riverbed sands were collected from the five south-western states of Nigeria, and electrical conductivity determined after samples have been contaminated with varying concentrations of petrol, cow dung, palm oil, manganese (IV) oxide, cement, cassava extract, wood ash, and washing soda. Impact of density was also studied for solid contaminant. Conductivity was largely proportional to contaminant concentration in all the samples in general. However with increase in concentrations of petrol and palm oil, decrease in electrical conductivity of samples was observed while increase in concentrations of other contaminants increased conductivity of samples. Conductivity was also largely influenced by the density of contaminated samples. At higher concentrations of contaminants, conductivity became constant especially when conductivity values of contaminants itself was achieved.

Keywords : Riverbed Sands, Electrical Conductivity, Contaminants, Density

Introduction

In urban environments, riverbed sands receive considerable pollution from industry, traffic and refuse, and since contaminated soil particles can easily be inhaled or ingested, there is a potential transfer of toxic pollutants to humans (El Khalil *et al.*, 2008). Electrical conductivity plays a determinant role in two complementary aspects of the exploration of the ground: surface measurements and well-logging measurements (Keller, 1988; Tabbagh *et al.*, 2002). This property exhibits the widest range of variation among all the properties used in exploration geophysics. Furthermore, the concept of electrical conductivity had found a very good use in Agriculture.

The conductivity meters measure the number of ions that are mobile within a liquid and the measurement of electrical conductivity is considered the most accurate way of determining the salinity of the soil. Plants absorb nutrients in form of ions *e.g.* nitrogen ions, phosphorous ions and potassium ions (NPK) and measuring a soil solution is a good indication of the total amount of nutrients available to the plants (although not the specific nutrients).

In petroleum engineering, it is frequently assumed that the electrical conductivity has contributions from both the liquid and solid phases (Patnode and Wylie, 1950; Cremers and Laudelout, 1965; Rhoades *et al.*, 1976). Electrical conductivity has also been used to estimate diffusion in porous media (Snyder, 2001; Garrouch *et al.*, 2001, Klein and Santamarina, 2003). Electrical conductivity depends on the concentration of the ions, temperature of the solution and the nature of the ions (Waxman and Smits, 1968; Bussian, 1983). Given the presence of industrial and agricultural activities around major rivers in south-western Nigeria, contamination has been a major issue. However, information about the electrical properties of contaminated Nigerian sands is very scarce despite their potential usage as ceramic tiles and bricks in building constructions. This work was therefore aimed at assessing the electrical properties of Nigerian riverbed sands which are highly impacted by human activities with increase in packing densities and local contaminants concentrations. The results of this study should therefore be sufficiently representative of the present situation to constitute guidance for future research which will help to develop management and remediation strategies.

Materials and Method

Riverbed sands were collected from the five south-western states of Nigeria as presented in Table-1 and Fig. 1. Samples were thoroughly washed to remove any hidden contaminant, oven-dried and sieved to ensure uniform grain sizes. Electrical conductivity was measured by a conductivity meter at a constant room temperature (25°C). The contaminants used were petrol, cow dung, palm oil, MnO_2, cement, cassava extract, wood ash, and washing soda. Most of the pollutants are agricultural and industrial wastes, with block making industries using cements, saw mills converting saw dust to wood ash, local palm oil industries, and cassava extract produced by cassava flour producers.

Petrol

Petrol is a fuel derived from petroleum crude oil and used in spark-ignited internal combustion engines. Conventional gasoline is mostly a blended mixture of more than 200 different hydrocarbon liquids ranging from those containing 4 carbon atoms to those containing 11 or 12 carbon atoms (Speight, 2008).

Cow dung

Cow dung is the waste of bovine animal species especially domestic cattle. Cow dung is the undigested residue of herbivorous matter, having passed through the animal's gut, and passed out as resultant faecal matter which is rich in minerals. Colour ranges from greenish to blackish, often darkening in colour soon after exposure to air. In many parts of the developing world, the banks of the river and dry river beds are contaminated with cow dung as cows are led to wander about and looking for rivers to drink water. Chemical composition of cow dung revealed that it contains organic matter, Nitrogen (N) and Manganese (Mn) (Kumar and Vishal, 2007).

Palm oil

Palm oil is obtained from the mesocarp of the *Elaeis guineensis* fruit, simply by cooking, mashing and pressing. Palm oil extraction is predominantly done by rural women around rivers especially because of the volume of water required during the production of oil. Seeds are separated and after cracking and removing the shell, the kernel processed to yield palm kernel oil and palm kernel cake. Refined, Bleached and Deodorized (RBD) palm oil is obtained from refining crude palm oil. Palm oil is a light yellow liquid and semi-solid at room temperature, melting to a clear yellow liquid on slight heating. It naturally contains carotenes (pro-vitamin A) which give red colour to the oil. Palm oil processing is predominantly the activity of most rural dwellers, and the disposed extract mixes with soils and ends up affecting water quality of streams and rivers especially during the rainy season.

Manganese (IV) Oxide

Manganese (IV) oxide is a blackish solid used in Lechlanche dry cell where it is contained in a muslin bag round the carbon pole to function as a depolarizer. MnO_2 is a poor conductor and therefore has to be mixed with powdered carbon, which is a good conductor. When a primary cell works continuously, it depolarizes owing to the hydrogen librated at the carbon pole. According to Oniawa and Fakayode, 2010, Lead levels measured in the vicinity of a battery factory in Nigeria were found to be elevated around the factory with average levels of about 2000mg kg^{-1} close to the fence but decreased gradually to about 50mg kg^{-1} some 750m away.

Cement

Portland cements are commonly characterized by their physical and chemical properties for quality control purposes. They are common building materials used for constructions of bridges and buildings and for plastering and concreting of blocks. The ability to easily mix with soils makes cement, which is mostly used for construction materials, end up in streams and rivers, thus affecting water quality which many rural populace drink in developing countries. The constituent components include Tricalcium Silicate ($3CaO.SiO_2$), Dicalcium Silicate ($2CaO.SiO_2$), Tricalcium Aluminate ($3CaO.Al_2O_3$), Tricalcium Aluminoferrite ($4CaO.Al_2O_3.Fe_2O_3$) and Gypsum ($CaSO_4.H_2O$). According to a research conducted by Olaleye and Oluyemi, 2009 in the vicinity of Ewekoro cement factory in Nigeria, high levels of total dissolved particulates and atmospheric deposition rates were recorded, with planktonic flora and fauna of the river systems draining the area very poor.

Cassava Extract

Cassava is annually cultivated as annual crops in tropical and subtropical regions of the world. It is the third largest source of carbohydrates for human food in the world, with Africa being the largest centre of production (Claude and Denis, 1990). Cassava contains cyanide (Arguedes and Cook, 1982), either in expressed juice or wash water spray (Cook and Maduagwa, 1987), and the effluent from cassava is usually disposed as waste water. According to a research conducted by Ehiagbonare *et al.*, 2009 on the impact of cassava effluent on the surrounding waters, it was observed that the colour, taste, odour and quality of the waters changed after cassava effluent have been discharged into it from the neighbouring cassava processing base. Environmental problems from cyanide may occur; for example growing stages of plants including vegetables and sensitive stages of fish and other aquatic animals may be negatively affected (Bengtsson and Trient, 1994).

Wood Ash

Wood ash is the inorganic and organic residue remaining after the combustion of wood or unbleached wood fibre. The physical and chemical properties of wood ash vary significantly depending on many factors. Hardwoods usually produced more ash than softwoods and the bark and leaves generally produce more ash than the inner woody parts of the tree. On the average, the burning wood results in about 6-10 per cent ash. Bush burning is a major trend in most African countries, especially as a means of clearing overgrown bushes, and also by hunters during dry seasons. It mixes easily with sands and is a major contaminant of water in rivers which are primary sources of drinking water in rural areas. It has been established that wood ash is an unusual cause of irritations and chemical burns when in contact with human skins (Kilic and Kilic, 2002).

Washing Soda

Sodium carbonate (also known as washing soda, Na_2CO_3) is a sodium salt of carbonic acid. It most commonly occurs as a crystalline heptahydrate, which readily effloresces to form a monohydrate white powder and is domestically well known for its everyday use as a water softener. Production of washing soda is very common in most Nigerian households, and soda residues are usually disposed in careless manners, making them contaminate the soils and shallow/surface waters especially during rainy season. Redness and burning of eyes, redness and swelling of skin, nose and/or throat irritation, nausea and/or vomiting are some of the known health hazards (USOSHA, 1985).

Sample Preparation

0.2 kg of each sample was moistened with 30 ml of distilled water. Several readings of electrical conductivity were taken at different points in the cylinder in order to obtain the average values. 0.000005m^3 of each contaminant was then added and mixed thoroughly, after which conductivity was measured. The concentration of the contaminant was increased by the same volume at each experimental set-up until dose of the concentration reached a level of $0.00003 m^3$. Measurements were made following the procedure explained above with 0.3, 0.4, 0.5, and 0.6 kg of the samples. For all the contaminants where the concentration of solid contaminants was by masses (kg) *i.e.* Cow dung, Manganese (IV) oxide, Cement, Wood ash and Washing soda, a sensitive scale was used to achieve the accuracy.

Fig. 1: Map showing sample locations.

In the second round of experiments, the concentration of cassava liquid waste was dosed with 0.3 kg of each sample step-by-step by increasing the concentration and conductivity was measured at each experimental step while sample dose was increased till the measured conductivity became constant.

In the third part of the work, 0.3 kg of the samples with 30 ml of the water was thoroughly mixed in a beaker with 0.00001 kg of MnO_2. The mixture was then transferred into a container and compressed to different densities and conductivity measured at each density. The process was repeated for the remaining two solid contaminants *i.e.* Cement and Wood ash.

Table-1 : Sample location parameters and natural conductivity values.

Sample	River Name	State	Elevation (m)	Porosity	Latitude	Longitude	Electrical Conductivity (ms/cm)
A	Lekki Beach	Lagos	5	0.432	6° 25′ N	3° 31′ E	58
B	Owena	Ondo	208	0.343	7° 10′ N	4° 43′ E	254
C	Water Works	Osun	263	0.396	7° 38′ N	4° 12′ E	65
D	Odoona	Oyo	160	0.401	7° 23′ N	3° 51′ E	386
E	Isolu	Ogun	64	0.394	6° 48′ N	3° 13′ E	146

Results and Discussion

Increase in the weight of samples while maintaining same amount of contaminants had little impact on sample (A) from Lekki Beach which had the smallest variations in conductivity values with regards to sample weight increment, as markedly opposed to other samples with pronounced variations in conductivities as weights of samples increase. Incidentally, Lekki Beach which is the most porous sample as shown in Table-1 also had the highest conductivity after contamination irrespective of sample weights. Samples with widest variations were those obtained from Odoona and Isolu.

Slope Analysis

Analysis of slopes for the plots of variations of conductivity with increment of contaminant concentrations was carried out as shown in Table-2 for Petrol, Cow dung, Palm oil, Manganese (IV) Oxide and Cement, while Table-3 shows the slope analysis for Cassava liquid waste, Wood ash and Washing soda. The slope analysis clearly revealed that values of gradients reduced with masses of samples, as equal amount of contaminants was being added irrespective of the samples.

Conductivity against Concentration

Conductivity variations with contaminants for the various samples were carried out but only the plots for Petrol (representing liquid contaminants) and Manganese (IV) oxide (representing solid contaminants) were shown. Conductivity decreased with increase in concentration of petrol in all the samples as shown in Fig. 2 to 6. Petrol as a covalent hydrocarbon is a poor conductor of electrical current. The electrical conductivity on the other hand increased with increase in the concentration of cow dung as shown in Table-2. The cow dung used in the study consisted of the following chemical composition: 1.6% N, 0.70% P, 0.53% K, 0.91% Mg, 2.71% Ca, 0.50% Na, 56.8% organic matter and C:N of 7.9 (Onwudike, 2010), which has resulted in the observed conductivity trend.

Table-2 : Slopes of Conductivity Vs Concentration of 5 contaminants for different samples.

Sample	0.2 kg	0.3 kg	0.4 kg	0.5 kg	0.6 kg
			Petrol (m^3)		
A	-3E +06	-3E +06	-2E +06	-2E +06	-2E +06
B	-2E +06	-1E +06	-972571	-694857	-720000
C	-2E +06	-2E +06	-1E +06	-8937	-769143
D	-2E +06	-2E +06	-2E +06	-1E +06	-1E +06
E	-2E +06	-2E +06	-2E +06	-1E +06	-982653
			Cow dung (kg)		
A	2611.7	2470.3	2161	2124.3	2053.1
B	1733.7	1455.4	1369.1	1249.4	932
C	1936	1604.6	1428.9	1212.9	1043.1
D	1472.3	1282.6	1054.9	921.71	756.57
E	2035.4	1706.3	1326	1344.3	931.43
			Palm oil (m^3)		
A	-2E +06	-2E +06	-2E +06	-2E +06	-2E +06
B	-2E +06	-1E +06	-883429	-690286	-472571
C	-2E +06	-1E +06	-882286	-800000	-739429
D	1E +06	-894857	-684571	-660571	-381714
E	-2E +06	-2E +06	-2E +06	-1E +06	-899429
			MnO_2 (kg)		
	0.2 kg	0.3 kg	0.4 kg	0.5 kg	0.6 kg
A	277629	237586	181414	164214	146829
B	59166	48583	33211	25040	14149
C	62674	49897	41817	32537	22069
D	45577	35360	29063	22903	10651
E	69383	50446	41257	24023	29349
			Cement (kg)		
	0.2 kg	0.3 kg	0.4 kg	0.5 kg	0.6 kg
A	67977	64160	57851	43303	36537
B	43703	40891	31600	28034	20469
C	50571	38777	38777	28960	24823
D	40023	36480	36046	23977	20366
E	50331	40823	36240	34617	25143

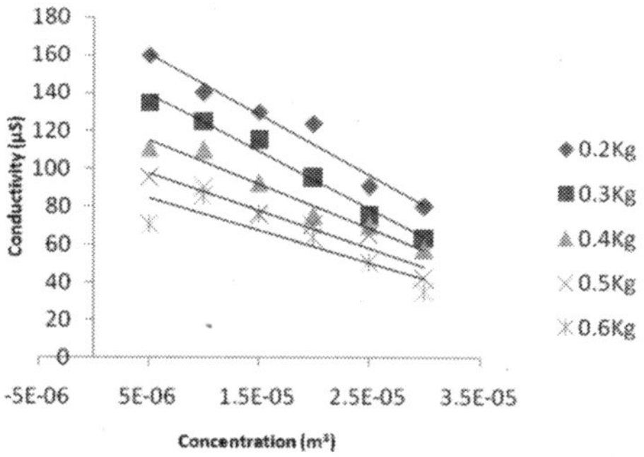

Fig. 2: Conductivity against concentrations of Petrol for various masses of Sample A

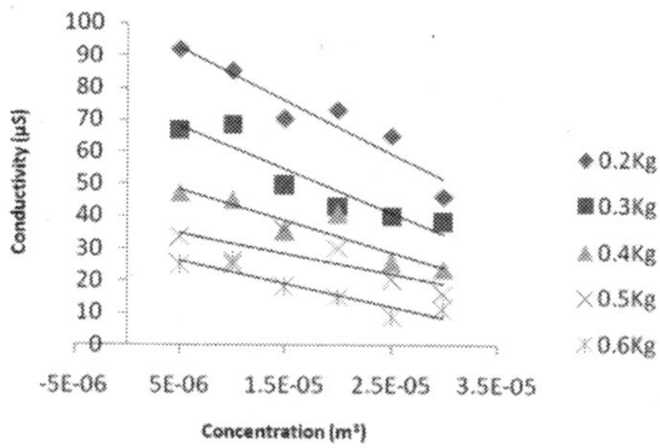

Fig. 3: Conductivity against concentrations of Petrol for various masses of Sample B

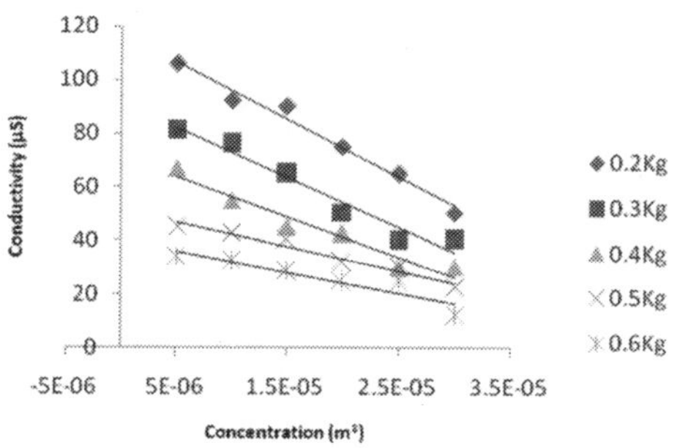

Fig. 4: Conductivity against concentrations of Petrol for various masses of Sample C.

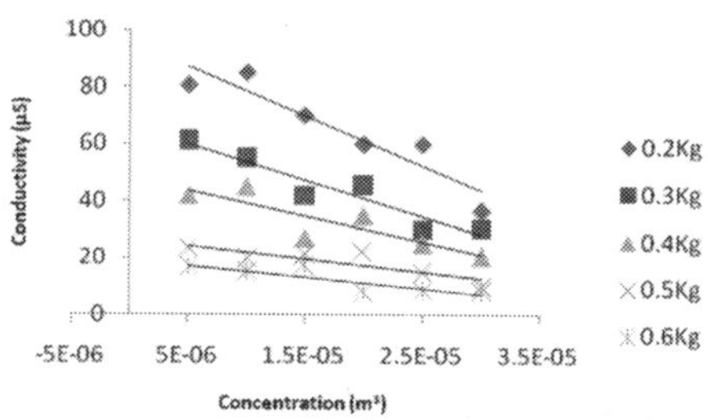

Fig. 5: Conductivity against concentrations of Petrol for various masses of Sample D.

Conductivity decreased with increase in the concentration of palm oil as observed in Table-2 with negative gradient as was the case with petrol. Palm oil is also made up of covalent bond, and thus behaved in a similar way as petrol. Conductivity increased with increase in density of MnO_2 as shown in Fig. 7 to 11. Although MnO_2 is a poor conductor, chemical composition analysis however revealed that the conductivity behaviour is due to MnO_2 being usually mixed with powdered carbon which is a good conductor. From the studies on cement dosages, the results revealed that the conductivity increased with increase in concentration of cement as shown in Table-2 with positive gradient. The increase in conductivity was probably contributed by calcium which is a very reactive element in the chemical composition of cement. In the activity series, calcium reacts with water to form hydroxide ($Ca(OH)_2$). Calcium comes third in the activity series, Potassium (K) and Sodium (Na) takes the first and second positions respectively.

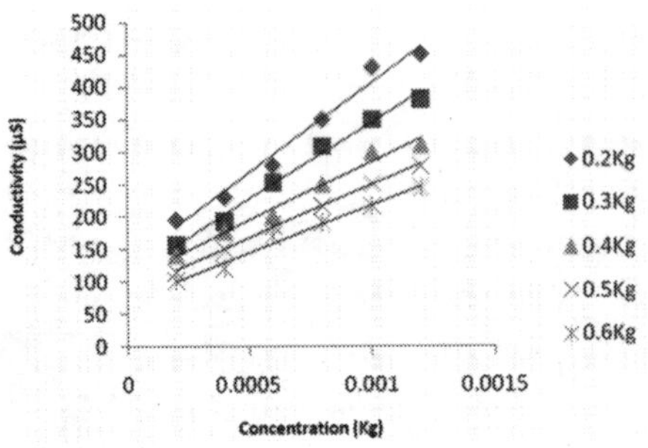

Fig. 6 : Conductivity against concentrations of Petrol for various masses of Sample E.

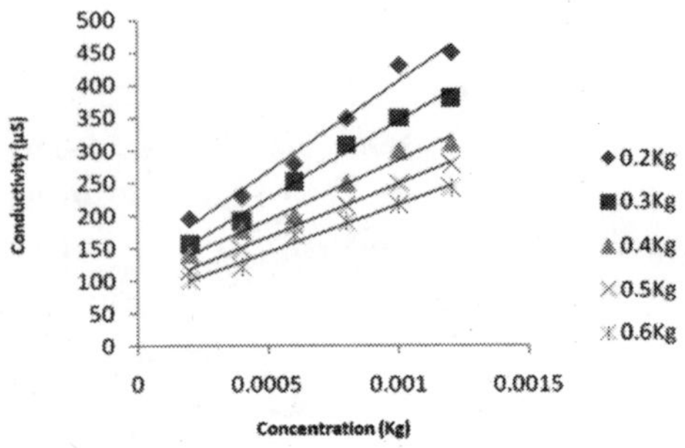

Fig. 7 : Conductivity against concentrations of Mn O_2 for various masses of Sample A.

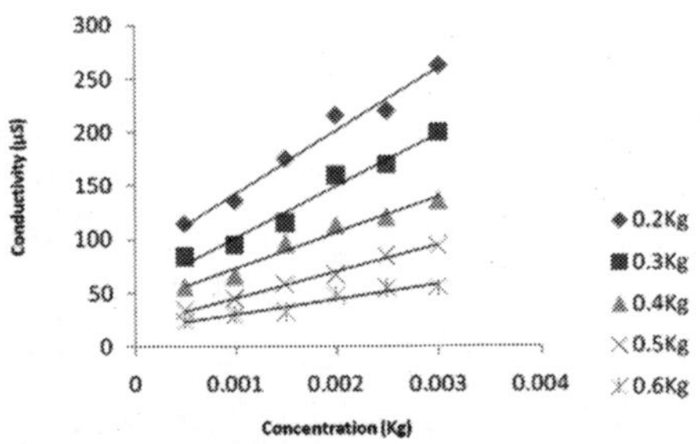

Fig. 8 : Conductivity against concentrations of Mn O_2 for various masses of Sample B.

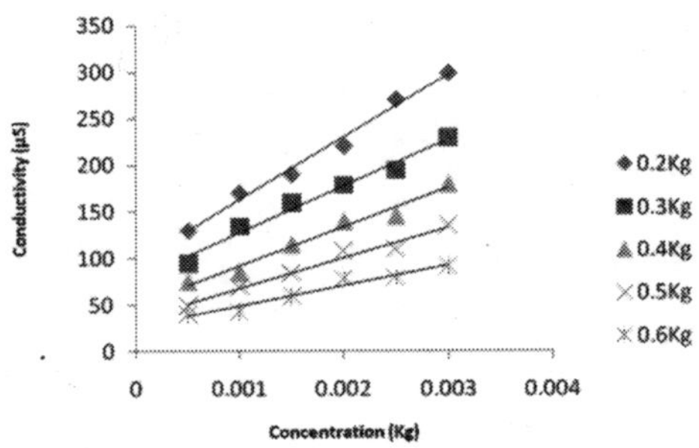

Fig. 9 : Conductivity against concentrations of Mn O_2 for various masses of Sample C.

For cassava liquid waste, conductivity increased sharply, with increase in its concentration as observed in the slopes analysis as shown in Table-3. It is characterized by 1,300 ppm of CN^- which had strong influence on the electrical conductivity behaviour of the samples dosed with cassava liquid. Same trend was observed with wood ash (Table-3) where conductivity also increased with increase in concentration, though not as sharply as cassava extract. Calcium is the most abundant element in wood ash, which also is a good source of potassium, phosphorus, magnesium and aluminium all of which are conductive ions. From Table-3, conductivity was observed to increase with concentration of washing soda. Electrical conductivity increase was also very sharp in soda, and the variation in conductivity values across the samples was very minimal. Sodium carbonate (Na_2CO_3), sodium bicarbonate ($NaHCO_3$), sodium hydroxide (NaOH) and sodium oxide (Na_2O) are all sodium salts that might have contributed for the conductivity behaviour.

Table-3 : Slopes of conductivity Vs Concentration of 3 contaminants for different samples.

			Cassava liquid waste (m^3)		
Sample	0.2 kg	0.3 kg	0.4 kg	0.5 kg	0.6 kg
A	3E +07	3E +07	3E +07	3E +07	2E +07
B	3E +07	2E +07	2E +07	2E +07	2E +07
C	3E +07	2E +07	2E +07	2E +07	2E +07
D	2E +07	2E +07	2E +07	2E +07	1E +07
E	3E +07	2E +07	2E +07	2E +07	2E +07
			Wood ash (kg)		
	0.2 kg	0.3 kg	0.4 kg	0.5 kg	0.6 kg
A	63840	63223	47840	43771	40514
B	48297	34857	26960	21874	18457
C	45131	35017	28594	24914	22034
D	43931	28971	22309	24343	16034
E	48160	34960	33989	31349	26983
			Washing soda (m^3)		
	0.2 kg	0.3 kg	0.4 kg	0.5 kg	0.6 kg
A	3E +07	3E +07	2E +07	2E +07	2E +07
B	2E +07	2E +07	2E +07	1E +07	1E +07
C	2E +07	2E +07	2E +07	1E +07	1E +07
D	2E +07	2E +07	1E +07	1E +07	1E +07
E	2E +07	2E +07	2E +07	2E +07	1E +07

Furthermore, range of conductivity values was analysed for the investigated contaminants with sample weight 0.2 kg as shown in Table-4. Samples dosed with Petrol and Palm oil contaminants fall within the lowest conductivity range, while Washing soda and Cassava extract fall within the highest conductivity range. However, at some instances, Washing soda and Cassava extract also fall within the middle range of conductivity.

Comparing the electrical conductivity values at natural locations in Table 1 with Table-4 revealed that samples B, D and E were more contaminated than A and C. This was expected from the point of view of human activities around B, D and E as opposed to A and C that were not directly within residential areas. For instance, Sample E was taken from Isolu river which is very close to the University of Agriculture, Abeokuta off campus residential area characterized with a lot of human activities including new building constructions.

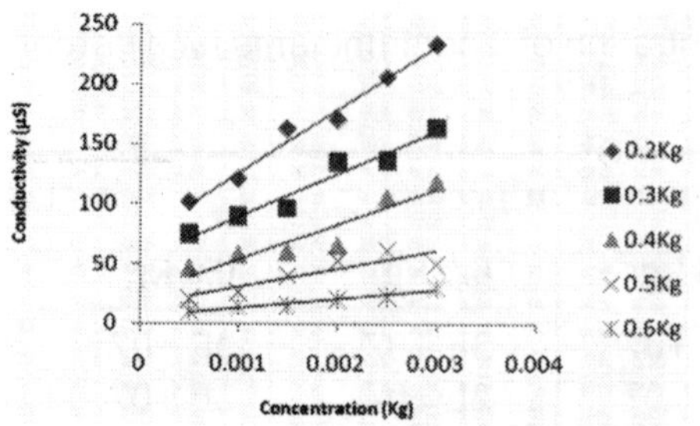

Fig. 10 : Conductivity against concentrations of Mn O_2 for various masses of Sample D

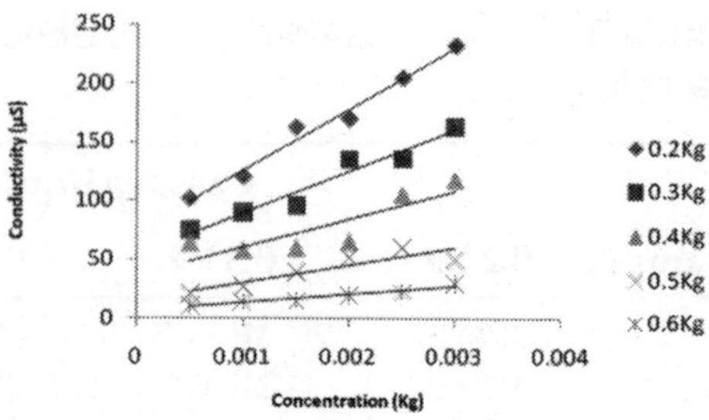

Fig. 11 : Conductivity against concentrations of MnO_2 for various masses of Sample E.

Table-4 : Contaminant Type and Range of Conductivity from Experiment.

Electrical Conductivity Range (µs/cm)	Possible Contaminant Type
55 – 150	Petrol, Palm Oil
150 – 500	Cement, Washing Soda, Cassava Extract, MnO_2, Wood Ash, Cow Dung
500 – 1000	Washing Soda, Cassava Extract

Conductivity at Higher Concentrations

Conductivity of sample A was observed to be highest as compared to other samples at lower concentrations. It was also already observed for all the samples investigated that conductivity varied linearly with the increase in concentration. However, behaviour of conductivity at very high concentrations was investigated with cassava liquid waste as shown in Fig. 12. Cassava liquid waste was selected for the test because its slope analysis in Table-3 revealed very steep slopes and close values of gradients. At very high concentrations of cassava liquid waste, conductivity was largely constant for the samples and finally converged. The observed convergence of conductivity irrespective of the samples was most probably due to ionic interference which is restricting the mobility of the ions.

Conductivity against Density

The graphs of conductivity against density of the samples dosed with Manganese (IV) Oxide, Cement and Wood ash were shown in Fig. 13 to 15. Sample A consistently remained the highest conductive, while the least conductive was sample D. It was also observed that for all the samples, the conductivity increased linearly with increase in the density of the sample. It is expected that due to increase in density, the particles of the porous media are brought more closely together along with the particles of the contaminant, thus making the ions to be locked in the pores.

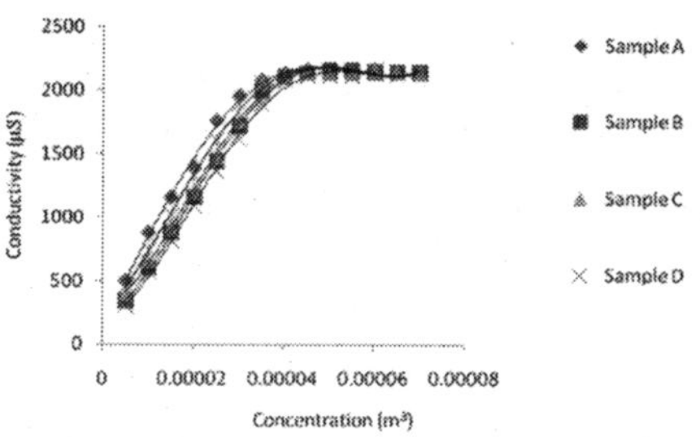

Fig. 12 : Conductivities of the samples at higher concentrations of cassava liquid waste.

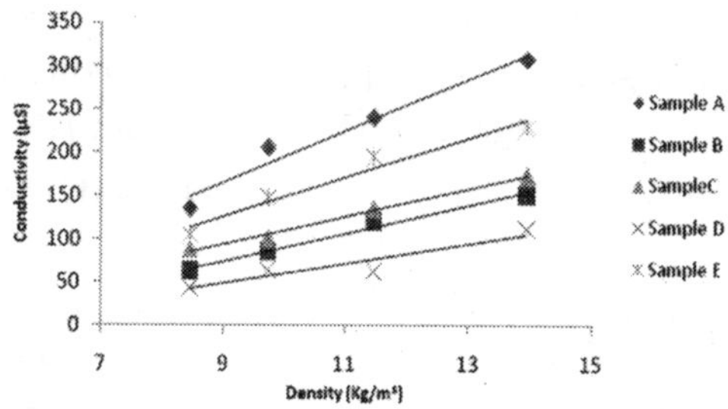

Fig. 13 : Conductivity against density of samples contaminated with MnO_2.

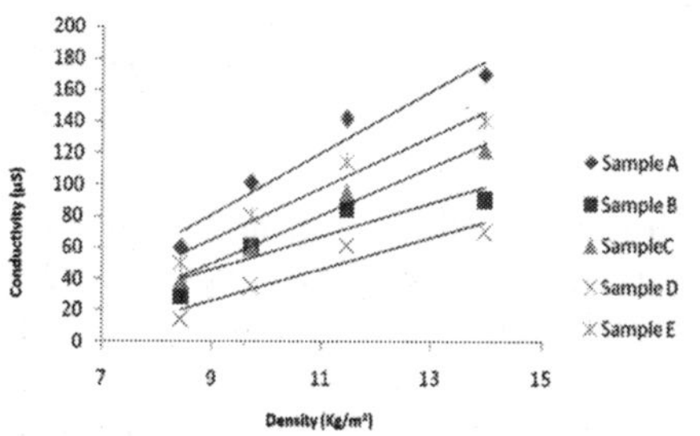

Fig. 14 : Conductivity against density of samples contaminated with Cement.

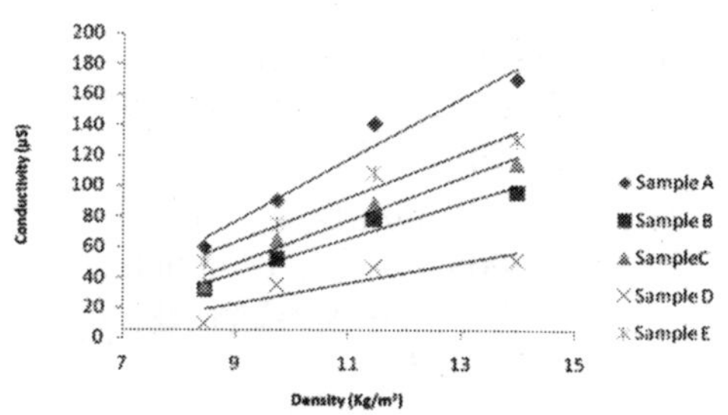

Fig. 15 : Conductivity against density of samples contaminated with Wood Ash.

Conclusion

The presence of contaminants greatly influences the conductivity of porous media. A contaminant with lower or zero Total Dissolved Solids value *e.g.* palm oil reduced conductivity while that with higher Total Dissolved Solids value *e.g.* cement and cassava liquid waste increased conductivity of dosed samples. This is similar to what obtains in doped extrinsic semiconductors where the degree of doping makes a large difference in the conductivity with more doping implying higher conductivity.

In moderate concentration ranges, conductivity was proportional to concentration and largely dependent on the density (of the samples), the denser the sample the more closely packed the particle and the more conductive it is. At higher concentrations of contaminant like cassava liquid waste, the conductivity levelled off despite the increase in the concentration, and converged irrespective of the samples.

References

Arguedes, P. and Cooke, R. D. (1982) Cyanide Concentrations during Extraction of Cassava Starch. Food Technol. v. 17, pp. 251-262.

Bengtsson B. E. and Trient, T. (1994) Tapioca-starch waste water toxicity characterized by microtox and duck week tests. Ambio., v. 23, pp. 473-477.

Bussian, A. E. (1983) Electrical Conductance in a Porous Medium. Geophysics, v. 48(9), pp. 1258 1268.

Claude, F. and Denis, F. (1990) African Cassava Mosaic Virus: Etiology. Epidemiology and control plant disease, v. 64(6), pp. 404-411.

Cooke, R.D. and Maduaguwu, E. N. (1987) The effort of simple processing on the cyanide content of cassava chips. Food Technol., v. 13, pp. 299-306.

Cremers, A. and Laudelout, H. (1965) Note on the 'Isoconductivity Value' of clay gels. Soil Science. v. 100, pp. 298 -299.

Ehiagbonare, J. E., Adjarhore, R. Y. and Enabulele, S. A. (2009) Effect of cassava effluent on Okada natural water. African Journal of Biotechnology, v. 8 (12), pp. 2816-2818.

El Khalil, H., Schwartz, C., Elhamiani, O., Kuniniok, J., Morel, J. L. and Boularbah, A. (2008) Contribution of technic materials to the mobile fraction of metals in urban soils in Marrackech (Morocco). Journal Soils Sediments, v. 8 (1), pp. 17 - 22.

Garrouch, A. A., Ali, L. and Qasem, F. (2001) Using diffusion and electrical measurements to assess tortuosity of porous media. Industrial Eng. Chem. Res,. v. 40, pp. 4363 – 4369.

Keller, G. V. (1988) Rock and mineral properties. In: *M. N. Nabighian (ed.)* Electromagnetic methods in Applied Geophysics, I, Theory, pp. 13 – 51.

Kumar, G. A. and Vishal, M. (2007) Organic and mineral composition of *Gomeya* (cow dung) from *Desi* and crossbred cows – A comparative study. International J. of C. Sci. v. 3(1&2).

Kilic, A. and Kilic, A. (2002) Wood ash: an unusual cause of a chemical burn. J. International Society for Burns, v. 28(1), pp. 95 – 96.

Klein, K. A., and Santamarina, J. C. (2008) Electrical conductivity in soils: Underlying phenomena. J. Environmental Eng. Geophysics, v. 8, pp. 263 - 273.

Olaleye, V. F. and Oluyemi, E. A. (2000) Effects of cement flue dusts from a Nigerian cement plant on air, water and planktonic quality. Envr. Monitoring and Assessment. v. 162(1-4), pp. 153 – 162.

Oniawa, P. C. and S. O. Fakayode (2010) Lead contamination of topsoil and vegetation in the vicinity of a battery factory in Nigeria. Envir. Geochemistry and Health, v. 22, pp. 211 - 218.

Onwudike, S.U. (2010) Effectiveness of cow dung and mineral fertilizer on soil properties, nutrient uptake and yield of sweet potato (Ipornoea batatas) in Southeastern Nigeria. Asian J. Agric. Res., v. 4, pp. 148-154.

Patnode, W. H., and Crowder, M. R. J. (1950) The presence of conductive solids in reservoir as factor in electric log interpretation. Petroleum Transactions AIME, v. 189, pp. 47 - 52.

Rhoades, J., Raats, P. A. C. and Prather, R. J. (1976) Effects of liquid-phase electrical conductivity, water content and surface conductivity on bulk soil electrical conductivity. Soil Sci. Soc. Am. J., v. 40, pp. 651 - 655.

Snyder, K. A. (2001) The relationship between the formation factor and the diffusion coefficient of porous materials saturated with concentrated electrolytes: Theoretical and experimental considerations. Concrete Sci. Eng., v. 3, pp. 216 - 224.

Speight, J. (2008) Synthetic Fuels Handbook, 1st Edition. McGraw-Hill, pp. 377 - 378.

Tabbagh, A., Panissod, C., Guerin, R. and P. Cosenza (2002) Numerical modelling of the role of water and clay content in soils' and rocks' bulk electrical conductivity. J. Geophysical Research, v. 107(B11), pp. 2318 - 2322.

United States Department of Occupational Safety and Health Administration (1985) Material Safety Data Sheet: Washing Soda. OBM no. 1218-0072.

Waxman, M. H., and J. M. Smits (1968) Electrical conductivities in oil-bearing shaly sand. Soc. Petroleum Eng. J., v. 243, pp 107 - 122.

